Federated Learning for Smart Communication using IoT Application

The effectiveness of federated learning in high-performance information systems and informatics-based solutions for addressing current information support requirements is demonstrated in this book. To address heterogeneity challenges in Internet of Things (IoT) contexts, *Federated Learning for Smart Communication using IoT Application* analyses the development of personalized federated learning algorithms capable of mitigating the detrimental consequences of heterogeneity in several dimensions. It includes case studies of IoT-based human activity recognition to show the efficacy of personalized federated learning for intelligent IoT applications.

Features:

- Demonstrates how federated learning offers a novel approach to building personalized models from data without invading users' privacy.
- Describes how federated learning may assist in understanding and learning from user behavior in IoT applications while safeguarding user privacy.
- Presents a detailed analysis of current research on federated learning, providing the reader with a broad understanding of the area.
- Analyses the need for a personalized federated learning framework in cloud-edge and wireless-edge architecture for intelligent IoT applications.
- Comprises real-life case illustrations and examples to help consolidate understanding of topics presented in each chapter.

This book is recommended for anyone interested in federated learning-based intelligent algorithms for smart communications.

Chapman & Hall/CRC Cyber-Physical Systems

Series Editors:
Jyotir Moy Chatterjee
Lord Buddha Education Foundation, Kathmandu, Nepal
Vishal Jain
Sharda University, Greater Noida, India

Cyber-Physical Systems: A Comprehensive Guide
By: Nonita Sharma, L K Awasthi, Monika Mangla, K P Sharma, and Rohit Kumar

Introduction to the Cyber Ranges
By: Bishwajeet Pandey and Shabeer Ahmad

Security Analytics: A Data Centric Approach to Information Security
By: Mehak Khurana and Shilpa Mahajan

Security and Resilience of Cyber Physical Systems
By: Krishan Kumar, Sunny Behal, Abhinav Bhandari, and Sajal Bhatia

Cyber Security Applications for Industry 4.0
By: R Sujatha, G Prakash, and Noor Zaman Jhanjhi

Cyber Physical Systems: Concepts and Applications
By: Anupam Baliyan, Kuldeep Singh Kaswan, Naresh Kumar, Kamal Upreti, and Ramani Kannan

Intelligent Security Solutions for Cyber-Physical Systems
By: Vandana Mohindru Sood, Yashwant Singh, Bharat Bhargava, and Sushil Kumar Narang

Federated Learning for Smart Communication using IoT Application
By: Kaushal Kishor, Parma Nand, Vishal Jain, Neetesh Saxena, Gaurav Agarwal, and Rani Astya

For more information on this series please visit: https://www.routledge.com/Chapman--HallCRC-Cyber-Physical-Systems/book-series/CHCPS

Federated Learning for Smart Communication using IoT Application

Edited by

Kaushal Kishor, Parma Nand, Vishal Jain, Neetesh Saxena, Gaurav Agarwal, and Rani Astya

CRC Press
Taylor & Francis Group
Boca Raton London New York

CRC Press is an imprint of the
Taylor & Francis Group, an **informa** business
A CHAPMAN & HALL BOOK

Front cover image: mijiworld/Shutterstock

First edition published 2025
by CRC Press
2385 NW Executive Center Drive, Suite 320, Boca Raton FL 33431

and by CRC Press
4 Park Square, Milton Park, Abingdon, Oxon, OX14 4RN

CRC Press is an imprint of Taylor & Francis Group, LLC

ISBN: 978-1-032-78812-8 (hbk)
ISBN: 978-1-032-78813-5 (pbk)
ISBN: 978-1-003-48936-8 (ebk)

DOI: 10.1201/9781003489368

Typeset in Times
by codeMantra

Contents

Preface

This book describes how federated learning may assist in understanding and learning from user behavior in Internet of Things (IoT) applications while safeguarding user privacy. The authors begin by demonstrating how federated learning offers a novel approach to building personalized models from data without invading users' privacy. The authors then present a detailed analysis of current research on federated learning, providing the reader with a broad understanding of the area. The book also analyses the need for a personalized federated learning framework in cloud-edge and wireless-edge architecture for intelligent IoT applications. To address heterogeneity challenges in IoT contexts, the book analyses developing personalized federated learning algorithms capable of mitigating the detrimental consequences of heterogeneity in several dimensions. The book includes case studies of IoT-based human activity recognition to demonstrate the efficacy of personalized federated learning for intelligent IoT applications, as well as a variety of controller design and system analysis tools such as model predictive control, linear matrix inequalities, optimal control, and so on. This one-of-a-kind and comprehensive co-design framework will assist control theory and engineering researchers, graduate students, and engineers. According to recent advancements in communication computing, technology has grown more adaptable and scalable than ever before. This provides industries with greater control over their data. It also aids in offering higher levels of security at each data center. The most important components of the most recent communication computer technology are well structured. The purpose of this book is to showcase federated learning high-performance information systems and informatics-based solutions for validating current information support needs. Cloud computing is a new trend in information technology in which computer-based services are provided over the Internet or in the cloud and are accessible to consumers from any place and on any device. Software, hardware (servers), databases, storage, and other services are examples of services. Cloud computing for smart societies is here to stay. Businesses must embrace and flourish with this technology. Technology is both effective and exciting. In the long term, it shows to be a cost-effective method of service execution for many firms, both large and small. Cloud computing has a bright future since it has a broader scope, particularly in terms of reach. Customers and host service providers will both profit. However, it all depends on the technological decisions made by business product owners. It covers important topics such as computational intelligence in wireless networks and communications, artificial intelligence and wireless communication security, security risk scenarios in communications, security/resilience metrics and their measurements, cyber-crime data analytics, modeling for wireless communication safety risks, breakthroughs in cyber threats and computer crimes, adaptive and learning techniques for secure estimation and control, decision support, and decision support systems.

Editor Biographies

Kaushal Kishor He received his PhD in Computer Science and Engineering from AKTU Lucknow, in the domain of Mobile Ad hoc Network and MTech & BTech in Computer Science and Engineering from UPTU Lucknow. Currently, he is working at the ABES Institute of Technology, Ghaziabad as Professor of Information Technology. He has supervised more than 50 projects for graduate and postgraduate students. He has more than 19 years of experience in teaching. He has a Gate Qualified 2003 score of 94.5 percentile. He has published and edited books (1) *Cloud-based Intelligent Informative Engineering for Society 5.0*, pp. 1–234. Chapman and Hall/CRC., 2023, eBook ISBN: 9781003213895, (2) *Design and Analysis Algorithms* (ISBN NO.: 978-93-81695-20-3), (3) *Computer Networks* (ISBN: 978-93-81695-27-2), (4) *Compiler Design* (ISBN: 938169530 x and ISBN-13 9789381695302), (5) *Design and Analysis of Algorithms: Techniques and Control Management* (ISBN: 978-81-8220-516-1), (6) *Computer Networks a System Approach* (ISBN: 978-81-8220-516-3), (7) *Compiler Design Principles, Techniques, and Tools* (ISBN: 978-81-8220-626-7) for various engineering field like BTech and MCA student. He has published 40 papers in peer-reviewed international/national journals and conferences. His research interests include artificial intelligence, computer networks, algorithms, compiler design wireless, and sensor networking.

Parma Nand He earned a PhD in Computer Science and Engineering from IIT Roorkee and an MTech and BTech in Computer Science and Engineering from IIT Delhi. Dr. Parma Nand has more than 29 years of experience both in industry and academia. He has received various awards such as Best Teacher from Union Minister, Best Students Project Guide Award from Microsoft in 2015, and Best Faculty Award from Cognizant in 2016. He has successfully completed government-funded projects and spearheaded the last five IEEE International conferences on Computing, Communication & Automation (ICCCA), IEEE students chapters, Technovation Hackathon 2019, Technovation Hackathon 2020, International Conference on Computing, Communication, and Intelligent Systems (ICCCIS-2021). He is a member of the Executive Council of IEEE UP section (R-10), member of the Executive Committee IEEE Computer and Signal Processing Society, member of the Executive India Council Computer Society, member of the Executive Council Computer Society of India, Noida section, and has acted as an observer in many IEEE conferences. He also has active memberships in ACM,

IEEE, CSI, ACE, ISOC, IAENG, and IASCIT. He is a lifetime member of the Soft Computing Research Society (SCRS) and ISTE.

He has delivered many invited/keynote talks at international and national conferences/workshops/seminars in India and abroad. He has published more than 150 papers in peer-reviewed international/national journals and conferences.

Vishal Jain He is currently working as an Associate Professor in the Department of Computer Science and Engineering, Sharda School of Engineering and Technology, Sharda University, Greater Noida, UP, India. Before that, he worked for several years as an Associate Professor at Bharati Vidyapeeth's Institute of Computer Applications and Management (BVICAM), New Delhi. He has more than 16 years of experience in academics. He obtained a PhD (CSE), MTech (CSE), MBA (HR), MCA, MCP, and CCNA. He has more than 1,130 research citation indices with Google Scholar (*h*-index score 17 and *i*-10 index 27). He has authored more than 95 research papers in reputed conferences and journals, including Web of Science and Scopus. He has authored and edited more than 45 books with various reputed publishers, including Elsevier, Springer, DeGruyter, IET, River Publishers, Apple Academic Press, CRC, Taylor and Francis Group, Scrivener, Wiley, Emerald, NOVA Science, Bentham Books, and IGI-Global. He is the series editor of 10 book series. He is a life member of CSI, ISTE, and a senior member of IEEE. His research areas include information retrieval, semantic web, ontology engineering, data mining, ad hoc networks, and sensor networks. He received a Young Active Member Award for the year 2012–2013 from the Computer Society of India, the Best Faculty Award for the year 2017, and the Best Researcher Award for the year 2019 from BVICAM, New Delhi.

Neetesh Saxena He is currently an Associate Professor (Lecturer) with the School of Computer Science and Informatics at Cardiff University, UK with more than 14 years of teaching/research experience in academia. Before joining CU, he was an Assistant Professor at Bournemouth University, UK. Prior to this, he was a postdoctoral researcher in the School of Electrical and Computer Engineering at the Georgia Institute of Technology, USA. He was also with the Department of Computer Science, The State University of New York (SUNY) Korea, South Korea as a Postdoctoral Researcher and a Visiting Scholar in the Department of Computer Science, Stony Brook University, USA. He earned his PhD in Computer Science and Engineering from the Indian Institute of Technology (IIT), Indore, India. He was a DAAD Scholar at Bonn-Aachen International Center for Information Technology (B-IT), Rheinische- Friedrich-Wilhelms Universität, Bonn, Germany, and also a TCS Research Scholar. He is a senior member of IEEE and a member of IEEE SMC, IEEE ComSoc, ACM, and Eta Kappa Nu.

Gaurav Agarwal He completed his PhD in Computer Science and Engineering from IIT (ISM), Dhanbad, MTech in Computer Science and Engineering, from Shobhit University, Meerut and BE in Computer Science and Engineering from North Maharashtra University, Jalgaon. He is an NBA Expert and has received two international awards and one national award. He received 6+ SCI & 6+ SCOPUS and holds four national patents (two granted) and has 123 citations with *h*-index 5. He is currently working as an Associate Professor at Galgotias University. He has 20+ years of teaching experience with leading engineering institutions and he is working for NBA Accreditation.

Dr. Rani Astya is an Associate Professor in the Computer Science & Engineering Department at School of Engineering & Technology, Sharda University. Holding degrees of (B.E.) in Computer Science & Engineering from Jiwaji University (Gwalior), M.Tech from (IIT Roorkee)and Ph.D. in Computer Science from Banasthali Vidyapith (Rajasthan). Dr. Astya has contributed to the academic community by publishing 45 research papers in esteemed international conferences and journals. Her dedication to innovation is evident, having filed over 40 patents, with 35 already published & 2 granted and more anticipated in the future.

Contributors

Gaurav Agarwal
Department of CSE
Invertis University
Bareilly, India

Gaurav Agarwal
Department of CSE
Galgotias University
Greater Noida, India

Rupanshi Agarwal
Department of CSE
Invertis University
Bareilly, India

Idaraesit Akpabio
Library and Information Science
University of Uyo
Uyo, Nigeria

Ifiokobong Akpabio
Civil Engineering
University of Uyo
Uyo, Nigeria

Edidiong Akpabio
Bachelor of computer Application
Datta Meghe Institute of Higher Education and Research,
Wardha, India

Sangeeta Arora
Department of Computer Applications
KIET Group of Institutions
Ghaziabad, India

Debashree Chakravarty
Department of Management
School of Management
KIIT University
Bhubaneswar, India

Alok Singh Chauhan
Department of Computer Science and Engineering
Galgotias University
Greater Noida, India

Saurabh Dave
Faculty of Computer Applications
Ganpat University
Mehsana, India

Antonino Galletta
Department of Computer Science
Faculty of Engineering: Universita degli Studi di Messina, University of Messina
Messina, Italy

Sachin A. Goswami
Faculty of Computer Applications
Ganpat University
Mehsana, India

Sachi Gupta
Department of Computer Science
Galgotias College of Engineering and Technology
Greater Noida, India

Vishal Jain
Department of Computer Science and Engineering
School of Engineering and Technology
Sharda University
Greater Noida, India

Christian Kaunert
Department- School of Law and Government
Dublin City University
Dublin, Ireland
and
University of South Wales
Newport, UK

Kaushal Kishor
Department of Information Technology
ABES Institute of Technology
Ghaziabad, India

Arun Kumar
Department of CSE
ITS Engineering College
Greater Noida, India

Kanchan Naithani
Department of Computer Science and Engineering
Hemvati Nandan Bahuguna Garhwal University (A Central University)
Srinagar, India

Supriya Narad
Faculty of Science & Technology
Datta Meghe Institute of Higher Education and Research

B.C.M. Patnaik
Department of Management
School of Management
KIIT University
Bhubaneswar, India

Kashyap C. Patel
Faculty of Computer Applications
Ganpat University
Mehsana, India

Sandeep Poddar
Department of Research & Innovation
Lincoln University College
Petaling Jaya, Malaysia

Roheen Qamar
Department of Information Technology
Quaid-e-Awam University of Engineering, Science and Technology
Nawabshah, Pakistan

Angeles Quezada
Department of Systems and Computing
Institute of Tijuana
Tijuana, Mexico

Dina Rajput
Manager (Solution Architect)
Deloitte LLC
Phoenix, USA

Y. P. Raiwani
Department of Computer Science and Engineering
Hemvati Nandan Bahuguna Garhwal University (A Central University)
Srinagar, India

Vrinda Sachdeva
Department of CSE
ITS Engineering College
Greater Noida, India

Akash Sanghi
Department of CSE
Invertis University
Bareilly, India

Ipseeta Satpathy
Department of Management
KIIT School of Management
KIIT University
Bhubaneswar, India

Rajasekaran Selvaraju
Department of Information Technology,
University of Technology and Applied Sciences–Ibri
Ibri, Oman

Swati Sharma
Computer Science & Engineering Department
KIET Group of Institutions
Ghaziabad, India

Saima Siraj
Department of Information Technology
Quaid-e-Awam University of Engineering, Science and Technology
Nawabshah, Pakistan

Bhupinder Singh
Department- Sharda School of Law
School of Engineering and Technology
Sharda University
Greater Noida, India

Aditya Singhal
Department of ENT
Teerthanker Mahaveer Medical College & Research Centre, Moradabad,India

Shrikant Tiwari
Department of Computer Science and Engineering
Galgotias University
Greater Noida, India

Vivek Tomar
Department of Computer Engineering
Graphic Era (Deemed-to-be) University
Dehradun, India

Shailja Varshney
Department of CSE
ITS Engineering College
Greater Noida, India

Raj Kishor Verma
Department of CSE-DS
ABES Institute of Technology
Ghaziabad, India

Satya Prakash Yadav
School of Computer Science Engineering and Technology (SCSET)
Bannet University
Greater Noida, India

1 Introduction to Federated Learning

Transforming Collaborative Machine Learning for a Decentralized Future

Akash Sanghi, Rupanshi Agarwal, Gaurav Agarwal, Sachi Gupta, and Rajasekaran Selvaraju

1.1 INTRODUCTION

Federated learning (FL) is leading the way in revolutionary developments in machine learning, transforming the traditional field of centralized model training. Fundamentally, FL is a novel technique that enables a network of dispersed devices to jointly train machine learning models. FL prioritizes privacy above central processing of raw data, as is the case with traditional approaches. Individual devices—such as cellphones, edge devices, or other endpoints—contribute to model training under this novel paradigm without disclosing private information. We will explore the fundamentals of FL, its uses, and its potential to revolutionize the ever-evolving field of artificial intelligence as we delve into its depths.

1.1.1 Evolution of Federated Learning

In the broad timeline of machine learning evolution, FL is emerging as a transformative paradigm rooted in the historical development of intelligent systems. The story begins in the mid-twentieth century with the first attempts to simulate human cognition using rule-based systems and symbolic reasoning. However, neural networks only gained popularity in the late 1950s and 1960s, but they faced computational limitations that led to the rise of symbolic artificial intelligence in the following decades [1].

The resurgence of neural networks in the late 1980s marked a renaissance, driven by advances such as backpropagation. The beginning of the twenty-first century saw a paradigm shift as the community focused on leveraging big data and increasing computing power, bringing deep learning to the forefront with open frameworks, and democratizing access. However, the great success of centralized machine learning models has led to significant challenges: privacy issues, communication bottlenecks, and the inability to centralize disparate datasets. In response to these challenges, FL has made significant progress [2].

DOI: 10.1201/9781003489368-1

1.1.2 Decentralization as a Core Principle

FL introduced a decentralized approach to model training. At the heart of FL is the principle of decentralization [3]. Unlike traditional models that aggregate and process data on a central server, FL distributes the learning process. Individual devices or nodes, often mobile phones, act as decentralized clients that store and process data locally. This approach not only addresses privacy concerns associated with centralized models but also enables learning at the edge and supports real-time adaptation to evolving data streams.

In a federated mechanism, it is generally assumed that participants have the same identity and support the establishment of shared data policies. Because the data are not directly transferred, it does not affect data specifications or compromise user privacy.

Let's explain the fundamental idea of FL: Define N FL candidates that wish to train a global model by combining all of their data ($\{D_1, D_2,\ldots, D_N\}$). One popular way to train a model M_{sum} with a P_{sum} performance is to combine all the data and use the entire data $D=(D_1 \text{U } D_2 \text{U } D_3 \text{ U}\ldots\text{U } D_N)$.

FL is a platform in which candidates co-train a model M_{fed} using a P_{fed} performance, wherein no candidate divulges their personal information to other parties. Let μ be some non-negative value, thus the performance loss of the model based on FL is expressed as given in Equation 1.1.

$$\left|P_{\text{fed}} - P_{\text{sum}}\right| < \mu \tag{1.1}$$

FL learns by minimizing a loss function that is computed for each client with the use of a weighted aggregation technique. FL aims to minimize the subsequent objective function $f(x)$ given in Equation 1.2.

$$\min f(x) = \sum_{i=1}^{N} \frac{n_k}{n} F_k(x) \tag{1.2}$$

where N is the number of clients, n_k is the k_{th} client's data, and $F_k(x)$ is the k_{th} client's local objective function.

1.1.3 Key Components of Federated Learning

To understand FL, you need to examine its key elements. Decentralized devices or clients form the first pillar. These devices store local datasets, each providing a unique perspective without sharing the raw data. The central server coordinates the collaborative model training process, collecting model updates from each client and combining them into an overall model. This integrated approach ensures that confidential information remains with the customer, thereby mitigating data protection risks. Another key component is communication protocols, the lifeline of FL. Federated Averaging, a widely used protocol, allows a central server to aggregate model updates without direct access to the raw data, preserving privacy during the collaborative learning process [4].

1.1.4 The Future Landscape

As FL continues to evolve, its impact on the future of machine learning will become more significant. This paradigm not only addresses privacy concerns but also represents a scalable and efficient model for learning from decentralized data sources. Continuous research into new algorithms, robust security measures, and standardization efforts signify a commitment to refining and improving FL and pave the way for an era in which intelligence is not only distributed but also fundamentally respects the private life of the individual [1]. In essence, FL embodies the evolution of machine learning paradigms and offers a glimpse of a future where decentralized collaboration is not just a preference but a necessity.

1.2 ARCHITECTURE OF FEDERATED LEARNING

FL's architecture is a complex system that prioritizes data privacy while enabling cooperative model training across decentralized devices.

1.2.1 Client–Server Architecture

The FL architecture, encapsulated in the client–server model, represents a dynamic structure in which a central server organizes the joint training of several clients. This architecture not only promotes a privacy-friendly training environment but also fits perfectly in scenarios where data decentralization is required [5]. We explore the complexity of the FL client–server model by examining the key roles of the central server, decentralized clients, and the overall model in this innovative machine learning approach. Figure 1.1 shows the basic architecture of FL.

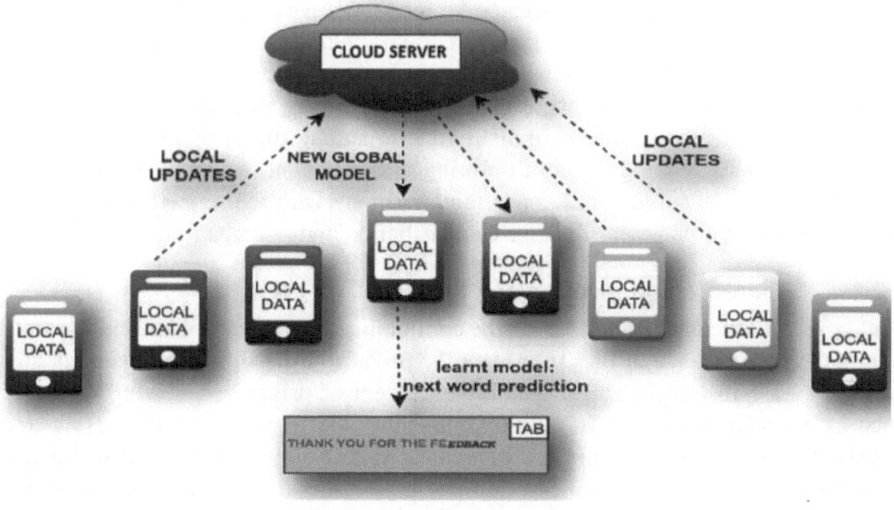

FIGURE 1.1 Basic architecture of federated learning.

1.2.1.1 System Bootstrapping

In order to engage with the client, the modeling task is first sent by the central server. The owner of the client data then suggests a collaborative modeling vision based on its own requirements. Each data holder begins the collaborative modeling process after coming to an agreement with other cooperative data holders and establishing the joint modeling vision. At last, the data holders receive the starting parameters from the central server.

1.2.1.2 Global Model

At the core of the FL architecture is the pivotal role played by the global model, serving as the comprehensive machine learning framework subjected to collaborative training. Its functions are integral to the fluid dynamics of FL. Initially, the global model is instigated into existence on the central server, marking the commencement of its collaborative learning journey. Subsequently, this model is disseminated selectively among chosen clients, each entrusted with a localized copy for individualized training on their unique datasets. The evolution of the global model unfolds through a nuanced process: the aggregation of model updates derived from the diverse insights and knowledge accumulated during local training across all participating clients. This iterative cycle of distribution, localized refinement, and amalgamation of updates ensures the continuous enhancement of the global model, embodying the collective intelligence harnessed from the entirety of the decentralized network. In essence, the global model encapsulates the amalgamated learning experiences, ensuring a robust and adaptive machine learning model refined collaboratively by the diverse perspectives of each contributing client.

1.2.1.3 Model Distribution and Aggregation

Within the FL choreography, the central server kindly distributes the most recent version of the global model to a selected group of clients during the distribution stage. This action prepares the audience for a cooperative symphony in which every client returns to the central server with its own melodies after completing careful local training. The central server then skillfully mixes these many model updates in the peaceful aggregation process that follows, creating an updated version of the global model. Federated averaging distinguishes among the symphony of aggregation techniques by leading a smooth mixing of participant insights. The decentralized ensemble's collective intelligence is in harmony with the refined global model that is ensured by this graceful dance of distribution and aggregation.

1.2.1.4 Secure Communication and Protocol

With a veil of efficiency and security, the communication protocol is the main attraction in the performance of FL. To protect the privacy of the sensitive model updates while they are in transit, an intricate arrangement of privacy-preserving methods, such as encryption, is used in the communication between the central server and clients. This secret waltz displays a sophisticated economy of movement while also guaranteeing the safety of confidential information. Like a well-rehearsed play, the communication protocol transmits only the refined ensemble of aggregated model changes, thereby minimizing data transfer. This graceful and secure movement of communication highlights the elegance and sophistication of the FL architecture [6].

1.2.1.5 Iterative Training Round

FL takes place over several iterations. A subset of clients is chosen, the global model is distributed, local training is conducted, model updates are generated, and aggregation is done for each round. The global model can change throughout subsequent rounds due to its iterative nature.

1.2.1.6 Model Update

Based on the aggregate result, the central server changes the global model once and sends it back to the data holders taking part in the modeling. The data holder updates the local model, initiates the subsequent partial computation, and assesses the updated model's performance simultaneously. The training is stopped and the joint modeling is finished when the performance is good enough. A global model that has been built will be stored on the central server for use in future classification or prediction tasks.

1.3 TYPES OF FEDERATED LEARNING

The three types of FL methods—horizontal federated learning (HFL), vertical federated learning (VFL), and federated transfer learning (FTL)—are often separated based on the extent of feature overlap in the client dataset and are intended to address distinct practical issues.

1.3.1 Vertical Federated Learning

A specific type of FL called VFL is intended to deal with situations in which data are divided horizontally among several groups or parties. While VFL works with data instances that have different features, HFL involves each participant holding data instances with the same features but different samples. When parties want to work together and train a machine learning model without giving up all of their feature sets, VFL is very pertinent [7].

1.3.1.1 Workflow of VFL

VFL's workflow carefully balances sensitive data protection with cooperative exchange. Figure 1.2 shows the architecture used in VFL. Each participating entity receives a global model first, and then distinct feature sets that are particular to that entity are recognized. Subsequently, each entity independently performs local training on its own feature-specific datasets, an important measure to protect the privacy of its data. After training, the resulting models are encrypted to protect the underlying data information and start a transfer focused on privacy.

The hub for collaboration is a central server that securely aggregates these encrypted model updates. Using a cycle of secure aggregation, local refinement, and dissemination, the global model can progressively advance and incorporate various

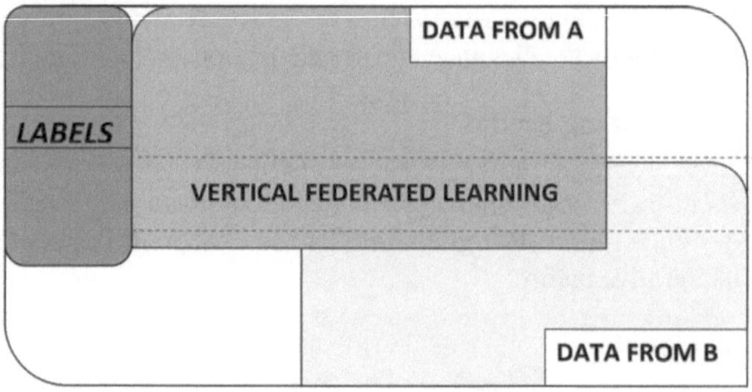

FIGURE 1.2 Vertical federated learning.

perspectives from all involved parties. This well-orchestrated process emphasizes the nuanced balance achieved in VFL, enabling collaborative model training without compromising the confidentiality of individual datasets.

1.3.2 Horizontal Federated Learning

A novel method of collaborative machine learning called "horizontal federated learning" was created to handle situations in which data are divided horizontally among several companies or parties [8]. Every participating entity in this paradigm has datasets with the same attributes, but different samples. The process makes sure that training models are done in a group setting while protecting each device's raw data privacy.

1.3.2.1 Workflow of HFL

HFL is a methodical process that carefully balances privacy protection with cooperative intelligence. Starting with a global model initialized by the central server, this paradigm distributes the model to all participating entities while preserving feature consistency across all devices. After that, each entity individually uses its dataset with the same features for local training, guaranteeing that the raw data's privacy is preserved on the local device. Figure 1.3 shows the architecture used in HFL.

After being encrypted, these updates are sent back to the central server, where they are combined to create an improved global model. The distribution, local refinement, and secure aggregation cycle that is iterative in nature allows the global model to gradually develop while incorporating knowledge from many datasets. The process highlights the intricate equilibrium attained in HFL, where cooperative model training flourishes without compromising the confidentiality of individual datasets.

1.3.3 Hybrid Federated Learning

The goal of hybrid FL is to combine the best features of FL and centralized learning strategies, minimizing the drawbacks of each while maximizing their respective benefits. In this hybrid architecture, a machine learning model is jointly trained by a central server and decentralized clients. The key differentiator is the

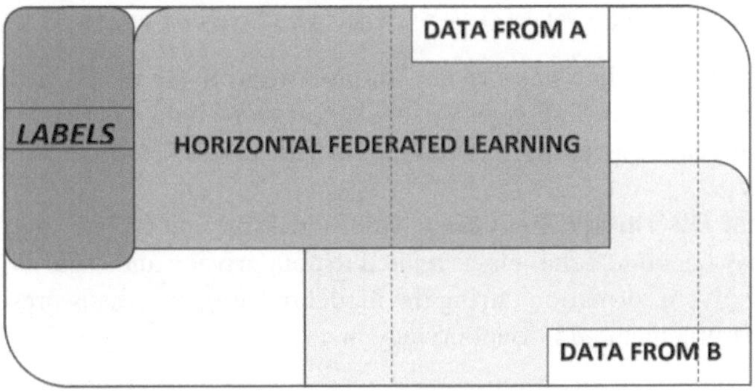

FIGURE 1.3 Horizontal federated learning.

deliberate fusion of periodic centralization of model updates with local training on individual devices.

1.3.3.1 Workflow

A global model is first initialized by the central server and then distributed to specific clients. These clients use their own private datasets to autonomously do local training, allowing them to make unique insights-based model improvements. Nevertheless, a portion of the model parameters or updates is routinely sent back to the central server for aggregation, as opposed to only depending on decentralized updates. The global model gains from the variety of insights across devices through this combination of local and centralized updates, all the while retaining a degree of coordination made possible by the central server.

1.4 FEDERATED LEARNING VS. DISTRIBUTED MACHINE LEARNING

Two major paradigms have arisen in the ever-changing field of machine learning, which are changing how data are used and the models are trained. Traditional distributed learning and FL are two different techniques, each with special traits, benefits, and applications. In this section, we embark on a comprehensive exploration of the differences between these two methodologies:

1.4.1 Data Decentralization vs. Centralized Training

FL trains models in a decentralized manner. The training process is dispersed among several local devices or servers. By using its own dataset, each device trains the model independently, involving local entities in the learning process. By combining model updates, the central server makes sure that the global model is improved without disclosing raw data [4].

Traditional distributed learning, on the other hand, uses a centralized training structure. Usually, a centralized server maintains the data that is then used to train the model. During the training phase, the central server is essential for sending complete models or large chunks of models across devices.

1.4.2 Privacy-Centric Approach vs. Centralization Challenges

FL places a strong emphasis on privacy. In order to mitigate the hazards associated with data exposure, it accomplishes this by keeping raw data localized on particular devices. FL appeals especially in situations where privacy and data sensitivity are critical issues [9].

Traditional Distributed Learning: Traditional distributed learning, by centralizing data, may encounter challenges related to data privacy and security. The risk of exposing sensitive information during the model training process is present, making it essential to implement additional privacy measures.

1.4.3 Optimized Communication vs. Communication Overhead

FL transmits only the most important model updates, optimizing transmission. This method lessens the need for continuous data transfer, which improves efficiency—particularly in settings with limited bandwidth.

Traditional distributed learning has a feature called communication overhead, which is the constant connection between devices and the central server. In contexts with limited network resources, this can result in higher bandwidth utilization and inefficiency.

1.4.4 Application and Use Cases vs. Common Usage

FL has a wide range of uses in industries including healthcare, banking, and personalized services where data privacy is a major concern. Its flexibility to adjust to decentralized cooperation makes it essential in developing domains.

Conventional distributed learning is frequently used in situations where the privacy of data is not as important. Based on well-established procedures, it remains applicable in situations when a centralized method complies with current standards.

To sum up, there are two different paradigms in machine learning approaches: FL and traditional distributed learning. FL is a shining example of innovation because of its decentralized, privacy-centric, and effective communication methods. In the meantime, traditional distributed learning—which is based on centralized procedures—continues to be relevant in situations when a centralized strategy is appropriate. The particular needs and goals of a particular application or industry determine which of these approaches is best.

1.5 PRIVACY IN FEDERATED LEARNING

One important aspect of FL that tackles the difficulties of decentralized collaboration in machine learning is security. Safeguarding sensitive data's security, integrity, and privacy becomes critical as we negotiate the complex terrain of FL. The development of secure FL is motivated by the desire to reduce potential risks and foster confidence in the cooperative learning process. This includes using cryptographic techniques to secure communication channels and putting in place privacy-preserving algorithms. This exploration of FL security reveals the creative approaches and tools used to protect private data, opening the door for responsible and safe applications in a variety of industries, including banking and healthcare. Here are some security aspects of FL.

1.5.1 Decentralized Model Training

The fundamental idea of FL is decentralized model training. This method relies heavily on local devices, like edge or smartphone devices, to train machine learning models. Take a look at a healthcare application, for example, where patient data are housed on individual devices. FL eliminates the need to centralize sensitive patient data by enabling direct model training on these devices. Individual health records are kept private and secure because the central server only gets aggregated model updates.

Considering the example of healthcare, because FL centralizes patient data for model training, it presents a serious challenge for healthcare applications in the lack of privacy-preserving techniques. Due to the centralized method, there is a vulnerability that exposes sensitive data that is part of each patient's medical record [7]. The result of this centralization is an increased possibility of these private and sensitive documents being accessed without authorization.

1.5.2 Secure Model Aggregation

The privacy of FL depends on secure aggregation techniques. Federated averaging with encryption is one noteworthy method. Before being transferred to the central server, model updates from local devices are encrypted in this process. These encrypted updates are combined by the central server, which then decrypts the outcome. This provides an additional degree of security to the collaborative learning process by preventing the central server from accessing the specifics of individual model changes, even throughout the aggregate phase. This is particularly important for financial applications where model aggregation requires the protection of transaction data [10].

FL programs face a problem when model updates are aggregated without encryption as a precaution. This situation presents a significant danger since it leaves participating devices' individual contributions open to possible unauthorized access. The potential for sensitive data to leak during the aggregation phase arises from aggregating model updates without encryption.

1.5.3 Homomorphic Encryption

A cryptographic technique called homomorphic encryption makes it possible to do calculations on encrypted material without requiring its decryption. This means that, in the context of FL, local devices can compute model updates based only on the encrypted data they get from the central server. This sophisticated privacy-preserving technique is essential in situations such as collaborative research where data are contributed by several universities [2,3]. Throughout the FL process, the specifics of each institution's contributions are kept private thanks to homomorphic encryption.

The privacy of cooperative efforts may be at risk when computations are carried out without encryption due to the increased susceptibility to data disclosure. This greater vulnerability to data disclosure highlights the critical requirement for strong privacy safeguards in cooperative projects. The integrity of collaborative data-driven processes is jeopardized in the absence of encryption, highlighting the crucial role privacy-preserving strategies play in strengthening the security of FL initiatives.

1.5.4 Differential Privacy

To avoid the extraction of specific data points, controlled noise is injected into computations as part of differential privacy. This method is used in FL to safeguard individual model updates. Think of a teaching application that uses student performance information to facilitate group learning. Differential privacy protects each participant's privacy by preventing an opponent from identifying the precise contribution of any one student, even if they manage to obtain access to the global model.

The problem arises when individual data points are not protected in the FL model updates environment. This lack of protection for individual data points presents a danger because it allows for the possible identification of participant contributions. As a result of this vulnerability, individuals' privacy is jeopardized because sensitive information may be identified through the disclosure of particular contributions. To reduce these dangers and guarantee participant privacy and the confidentiality of individual contributions in the collaborative environment of FL initiatives, strong privacy safeguards must be put in place.

1.5.5 Federated Averaging with Weighted Updates

Weighted updates are made possible by federated averaging, a crucial FL technique. In other words, without disclosing the details of their own datasets, certain local models contribute more than others to the global model update. Federated averaging with weighted updates guarantees that, in a manufacturing context where numerous plants contribute data, each plant's performance is recognized in terms of improving the overall model. On the other hand, proprietary information regarding the production procedures of each plant is kept private.

The problem occurs when updates in FL are weighted equally, which could lead to the disclosure of confidential data. The reason for this problem is that individual participant contributions are not given varied weights. This equal weighting has the unintended consequence of increasing the risk of divulging private production techniques that are unique to each participant.

1.6 SECURITY CHALLENGES AND SOLUTIONS IN FEDERATED LEARNING

Decentralized model training presents distinct security problems when navigating the FL ecosystem. Innovative methods are required to protect sensitive data on distributed devices. Strengthening FL against possible attacks requires addressing cryptographic flaws, providing secure communication routes, and putting strong privacy-preserving algorithms in place. Below are some security challenges in FL and give some solutions to these challenges.

1.6.1 Security Challenges

FL presents a number of complex security concerns that require thorough solutions, despite its promise of being decentralized and privacy-preserving. Several challenges are:

1.6.1.1 Eavesdropping Attack

The potential for eavesdropping attacks, in which attackers could intercept communications between devices and the central server while the model is updating, is a major cause for concern. The necessity of putting strong encryption techniques in place, like secure multi-party computation (SMPC), to protect against unauthorized access is highlighted by the possible exposure of sensitive data.

1.6.1.2 Sybil Attack

Another difficult obstacle is the possibility of Sybil attacks, in which malevolent actors can fabricate several false identities in order to influence the FL procedure. Such attacks have the potential to create privacy breaches and impair the trained model's quality. Advanced participant verification methods, reputation systems, and trust-based algorithms become essential to identify and stop malicious entities in order to reduce this risk [8].

1.6.1.3 Byzantine Attack

While a byzantine assault focuses on the cooperation of numerous users in a distributed learning environment, a poisoning attack primarily considers an attack on a single user or a server. Byzantine users in FL are malicious attackers who have control over many clients. Because of unstable communication links, faulty hardware, or even malicious attacks, byzantine users are able to upload bogus data, which might cause attackers to modify the global model and prevent it from converging. Byzantine attacks are a type of active internal attack technique that can readily disrupt regular client–server communication in FL. Thus, the main concerns in FL security research have been the prevention and identification of byzantine attacks.

1.6.1.4 Poisoning Attack

A serious security risk is represented by "model poisoning" attacks, in which attackers introduce malicious data into a model while it is being trained in order to alter its behavior. Inaccurate forecasts resulting from the breach of model integrity give rise to apprehensions over possible exploitation. Attackers who engage in data poisoning deliberately alter source tags or other features of local client data in order to affect how well-federated models train. Detecting data poisoning is a difficult issue because of the distributed nature of FL deployment, which makes it difficult to ascertain if a client is engaging in FL in good faith. Compared to model poisoning, data poisoning is a more subtle form of assault. The central server is unclear about participants' personal information and the local training procedure, despite the idea of protecting participants' privacy and security. Such attacks are frequently challenging to identify since the attacker can only interfere with the FL system client's data-gathering process without having the ability to directly damage the FL-trained model.

1.6.2 Security Solutions

Two techniques are used to prevent this attack:

1.6.2.1 End-to-End Encryption

Encrypting every communication channel between the central server and the participating devices is a necessary step in implementing end-to-end encryption. This refers to encrypting data as it travels from local devices to the central server and back in FL. By ensuring that only the intended recipients—devices and the server—possess the decryption keys, this encryption keeps any malevolent eavesdroppers from deciphering the data being transferred. Sophisticated encryption techniques, such as symmetric or asymmetric key cryptography, aid in creating a private and secure communication channel [11].

1.6.2.2 Secure Multi-Party Computation

SMPC is still another essential element in the defense against monitoring others. By using this method, computation on encrypted data is possible without disclosing the actual data to any parties. SMPC guarantees that model updates from several devices can be combined in FL without disclosing the individual updates. The central server can aggregate the model updates without knowing the specifics of each update by executing computations on encrypted data. This provides an extra degree of privacy protection, making it impossible for any potential eavesdropper to obtain insightful information even if they manage to intercept the conversation.

In FL, end-to-end encryption and SMPC work together to create a strong defense against eavesdropping attempts. By protecting the privacy of the collaborative learning process in addition to the secrecy of the sent data, these cryptographic techniques greatly increase the process' resistance to outside eavesdropping and unauthorized access.

1.6.2.3 Solution to Sybil Attack

A complex fusion of participant verification techniques is needed to mitigate Sybil assaults in FL and guarantee the authenticity of the devices taking part in the cooperative learning process.

The first line of defense against Sybil attacks is the implementation of strong device identity checks. A unique identifier is issued to each participating device, and this identifier is thoroughly checked throughout the FL process. To make sure the device is authentic, methods like hardware-based attestation or cryptographic signatures can be used. The FL system may identify and reject attempts by malevolent actors to introduce numerous falsified identities by verifying the unique identification of each contributing device.

1.6.2.4 Solution of Byzantine Attack

Mitigating byzantine attacks in FL involves strategies to detect and handle malicious participants who intentionally provide incorrect or malicious updates.

To improve the security of the FL model aggregation, use threshold cryptography. In order to reconstruct the cryptographic keys, a minimum number of participants must be involved in the distribution of the keys among numerous parties. It is a more secure aggregation procedure since byzantine attackers would have to compromise a sizable fraction of the participants to alter the model update.

1.6.2.5 Solution to Poisoning Attack

Using techniques to identify and offset the effects of malevolent actors injecting hostile data is part of mitigating poisoning threats in FL.

Examine how homomorphic encryption can be used to confirm model updates while keeping the real data hidden. Because homomorphic encryption permits computations on encrypted data, changes can be validated by the central server without requiring access to the raw data. This method helps identify and manage tainted contributions while strengthening the FL process' security.

1.7 APPLICATIONS OF FEDERATED LEARNING ACROSS INDUSTRIES

Applications for FL span a diverse range of domains and scenarios, showcasing its versatility and relevance in various industries.

1.7.1 Federated Learning in Intelligent Healthcare System

In the medical field, patients' privacy is usually more important to people. Since patient information cannot be exposed and hospitals are unable to transmit it secretly, using the FL approach to secure pathological information is a good idea. When using FL for healthcare applications, the server can be the government agency's data center, and each hospital represents a unique client. Each hospital downloads the initial common model from the server in this manner and then uses local data to train it. Once finished, the model data are encrypted and transferred to the server, whereupon the FL algorithm updates the global model.

Models for detecting different diseases have been constructed using FL. It demonstrated strong detection performance and resolved issues with COVID-19 detection, DNA sequencing, and other issues in the context of the current, severe epidemic [12].

In the field of cancer rehabilitation medicine, this model serves as a kind of guide. Patient data is sensitive information in the medical industry, and many patients are hesitant to give up their personal information for services like intelligent testing. The security and efficacy of FL technology in this case offer workable recommendations for the use of intelligent services in the medical industry. Combining FL with the medical sector allows for the development of a disease detection model that performs well without compromising patient privacy, which advances the medical industry's intelligence process.

1.7.2 Federated Learning in Recommendation System

In order to recommend content or services, traditional recommender systems need to gather and evaluate a lot of user data [13]. However, users frequently do not consent to this operation because of privacy security concerns and the data island issue. The advent of FL offers a viable solution to this issue, particularly in the area of cross-domain recommendation, which includes social media, online shopping, short video recommendations, and other platforms. Federalization of the recommendation algorithm based on collaborative filtering, federalization of the recommendation algorithm based on deep learning, and federation of the recommendation algorithm based on meta-learning are common components of federated recommendation systems. To deliver high-quality material to users, the vertical federated recommendation architecture allows the server to access user data from other domains in addition to the data created by users in this field.

1.7.3 Federated Learning Empowers Smart Cities

Governmental organizations, commercial businesses, and private citizens all work together to develop the city, and when data privacy laws are strengthened, data centers are no longer allowed to freely transmit data to other parties. This restricts data exchange on several levels. With the advent of FL, the issue of data isolation is effectively resolved, and data integration occurs throughout the city to improve urban applications and citizen mobility services. When FL is used in the development of smart cities, both public and commercial organizations will function as clients or data owners. And the service will be the urban data management center. To address a range of issues related to the development of smart cities, the service will collaborate with various municipal levels to train a global model. It will handle numerous tasks in order to offer citizens more convenient urban services.

1.7.4 Federated Learning in Finance

In the banking industry, FL is used to improve risk assessment and fraud detection models. Financial institutions can work together to enhance fraud prevention algorithms without disclosing private client information, maintaining security protocols, and maintaining confidentiality [14].

1.7.5 Federated Learning in Edge Computing

Personal mobile devices' increasingly sophisticated functionalities are a result of the information technology industry's rapid evolution, and thus effectively ensures the secure processing of distributed edge data in Florida. Edge devices may train a similar machine learning model more effectively by combining FL and edge computing, all without transferring source data to cloud devices.

1.7.6 Federated Learning in Intrusion Detection

One often used technology for network security defense is intrusion detection. In order to detect network intrusions more accurately these days, deep learning techniques are typically employed. However, a lot of data is typically needed for deep learning of machine learning model training; in fact, some universities do not have access to network intrusion detection datasets, which is detrimental to the model's training. Privacy concerns will also arise from uploading data from all institutions for centralized in-depth training and learning. When FL is used in intrusion detection, it not only guarantees a high detection rate but also safeguards the privacy of diverse institutions' data and reduces the issue of data islands that intrusion detection faces.

1.7.7 Federated Learning in Education

FL aids in the creation of customized learning models in the field of education. Student devices work together to improve overall educational outcomes, customize instructional content recommendations, and accommodate different learning styles.

1.8 QUANTUM FEDERATED LEARNING

The novel intersection of quantum computing and FL is represented by quantum FL. An explanation of the idea is given below:

1.8.1 Integrating Federated Learning with Quantum Computing

Utilizing the ideas of quantum physics, quantum computing enables computations. Due to their ability to exist in several states at once, quantum bits, or qubits, offer special computational advantages over classical computing and allow for parallel processing [15].

By incorporating ideas from quantum computing, quantum FL builds on conventional FL and improves specific areas of the learning process. In traditional FL, only aggregated updates are communicated with a central server after machine learning models are trained locally on each device. Through the introduction of quantum features, quantum FL incorporates aspects of quantum computing into this decentralized architecture, potentially enhancing efficiency and security. Several crucial elements help to understand the integration:

1.8.1.1 Quantum Superposition in Model Exploration

Quantum bits, or qubits, can exist in several states at once thanks to quantum superposition. This characteristic allows several model parameters to be explored simultaneously by quantum devices in the context of quantum FL. Quantum superposition enables concurrent exploration, potentially speeding up the convergence of the learning process, as opposed to processing one set of parameters at a time [16].

1.8.1.2 Entanglement for Synchronized Collaboration

Regardless of the physical distance separating two particles, entanglement is a unique quantum phenomenon in which particles become correlated such that the state of one particle affects the state of its entangled partner. Entanglement can be used in quantum FL to synchronize updates among dispersed quantum devices. This makes sure that every device makes a consistent and logical contribution to the model's evolution, which improves learning efficiency overall.

1.8.1.3 Quantum Properties Improve Privacy Preservation

The quantum features included in quantum FL further strengthen the privacy-preserving aspect of FL. A certain amount of uncertainty is introduced by quantum computing, which makes it harder for individual devices to distinguish the precise contributions or information from others. This enhances security and complies with FL's need for secrecy [17].

1.8.1.4 Hybrid Classical-Quantum Systems

Hybrid classical quantum systems are frequently used in quantum FL applications in practice. Some functions, such as investigating model parameters or creating updates, are performed by quantum computing, while classical components oversee the entire FL process, including device coordination and communication.

1.8.1.5 Possibility of Quantum Speedup

In certain situations, quantum computing may be able to solve problems more quickly. Quantum FL investigates how quantum processing might speed up specific parts of model training and optimization tasks inside the FL framework, albeit the amount of this speedup depends on the specific problem and is influenced by factors like the quantum algorithm utilized.

1.8.2 Security Enhancement by Integrating Quantum Computing

FL can benefit from the integration of quantum computing principles in a number of ways that will improve security. Quantum computing introduces an inherent amount of uncertainty because of the concepts of superposition and measurement, which amplifies privacy concerns. In FL, privacy can be enhanced by making use of this ambiguity. Quantum characteristics can be used by devices participating in quantum FL to mask their unique contributions to the model updates, hence increasing the difficulty for hostile actors to retrieve sensitive data. The following main ideas clarify how more security is provided by combining quantum FL with traditional FL [17].

1.8.2.1 Quantum Key Distribution

Through the use of quantum key distribution (QKD) protocols, quantum communication presents the possibility of secure key distribution. QKD uses the fundamentals of quantum mechanics to generate cryptographic keys that are immune to interception efforts. By incorporating QKD into FL's communication channels, the danger of unauthorized access to private model updates can be decreased and key exchange security can be improved.

1.8.2.2 Quantum-Secure Cryptography

The traditional cryptographic algorithms used in secure communication may be threatened by quantum computing. The FL framework's integration of quantum-safe or post-quantum cryptography approaches guarantees the system's security against assaults that utilize quantum algorithms. By doing this, the FL system is future-proofed against hypothetical quantum computing breakthroughs that would jeopardize established cryptographic techniques.

1.8.2.3 Quantum-Resistant Algorithms for Secure Computing

FL computing can be enhanced by including quantum-resistant algorithms, which are engineered to withstand attacks from quantum computers. These algorithms offer an extra degree of protection against future attacks on traditional cryptography techniques as quantum computers gain power.

1.8.2.4 Decentralized Trust with Quantum Verification

The FL network can be made to function with decentralized trust by utilizing the concepts of quantum computing. Devices can demonstrate the accuracy of their calculations using quantum verification techniques without disclosing the underlying data. This preserves the privacy of individual datasets while improving the reliability of participants in the FL process.

Quantum FL improves the system's overall robustness and privacy by using these quantum security mechanisms. It offers a more secure basis for cooperative machine learning tasks, safeguarding private data throughout the FL process and guarding against the risks brought by developments in quantum computing.

1.9 FEDERATED LEARNING'S FUTURE DIRECTIONS

The future of FL is multidimensional. This section explores the various possible future directions of FL.

1.9.1 Distributed AI Environments

It is anticipated that FL would aid in the growth of decentralized AI ecosystems. Across many platforms and devices, collaborative model training will enable the development of networked, intelligent systems that can adapt and learn in real-time.

1.9.2 Advances in Edge Intelligence

The progression of FL and edge computing integration is expected to be substantial. By bringing intelligence closer to the data source, this trend will improve real-time decision-making, lower latency, and make better use of edge devices [10].

1.9.3 Improved Methods for Preserving Privacy

Prospective advancements in FL are anticipated to center around optimizing privacy-preserving methodologies. It is anticipated that more sophisticated cryptographic techniques, differential privacy, and FL with homomorphic encryption will develop to offer even higher guarantees of security and confidentiality.

1.9.4 Collaboration across Domains

FL is expected to promote more collaboration across domains. Collaborative model training can be utilized by industries, research institutions, and organizations from various sectors to capitalize on insights from heterogeneous datasets, hence promoting innovation and resolving intricate problems.

1.9.5 Federated Learning in 6G Networks

As communication networks advance to 6G, FL will probably be essential to maximizing network performance. This will allow devices to work together to enhance user experience, efficiency of the network, and communication protocols [2,3].

1.9.6 Modular Learning Frameworks

Future developments in FL frameworks might see the creation of more adaptive learning models. There will be a growing number of models that can adapt dynamically to variations in data distribution, device capabilities, and environmental conditions.

1.10 CONCLUSION

This new method of machine learning, with its fundamental concepts, unique features, and possible applications, has been thoroughly covered in the introduction to FL. We have illustrated the distinct benefits of FL in decentralized and privacy-preserving contexts through a thorough comparison with conventional machine learning techniques.

We have examined different forms of FL, including vertical and horizontal FL, and have discussed their unique benefits and use cases. With an emphasis on its versatility and potential for revolutionary impact, the conversation has expanded to include FL's real-world applications in finance, healthcare, and other industries. The necessity of privacy preservation and safe communication in FL systems has been emphasized by security issues. A notable investigation on the potential combination of FL and quantum computing has expanded the field and hinted at the fascinating opportunities that result from combining these two state-of-the-art technology. Unprecedented gains in computational efficiency and safe data sharing may result from the combination of quantum computing and FL.

To sum up, this chapter's exploration of the many facets of FL has revealed its core ideas, many applications, security concerns, and forward-thinking endeavors. The development of FL points to a dynamic future that will be influenced by further study, advancements in technology, and cooperative efforts. The coming together of quantum computing with FL is evidence of machine learning's ongoing pursuit of innovation. The chapter lays the groundwork for an era where FL takes center stage in redefining the landscape of privacy-aware and collaborative artificial intelligence by identifying problems and seeking solutions.

REFERENCES

1. Kishor, K. (2022). Communication-efficient federated learning. In: Yadav S. P., Bhati B. S., Mahato D. P., Kumar S. (eds) *Federated Learning for IoT Applications*. EAI/Springer Innovations in Communication and Computing. Springer, Cham. https://doi.org/10.1007/978-3-030-85559-8_9.
2. Kishor, K. (2022). Personalized federated learning. In: Yadav S. P., Bhati B. S., Mahato D. P., Kumar S. (eds) *Federated Learning for IoT Applications*. EAI/Springer Innovations in Communication and Computing. Springer, Cham. https://doi.org/10.1007/978-3-030-85559-8_3.
3. McMahan, H. B., Moore, E., Ramage, D., Hampson, S., & Arcas, B. A. (2017). Communication-efficient learning of deep networks from decentralized data. In: *Proceedings of the 20th International Conference on Artificial Intelligence and Statistics (AISTATS)*, Fort Lauderdale, Florida, vol. 54, pp. 1273–1282.
4. Konečný, J., McMahan, H. B., Yu, F. X., Richtárik, P., & Suresh, A. T. (2016). Federated learning: Strategies for improving communication efficiency. arXiv preprint arXiv:1610.05492.
5. Yang, Q., Liu, Y., Chen, T., & Tong, Y. (2019). Federated machine learning: Concept and applications. *ACM Transactions on Intelligent Systems and Technology (TIST)*, 10(2), 12.
6. Kishor, K., Nand, P., & Agarwal, P. (2018). Secure and efficient subnet routing protocol for MANET. *Indian Journal of Public Health*, 9(12), 200. https://doi.org/10.5958/0976-5506.2018.01830.2.
7. Bonawitz, K., Eichner, H., Grieskamp, W., Huba, D., Ingerman, A., Ivanov, V., Kiddon, C., Konečný, J., Mazzocchi, S., McMahan, B., & Van Overveldt, T. (2019). Towards federated learning at scale: System design. arXiv preprint arXiv:1902.01046.
8. Li, T., Sahu, A. K., Talwalkar, A., & Smith, V. (2020). Federated learning: Challenges, methods, and future directions. *IEEE Signal Processing Magazine*, 37(3), 50–60.
9. Kishor, K. (2023). Impact of cloud computing on entrepreneurship, cost, and security. In: Kishor K., Saxena N., Pandey D. (eds) *Cloud-Based Intelligent Informative Engineering for Society 5.0*, 1st edition, pp. 171–191. CRC Press, New York. ISBN: 9781003213895. https://doi.org/10.1201/9781003213895-10.
10. Kishor K., & Nand, P. (2023). Wireless networks based in the cloud that support 5G. In: Kishor K., Saxena N., Pandey D. (eds) *Cloud-Based Intelligent Informative Engineering for Society 5.0*, 1st edition, pp. 23–40. Chapman and Hall/CRC, New York. ISBN: 9781003213895. https://doi.org/10.1201/9781003213895-2.
11. Shokri, R., & Shmatikov, V. (2015). Privacy-preserving deep learning. In: *Proceedings of the 22nd ACM SIGSAC Conference on Computer and Communications Security*, pp. 909–990.
12. Sharma A., Jha N., & Kishor K. (2022). Predict COVID-19 with chest X-ray. In: Gupta D., Polkowski Z., Khanna A., Bhattacharyya S., Castillo O. (eds) *Proceedings of Data Analytics and Management*. Lecture Notes on Data Engineering and Communications Technologies, vol. 90. Springer, Singapore. https://doi.org/10.1007/978-981-16-6289-8_16.
13. Li, T., Sanjabi, M., & Smith, V. (2021). Federated learning: Challenges, methods, and future directions. arXiv preprint arXiv:2104.07535.
14. Jain, A., Sharma, Y., & Kishor, K. (2020). Financial supervision and management system using Ml algorithm. *Solid State Technology*, 63(6), 18974–18982. https://solidstatetechnology.us/index.php/JSST/article/view/8080.

15. Kishor, K. (2023). Chapter: 17 Application of quantum computing for digital forensic investigation. In: Yadav S. P., Singh R., Yadav V., Al-Turjman F., Kumar S. A. (eds) *Quantum-Safe Cryptography Algorithms and Approaches: Impacts of Quantum Computing on Cybersecurity*, pp. 231–248. De Gruyter, Berlin, Boston. https://doi.org/10.1515/9783110798159-017.
16. Kishor, K. (2023). Chapter: 10 Study of quantum computing for data analytics of predictive and prescriptive analytics models". In: Yadav S. P., Singh R., Yadav V., Al-Turjman F., Kumar S. A. (eds) *Quantum-Safe Cryptography Algorithms and Approaches: Impacts of Quantum Computing on Cybersecurity*, pp. 121–146. De Gruyter, Berlin, Boston. https://doi.org/10.1515/9783110798159-010.
17. Kishor, K. (2023). Chapter: 12 Review and significance of cryptography and machine learning in quantum computing. In: Yadav S. P., Singh R., Yadav V., Al-Turjman F., Kumar S. A. (eds) *Quantum-Safe Cryptography Algorithms and Approaches: Impacts of Quantum Computing on Cybersecurity*, pp. 159–176. De Gruyter, Berlin, Boston. https://doi.org/10.1515/9783110798159-012.

2 Applications, Challenges, and Opportunities for Federated Learning in 6G

Roheen Qamar and Saima Siraj

2.1 INTRODUCTION

Federated learning (FL), often called collaborative education, a subfield of machine learning, concentrates on situations in which a number of entities (commonly called clients) work together to train a model while maintaining the decentralization of their data. In contexts involving machine learning, this is not the case. Where data are centralized. Data heterogeneity is one of the main features that set FL apart. There is no assurance that the data samples kept by each client are disseminated independently and uniformly since the customers' data are decentralized. FL, often called collaborative learning, is a branch of machine learning that focuses on scenarios where a number of entities (commonly called clients) work together to train a model while maintaining the decentralization of their data [1]. FL is typically driven by and concerned with concerns like data access rights, data reduction, and privacy. Applications for it may be found in many different research domains, such as the Internet of Things (IoT), military, telecommunications, and medicines.

FL aims to train a machine learning algorithm (deep neural networks, for instance) without requiring explicit data sample sharing, using a large number of local datasets maintained on local nodes. The basic idea is to train local models on local data samples and regularly exchange parameters (e.g., weights and biases of a deep neural network) across these local nodes in order to produce a global model shared by all nodes. As 5G communication networks are being widely installed worldwide, researchers and business leaders are starting to look into 6G communications as a possible replacement for 5G. Most people concur that pervasive artificial intelligence (AI) will form the basis of 6G. This will make data-driven machine learning (ML) solutions possible in networks with vastly dispersed and diverse users. However, traditional ML techniques need centralized data collection and processing by a single server due to the increasing privacy concerns; this is preventing large-scale implementation in day-to-day life. A cutting-edge distributed AI method that puts privacy protection first is called FL. Given its potential to allow pervasive AI in 6G, it is considered vital for a variety of wireless applications.

Figure 2.1 shows FL architecture. The central server, decentralized clients, and model of FL's client–server paradigm are examined to comprehend this unique ML approach.

DOI: 10.1201/9781003489368-2

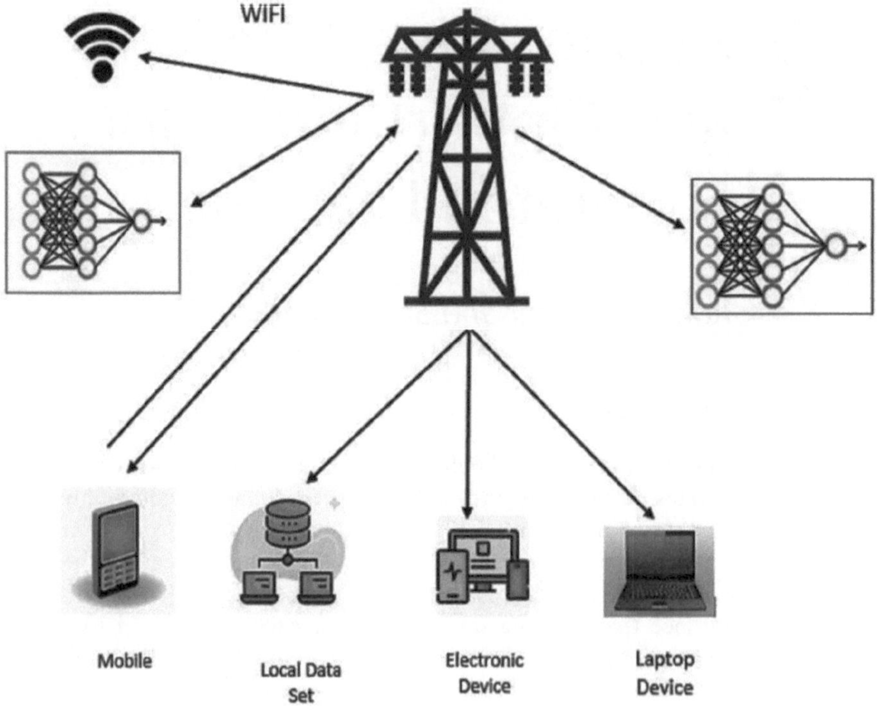

FIGURE 2.1 Federated learning architecture.

ML often stores training data in a data center for analysis and inference using centralized algorithms. However, privacy issues and wireless network limitations limit the devices' ability to submit data to a parameter server. FL may solve these challenges. Devices can train a shared ML model without sharing data. FL is extensively examined in 6G wireless networks in this article. The first section of the article covers wireless frequency modulation circumstances. FL's wireless communications applications are then discussed. Novel media, new services, and advanced infrastructures will be transformed by 6G. In various industries, the 6G system is expected to use advanced AI. The system will quickly gather, communicate, and learn to create cutting-edge applications and intelligent services [2].

6G stands for sixth-generation mobile connection, as the name would imply. Although the precise shape of 6G remains unknown until it is standardized, it is not too soon to make assumptions about the technology it will contain and the features it will possess. 6G networks will operate using radio waves at higher frequencies. Dr. Mahyar Shirvanimoghaddam, senior lecturer at the University of Sydney, says it's too early to anticipate 6G data speeds, even if wireless communications can reach 1 terabyte per second. This approximation applies to short-duration data transmissions across limited distances. South Korean company LG introduced adaptive beam shaping in 2021.

It's clear that the backend modifications done to mobile networks to support 5G will help 5G. More antennas on radio networks have been added by operators to improve signal strength, particularly inside. Additionally, edge computing and cloud technologies allow data to be processed closer to customers, even at the mast level, resulting in much-reduced latency. This base will be strengthened by 6G, which will bring additional features much beyond 5G's limitations. Considering the emergence of mobile connection as a front in geopolitical conflicts, it is hardly surprising that governments everywhere want to be at the forefront of the developing area of 6G technology. Numerous research initiatives, both government and privately sponsored, are underway globally. Notable among them is the €251 million "6Genesis" project located in Oulu, Northern Finland, a city historically linked to the advancements in mobile network technology. A 6G satellite has already been sent into orbit by China as a result of its research activities, while Samsung and Nokia are spearheading initiatives in South Korea and Europe. The main project in the United Kingdom is located at the University of Surrey's 6G Innovation Centre (6GIC) [2].

Figure 2.2 shows 6G services, which aim to revolutionize communication between consumers, networks, and devices by enhancing joint communication, sensing, and positioning, enabling new use cases, and optimizing industry operations.

More specifically, three areas—new media, new services, and new infrastructures—are where 6G will transform technology. The 6G system is expected to leverage advanced AI technology across several domains. It will quickly and efficiently collect, transmit, and learn data to offer a variety of innovative applications and services that are intelligent.

It is conceivable that 6G and 5G will cooperate for a while similar to how 5G and 4G will coexist for some time as they both use the same core network. Although there is still much to be done in the development of 5G technology, the 6GIC estimates that it will take 20 years to mature, suggesting it won't go extinct before at least 2040.

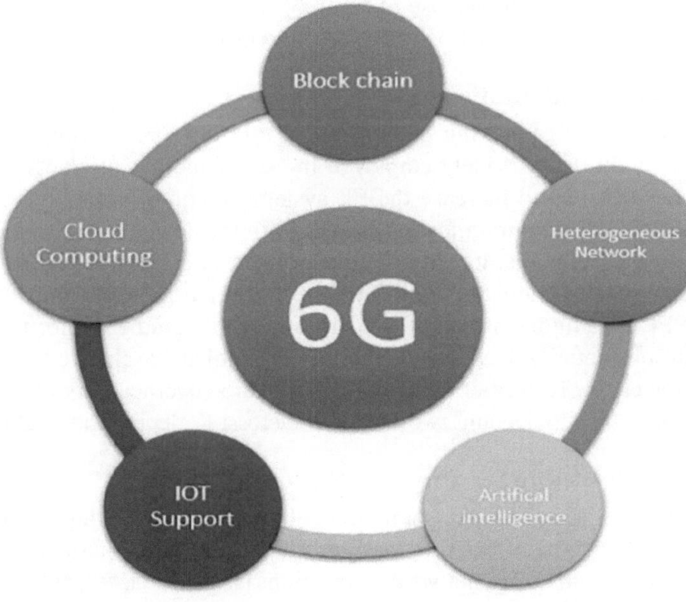

FIGURE 2.2 6G services.

In FL, localized data are maintained while statistical models are trained across dispersed devices or compartmentalized data centers, like cell phones or hospitals. Training in heterogeneous and possibly large-scale networks presents new difficulties that call for a fundamental shift from conventional methods for distributed optimization, extensive ML, and data analysis that protects privacy. In this paper, we discuss the unique characteristics and challenges of FL, provide an overview of current approaches, and propose many directions for future study that will be of interest to several teams of researchers [3].

Smart watches, wearable technology, and driverless cars are just a few examples of the contemporary dispersed networks that produce vast amounts of data every day. It is becoming more and more appealing because of the device's ability to store data locally and push network processing to the limit raising computational capability and worries about sending sensitive information [4].

The concept of edge computing is not new. In fact, fog computing, computing at the edge, and query processing in sensor networks have all been areas of study for decades, with a focus on calculating basic questions across dispersed, low-powered sensors. However, it is feasible to make use of improved local resources on each device as the computing and storage capacities of the devices inside dispersed networks increase. In addition, user-generated data must stay on local devices due to privacy issues around the transmission of raw data. Due to this, FL is becoming more and more popular [5].

2.1.1 Fundamental Difficulties for Federated Learning in 6G

This section presents the fundamental obstacles of FL, which serve as the primary roadblock prior to the widespread use of FL in 6G applications [6].

Figure 2.3 shows FL in 6G FL issues and concerns.

2.1.1.1 Costly Communication

With thousands of devices taking part in model training for FL, connectivity is a major barrier for FL to be widely employed in 6G. Numerous attempts have been made in earlier research to increase the FL system's communication effectiveness. Furthermore, achieving communication in sync with the device's local calculation is a challenge for FL networks. It will take the creation of a communication-efficient method to convince 6G networks with massively diverse networks and devices to embrace the FL paradigm. Instead of transmitting whole gradient information, this approach can substantially decrease the number of gradients exchanged between the devices and the cloud. To further reduce communication overhead in this scenario, it is imperative to take into account two essential factors: Reducing the total [7, 7a, 8].

2.1.1.2 Issues Related to Security

In the context of 6G networks, the communication capabilities of individual devices inside the network can be affected by various factors, including hardware improvements (such as CPU and GPU), network connections (such as 4G, 5G, 6G, and Wi-Fi), and the availability of energy (battery level). Over a broader geographic region.

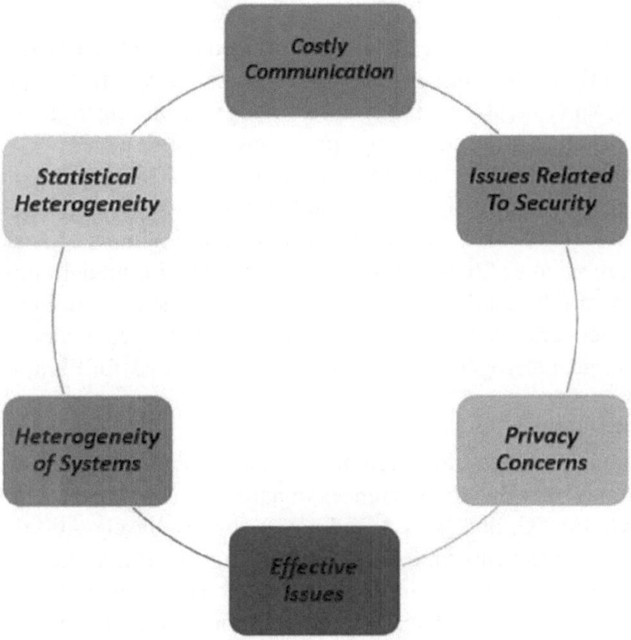

FIGURE 2.3 Federated learning in 6G.

It goes without saying that the heterogeneity of the devices' systems will cause some misunderstandings and errors in the FL model and 6G network. Furthermore, the FL can contain malfunctioning equipment, which might lead to the system's Byzantine failure. Similar to this, attackers may utilize active learning-based assaults on heterogeneous devices (such as backdoors and poisoning attempts) in order to introduce mistakes into the FL system. These FL systems' security flaws significantly exacerbate issues with mitigating assaults, tolerance, and shortcomings. Thus, creating a safe and reliable FL requires achieving three things: (i) strong aggregation techniques; (ii) tolerance for diverse hardware; and (iii) defense against attackers [6,9].

2.1.1.3 Privacy Concerns

Although FL protects each device's privacy by exchanging, when the device and cloud interact, private information will still be disclosed even though the model updates—such as gradient information—rather than raw data. For instance, in order to get local training data from the devices, attackers will use gradient leakage attacks or membership inference. Prior research has concentrated on utilizing technologies like homomorphic encryption (HE) and secure multi-party computing (SMC) to improve FL's privacy but these techniques are unable to defend against the aforementioned harmful assaults. SMC and HE are only capable of stopping data breaches; they are unable to fend off gradient leaks and member inference assaults. Thus, to mitigate or stop the above-indicated hostile attacks, the FL system has to develop additional privacy-enhancing techniques as soon as possible [10,11].

2.1.1.4 Effective Issues

In order to counteract or prevent the aforementioned malicious assaults, the FL system must promptly create supplementary mechanisms that enhance privacy for the device. . Users won't be able to make use of real-time intelligent services if model training and reasoning proceed at a comparatively sluggish pace.

Consequently, FL systems will face the following difficulties upon widespread deployment in 6G networks: (i) The FL model's size makes it impossible for it to be squeezed into a single device, (ii) The FL model's training moves too slowly to meet the latency requirements of the 6G network, and (iii) The FL model's inference moves too slowly to satisfy the user's real-time need. It takes efficient training and inference for FL and 6G networks to perform together flawlessly. Effective model training and inference in a large, heterogeneous network is a difficult task for FL systems [12].

2.1.1.5 Heterogeneity of Systems

The computational, storage, and communication capacity of each device in federated networks may vary due to differences in hardware (CPU and memory), network connection (3G, 4G, 5G, and Wi-Fi), and power (battery level). Additionally, due to network capacity constraints and system-related limits on individual devices, only a tiny percentage of the devices are typically active at any given time; for instance, there may be just a few hundred active devices in a network with millions of devices. Furthermore, it is common for an active device to malfunction after a specific number of iterations. Because of connectivity or energy issues. Problems like fault tolerance and straggler mitigation are significantly made worse by these system-level characteristics. Because of this, developed FL techniques need to:

1. Account for limited participation;
2. Handle diverse hardware; and
3. Be resilient enough to drop devices in the communication network [12].

2.1.1.6 Variation in Statistics

Devices often create and gather data in a very non-uniform way throughout the network; for example, while performing a next-word prediction job, mobile phone users utilize different languages in different contexts. Furthermore, there could be a statistical structure behind the amount of data points that represent the link between the devices and their corresponding distributions, and these variations might be rather large. This data-generation paradigm may complicate issue modeling, theoretical analysis, and the empirical evaluation of solutions. It also defies widely held independent and identically distributed assumptions in distributed optimization. In fact, there are more options outside learning a single global model, as in the case of (1)'s canonical FL issue, such as simultaneously learning several local models using multitask learning [13].

2.1.2 Possibility of Fulfilling 6G Needs

By 2030, approximately 125 billion wireless devices will need to be supported by 6G networks. Therefore, developing an intelligent signal and data processing system is crucial to enabling edge learning. FL is a crucial technology that might potentially fulfill the following predicted 6G specifications [14].

2.1.2.1 Super-Reliable Low Latency Communications on a Large Scale

Upgrades to the fifth-generation (5G) ultra-reliable low latency communication (URLLC) metrics are required since more 6G wireless end-user devices are expected to be deployed. By using FL, a shared network model may be collaboratively learned by several edge computing units, reducing service latency and ensuring high dependability [15].

2.1.2.2 Scalable Architecture

Edge intelligence, like FL, is distributed and requires a decomposable and scalable architecture for future 6G communications. Such architectures enable concurrent. Processing over several edge servers is essential to the development of new wireless communication services.

2.1.2.3 Services Focused on People

6G focuses on human-centric services, requiring user experience quality based on physical movement. FL can predict user movements, improving user experience through BS [16].

2.3 DIRECTIONS AND DIFFICULTIES IN RESEARCH

For wireless networks, FL guarantees that the behavior prediction or resource allocation problem may be resolved in a distributed fashion. The five primary directions and issues associated with FL usage for wireless networks are as follows.

2.3.1 Capability to Grow

Large-scale learning networks require scalable FL systems to address complexity and memory issues, and it's crucial to investigate distributed training issues for optimal performance [17].

2.3.2 Safety and Confidentiality

In FL, only the locally acquired model is sent to the center, protecting raw datasets. Eavesdroppers, however, have the ability to approximate reconstructions and divulge personal data. Confidentiality can be global or local, with model generation invisible to unknown devices and aggregation confidential to third parties [18].

2.3.3 Asynchronous Communication

FL involves wireless device-based communication, but synchronous methods can introduce stragglers. Asynchronous solutions can alleviate these issues, but bounded delay assumptions may be impractical in federated schemes.

2.3.4 Non-Independent Identically Distribution (Non-IID) Devices

In data modeling and convergence trend analysis, joint model training from heterogeneous data across devices poses issues, especially when handling heterogeneous settings and conflicting decision-making contexts.

2.3.5 Joint Communication and Computation Design

Real-world wireless communication networks necessitate the implementation of FL for several cells and hops. The performance of FL learning schemes is negatively impacted by limited radio resources, hence cooperative control of communications and computer resources is essential for effective FL [19].

2.4 CURRENT ISSUES AND FUTURE GOALS

This section identifies open problems and future research directions in wireless communication and FL, despite extensive research, highlighting key issues still to be studied.

2.4.1 Convergence

Limited wireless resources in communication networks limit user upload scheduling, affecting convergence time and performance. The center aims to involve all users' local FL models to determine the best global model, affecting FL convergence investigations using convex loss functions [20]. Nonconvex loss functions in educational settings pose challenges in investigating the convergence rates of FL. Key problems include requiring accurate convergence formulations, consistent with real experiment data, and conducting multitask FL simultaneously. Large-scale systems require multi-cell and multi-hop FL, requiring more insights into the FL. Studying wireless device mobility for FL convergence is challenging due to dynamic channel gains, potentially causing devices to exit the process due to serious CSI [21].

2.4.2 Privacy and Security

Privacy and security in FL face challenges such as protecting each user, preserving the BS, and ensuring the complete algorithm. A trade-off between privacy and FL performance is provided by differential privacy. Traditional methods like encryption and recent developments like secure multi-party computation provide security in IoT situations [22].

2.4.3 Performance Evaluation

The study focuses on the impact of broadband communication and FL latency efficacy, highlighting the need for research on communication-efficient FL due to the limited bandwidth of wireless communication, despite increasing computing resources [23].

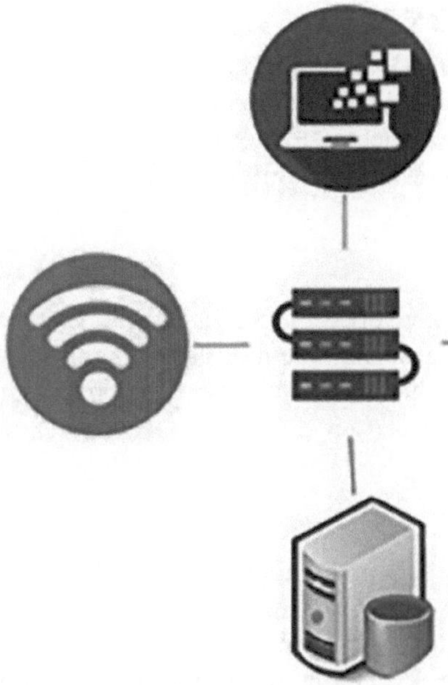

FIGURE 2.4 Emerging technologies.

2.4.4 FL for Emerging Technologies

FL's interaction with emerging technologies presents challenges, such as high propagation attenuation in the terahertz band, satellite beam optimization, and quantum key distribution parameter optimization in satellite communications [23].

Figure 2.4 shows emerging technologies the main server connects with Wi-Fi, CPU, computer system, and database. It increases IT operations.

2.5 APPLICATION FEDERATED LEARNING IN 6G

FL is an approach to ML in which local data samples are stored on decentralized devices or servers, and the model is trained across them without sharing them. It has drawn a lot of interest in the area of distributed learning and privacy-preserving ML. As we look ahead to 6G, which is envisioned to provide even faster and more reliable communication, there are several potential applications for FL:

2.5.1 Privacy-Preserving AI

Healthcare: FL can be employed for healthcare applications, allowing models to be trained across various medical organizations without having to exchange private patient information.

Finance: Financial institutions can collaborate on fraud detection models without sharing customer transaction details, enhancing security and privacy [24].

2.5.2 Edge Computing and IoT

Smart Cities: FL may be utilized on 6G networks to train models for traffic management, energy consumption prediction, and other urban planning applications.

Industrial IoT: Manufacturers can deploy FL for predictive maintenance across distributed edge devices, reducing downtime and improving efficiency [25].

2.5.3 Personalized Services

Content Recommendation: In 6G networks, FL can be used to personalize content recommendations on devices while ensuring that user data remains on the device.

Virtual Assistants: FL can enable more intelligent and personalized virtual assistants without compromising user privacy [26].

2.5.4 Network Optimization

Quality of Service Improvement: FL can be applied to optimize network parameters and improve QoS based on user experiences, device capabilities, and network conditions.

Dynamic Spectrum Allocation: FL can be used in wireless communication to dynamically assign spectrum resources according to local demand and interference circumstances [27].

2.5.5 Collaborative Learning

Educational Applications: FL can support collaborative learning platforms, allowing students to benefit from a collective model trained on data from various educational institutions without compromising privacy.

Research Collaboration: Researchers can collaborate on ML models without sharing sensitive datasets, fostering innovation across different organizations [28].

2.5.6 Security and Threat Detection

Cyber Security: FL can be utilized to enhance threat detection models by aggregating information from various sources without exposing vulnerabilities.

Anomaly Detection: In critical infrastructure, such as power grids, FL can help build robust anomaly detection models across distributed sensors [29].

As with any technology, the successful implementation of FL in 6G applications will require addressing challenges such as communication overhead, model aggregation, and security concerns. However, the potential for privacy-preserving collaborative learning makes FL a promising technique for the evolving landscape of 6G networks.

2.6 OPPORTUNITIES FOR FEDERATED LEARNING IN 6G

FL in 6G networks opens up numerous opportunities across various domains. Some key opportunities are as follows.

2.6.1 Enhanced Privacy and Security

Decentralized Data Processing: 6G networks can leverage FL to process data locally on devices, ensuring that sensitive information never leaves the local environment. This improves security and privacy, making it a good strategy for applications like healthcare, finance, and personal assistants.

Secure Model Updates: FL can facilitate secure model updates, where only the model parameters are transmitted, reducing the risk of exposing sensitive information during the training process [30].

2.6.2 Improved Quality of Service

Dynamic Network Optimization: FL can be applied to optimize network parameters dynamically, taking into account the unique characteristics and requirements of different devices. This can lead to improved quality of service (QoS) for users and better resource utilization.

Reduced Latency: By enabling local model updates, FL can contribute to lower latency in 6G networks, critical for applications like augmented reality, autonomous vehicles, and real-time communications [31].

2.6.3 Collaborative Learning across Devices

Cross-Device Learning: FL allows models to be trained collaboratively across devices, leading to more robust and generalizable models. This can be particularly beneficial for applications in smart homes, smart cities, and IoT ecosystems.

User-Centric Personalization: 6G can enable personalized services without compromising user privacy. FL allows models to be tailored to individual user preferences without centralized data collection [32].

2.6.4 Resource-Efficient Edge Computing

Edge Intelligence: FL can harness the requirement for centralized cloud processing by using edge devices' computing capabilities. Applications that need low latency responses and real-time processing would benefit most from this.

Energy Efficiency: FL can contribute to energy-efficient computing by distributing the computational load across devices, reducing the need for continuous data transmission to a centralized server [32].

2.6.5 Cross-Domain Collaboration

Interoperability: FL can facilitate interoperability between different domains and industries. For example, models trained on data from healthcare, finance, and telecommunications domains can collaborate to create more comprehensive and versatile solutions.

Global Collaboration: 6G networks can support global collaborations through FL, allowing organizations from different geographical locations to contribute to and benefit from shared ML models without compromising data privacy [32].

2.6.6 Adaptive Learning and AI Evolution

Adaptive Models: FL can enable models to adapt to changing environments and user behaviors by continuously learning from decentralized data sources. This is crucial for applications in rapidly evolving scenarios like social media trends or user preferences.

Continuous Improvement: FL can contribute to the continuous improvement of AI models by incorporating diverse data from various sources, ensuring that models remain relevant and effective over time [33].

2.6.7 Edge-to-Edge Communication

Communication between Edge Devices: 6G networks can facilitate efficient communication between edge devices, allowing them to collaboratively train models without the need for central servers. This is particularly valuable in scenarios where low latency communication is essential.

Decentralized Decision-Making: FL in 6G can enable decentralized decision-making processes, reducing reliance on centralized authorities for certain applications.

To fully capitalize on these opportunities, it's important to address technical challenges such as communication efficiency, model aggregation methods, and security considerations associated with FL in 6G networks. As the technology matures, these opportunities have the potential to revolutionize various industries and applications [33].

2.7 LITERATURE REVIEW

The purpose of this study is to offer a comprehensive assessment of the literature on the application of FL in EC circumstances with a taxonomy, taking into account advanced solutions and other outstanding difficulties [6]. The foundations of both EC and FL are examined in this investigation, which is followed by a discussion of pertinent work that has been done in FL within EC. We also describe the architecture, protocols, framework, and hardware requirements required to implement FL in the EC context. We also discuss applications, challenges, and pertinent contemporary solutions of the edge FL. Finally, we provide two relevant case studies of FL use in EC, highlight outstanding questions, and suggest future lines of inquiry. We think that this study will aid scholars in comprehending the relationship between FL and EC, enabling concepts and technology.

Federated learning involves training statistical models on distributed or isolated data centres, such as hospitals or mobile phones, while keeping the data localised. The training of heterogeneous and potentially large-scale networks presents new challenges that require a fundamental departure from traditional approaches to distributed optimisation, significant machine learning, and data analysis that ensures privacy protection [34].

This article discusses the unique characteristics and challenges of FLfederated learning, provides an overview of current approaches, and proposes several study areas that will be of interest to several research groups. Federated learning involves training statistical models on distributed or isolated data centres, such as hospitals or mobile phones, while keeping the data localised. The process of training in various and potentially extensive networks presents new challenges that require a fundamental change from traditional approaches to distributed optimisation, large-scale machine learning, and privacy-preserving data analysis [35].

This article discusses the possibility of this essential element to give rise to security vulnerabilities, breaches of privacy, and issues related to scalability. FL, short for federated learning, is a novel approach for DL algorithms aimed at safeguarding data privacy. Despite its ability to block client data transfers and limit privacy leaks, FL nevertheless encounters many challenges related to model defects and security breaches. The utilisation of blockchain and smart contracts, which are innovative technology, has the potential to safeguard FL across IoT ecosystems. This project aims to investigate the use of blockchain-based federated learning approaches to enhance the security of IoT systems. The text discusses the present state of blockchain research, the potential use of FL approaches in conjunction with it, the existing security challenges in the field of IoT, and proposes ways to emphasise the need of employing advanced techniques for ensuring security and privacy [36].

This research study aims to optimise the quantity of training samples for selected customers in each global iteration and perhaps reduce the time required for model training. To do this, we will investigate the combined approach of dynamic sample selection and client selection. Next, we will examine the method of identifying hazardous clients in order to assess the probability of a client being malicious. We will then establish client selection measures to prevent the system from selecting these undesirable clients. This is done to protect the federated learning process from malicious attacks by clients [37].

This research study provides a comprehensive explanation of the analytical approaches, procedures, and queries associated with potential functional analysis (FA) applications, domains, and types in 6G networks. In addition to examining FA, it explores supplementary approaches to enhance security and privacy in 6G networks and discusses the challenges and prerequisites for effectively adopting FA. Moreover, the text emphasises the distinctions between FL (Federated Learning) and FA (not defined) and showcases a network orchestration scenario to illustrate their potential for collaboration [38].

This article provides a comprehensive overview of the use of FL in intelligent healthcare. Firstly, we present the most recent advancements in FL, together with the rationales and prerequisites for implementing FL in intelligent healthcare. Next, we analyse the latest concepts in intelligent healthcare using federated learning (FL). These concepts encompass resource-conscious FL, incentive-focused FL, safe and privacy-conscious FL, and customised FL. Here, we present a comprehensive analysis of the unique uses of FL in important areas of healthcare, such as medical imaging, remote health monitoring, health data management, and COVID-19 detection. This paper examines many active smart healthcare initiatives based on FL and emphasises its important findings. To summarise, we discuss captivating research issues and possible avenues for future Federated Learning (FL) research in the field of smart healthcare [39].

This study paper provides a comprehensive analysis of the survey's utilisation of non-independent identical distribution (non-IID) data, specifically focusing on the use of both parametric and non-parametric models in horizontal and vertical federated learning. In addition, a comprehensive analysis of recent studies on the challenges of FLfederated learning with Nonnon-IID data is performed, and the advantages and disadvantages of various approaches are discussed. Ultimately, we conclude the work by providing some suggestions for future investigation [40].

FL is a novel distributed ML platform designed to safeguard privacy. However, models developed using FL usually perform worse than those learned using the traditional centralized learning mode when the training data are not independently and identically distributed (non-IID) on the local devices.

2.8 WHAT APPLICATIONS DOES 6G HAVE?

After the 2030s, how will life and our digital society be different? We start with hardware that can be used by people to access the network. There will probably be new interfaces between humans and machines that greatly simplify operations and information consumption, even if smartphones and tablets still exist. We expect that:

- We'll see more and more wearable electronics, including earphones and gadgets hidden in our clothes, and fewer and fewer bio-implants and skin patches. It's possible that we'll depend on novel brain sensors to control devices. We'll carry about a number of wearable's, and they'll all interact with one another naturally and intuitively to create various wearable interfaces.
- It's conceivable that touch-screen typing will become obsolete. We will become used to interacting with our devices through gestures and speech.
- Our gadgets will be completely aware of the environment in which they are used, and the network will get better at anticipating our requirements. Context awareness together with new HMIs will greatly improve the intuitiveness and efficiency of our interactions with the digital and physical worlds.
- Because of physical size and battery power concerns, it is unlikely that all of the computation required for these devices will be housed within the devices themselves. Instead, in order to finish activities, users might need to rely on locally accessible computer resources that go outside the edge cloud. Thus, a major part of the man-machine interaction of the future will be networks [41].

As customers, we should anticipate that the mainstream market will have access to today's self-driving concept automobiles by the year 2030. Even though they will primarily operate autonomously, there may be times when they still need assistance from a remote driver or the passenger. This will provide us a significant boost in the amount of time we can spend consuming internet-based information for learning, rich communications, or enjoyment. Additionally, the vehicles themselves will need a lot more bandwidth since they will download high-resolution maps, link directly to one another, and post vehicle sensor data to the network in real-time.

- The use of wireless cameras as sensors will be widespread. The camera is set to become a ubiquitous sensor, capable of detecting objects and people, thanks to advancements in artificial intelligence (AI) and machine vision. More broadly, cameras will be able to autonomously extract information from photos and videos. Limiting access to data and anonymizing information will allay privacy worries. In order to collect environmental data, radio and other sensing modalities like acoustics will also be employed.
- Security-screening methods will utilize advanced tools to cut down on security queues. Instead of only screening individuals at gates, a variety of people will be screened as they go through crowded locations using sensing modalities. In order to do this, radio sensing will be crucial, assisted by the communication systems of the future.

2.9 6G NETWORK TECHNIQUES

In the end, a new generation may be identified by how many innovative, crucial innovations it brings to the communication system. It usually takes 10 years or longer for really revolutionary new technology to be implemented in daily life. Because of this, the genuinely innovative technologies that will make up 6G need to be research ideas right now. Following the "six" concept for 6G, we have identified six further possible technological transformations that we believe will contribute to the development of the 6G framework: (i) designing and optimizing the air interface using AI/ML; (ii) expanding into new spectrum bands and using cutting-edge methods for cognitive spectrum sharing; (iii) integrating sensing and localization into the system's definition; (iv) satisfying strict latency and reliability standards. (v) Innovative methods for security and privacy; and (vi) fresh perspectives on network architecture that integrate sub-networks and RAN-Core converging [41].

2.10 WHAT CAN 6G WORK?

Wireless sensing devices operating on the 6G network have the capacity to monitor absorption levels across different frequencies and make adjustments to the frequencies as needed. 6G, the sixth generation of cellular networks, is now in the process of development. The objective of 6G cellular technology is to facilitate instantaneous augmented reality, virtual reality, and a forthcoming IoT framework with minuscule intelligent gadgets used both inside and outdoors. Furthermore, smartphones have the ability to access essential content at a quicker rate, such as streaming video. The efficacy of this technology is attributed to the fact that atoms and molecules emit and absorb electromagnetic radiation at predictable frequencies. The implementation of 6G technology will have a significant impact on various aspects of government and business, particularly in terms of security for critical assets and public safety strategies. This includes areas such as threat detection, health tracking, facial and feature recognition, social credit systems, law enforcement decision-making, air quality monitoring, gas and toxin sensing, and the development of realistic sensory interfaces. The progress in mobile network technology, smartphones, and upcoming technologies such as driverless cars, smart cities, and virtual and augmented reality will provide advantages [42].

2.11 CONCLUSION

In this study, the authors explored the fundamental difficulties of FL for 6G applications as well as the need for 6G communication. They also gave an outline of the integration of FL into 6G communications. Researchers have examined FL applications for wireless communications in this paper. There are now two primary FL categories available: FRL and FSL. We have also spoken about the reasons for employing FL in wireless communication applications. Moreover, researchers have determined a few of the methods needed to overcome the difficulties in applying FL in real-world wireless communication scenarios. Thus, it is anticipated that this research on FL for wireless communications will offer knowledge beneficial to FL-based wireless network operation, design, and optimization. Researchers provide a comprehensive examination of FL's applications and associated difficulties in this chapter discusses 6G wireless networks.

REFERENCES

1. Zhu, G., Y. Du, D. Gunduz, and K. Huang "One-bit over-the-air aggregation for communication-efficient federated edge learning: Design and convergence analysis," 2020. arXiv: 2001.05713.
2. Bonomi, F., R. Milito, J. Zhu, and S. Addepalli, "Fog computing and its role in the Internet of Things." In: *Proceedings of SIGCOMM Workshop on Mobile Cloud Computing*, Association for Computing Machinery, New York, NY, United States. 2012. https://doi.org/10.1145/2342509.2342513.
3. McMahan, H. B., E. Moore, D. Ramage, S. Hampson, and B. Aguera y Arcas, "Communication-efficient learning of deep networks from decentralized data." In: *Proceedings 20th International Conference on Artificial Intelligence and Statistics*, PMLR, Fort Lauderdale, FL, USA. 2017, pp. 1273–1282.
4. Liu, Y., X. Yuan, Z. Xiong, J. Kang, X. Wang, and D. Niyato. "Federated learning for 6G communications: Challenges, methods, and future directions." *China Communications* 17, no. 9 (2020): 105–118.
5. Yang, Z., M. Chen, K. K. Wong, H. V. Poor, and S. Cui. "Federated learning for 6G: Applications, challenges, and opportunities." *Engineering* 8 (2022): 33–41.
6. Li, T., A. K. Sahu, A. Talwalkar, and V. Smith. "Federated learning: Challenges, methods, and future directions." *IEEE Signal Processing Magazine* 37, no. 3 (2020): 50–60.
7. Kishor, K. "Communication-efficient federated learning." In: Yadav S. P., Bhati B. S., Mahato D. P., Kumar S. (eds) *Federated Learning for IoT Applications*. EAI/Springer Innovations in Communication and Computing. Springer, Cham, 2022. https://doi.org/10.1007/978-3-030-85559-8_9.

7a. Kishor, K. "Personalized federated learning." In: Yadav S. P., Bhati B. S., Mahato D. P., Kumar S. (eds) *Federated Learning for IoT Applications*. EAI/Springer Innovations in Communication and Computing. Springer, Cham, 2022. https://doi.org/10.1007/978-3-030-85559-8_3.

8. Y. Liu, J. Peng, J. Kang, A. M. Iliyasu, D. Niyato, and A. A. Abd El-Latif. "A secure federated learning framework for 5G networks," 2020. arXiv preprint arXiv:2005.05752.
9. Wang, Z., M. Song, Z. Zhang, Y. Song, Q. Wang, and H. Qi. "Beyond inferring class representatives: Userlevel privacy leakage from federated learning." In: *IEEE INFOCOM 2019-IEEE Conference on Computer Communications*. IEEE, Paris. 2019, pp. 2512–2520. DOI: 10.1109/INFOCOM.2019.8737416
10. Kishor, K. "Chapter 10 Study of quantum computing for data analytics of predictive and prescriptive analytics models". In: Yadav S. P., Singh R., Yadav V., Al-Turjman F., Kumar S. A. (eds) *Quantum-Safe Cryptography Algorithms and Approaches: Impacts of Quantum Computing on Cybersecurity*. De Gruyter, Berlin, Boston, 2023, pp. 121–146. https://doi.org/10.1515/9783110798159-010
11. Li, L., H. Xiong, Z. Guo, J. Wang, and C. Z. Xu. "Smartpc: Hierarchical pace control in real-time federated learning system." In: *2019 IEEE Real-Time Systems Symposium (RTSS)*, IEEE, Houston, TX, 2019, pp. 406–418.
12. Smith, V., C.-K. Chiang, M. Sanjabi, and A. Talwalkar. "Federated multi-task learning." In: Guyon, I., Von Luxburg, U., Bengio, S., Wallach, H., Fergus, R., Vishwanathan, S., and Garnett, R. (eds), *Proceedings of Advances in Neural Information Processing Systems*, Long Beach, CA, 2017, pp. 4424–4434.
13. Kishor, K., and P. Nand, "Wireless networks based in the cloud that support 5G." In: Kishor K., Saxena, N., and Pandey, D. (eds), *Cloud-based Intelligent Informative Engineering for Society 5.0*, 1st edition. Chapman and Hall/CRC, New York, 2023, pp. 23–40. ISBN: 9781003213895. https://doi.org/10.1201/9781003213895-2.
14. Park, J., S. Samarakoon, H. Shiri, M. K. Abdel-Aziz, T. Nishio, A. Elgabli, and M. Bennis. "Extreme URLLC: Vision, challenges, and key enablers," 2020. arXiv: 2001.09683.
15. Kishor, K. "Chapter 17 Application of quantum computing for digital forensic investigation". In: Yadav S. P., Singh R., Yadav V., Al-Turjman F., Kumar S. A. (eds) *Quantum-Safe Cryptography Algorithms and Approaches: Impacts of Quantum Computing on Cybersecurity*. De Gruyter, Berlin, Boston, 2023, pp. 231–248. https://doi.org/10.1515/9783110798159-017.
16. Kishor, K. "Chapter 12 Review and significance of cryptography and machine learning in quantum computing." In: Yadav S. P., Singh R., Yadav V., Al-Turjman F., Kumar S. A. (eds) *Quantum-Safe Cryptography Algorithms and Approaches: Impacts of Quantum Computing on Cybersecurity*. De Gruyter, Berlin, Boston, 2023, pp. 159–176. https://doi.org/10.1515/9783110798159-012.
17. Luo, S., X. Chen, Q. Wu, Z. Zhou, and S. Yu. "HFEL: Joint edge association and resource allocation for cost-efficient hierarchical federated edge learning." *IEEE Transactions on Wireless Communications*, 19, no. 10 (2020): 6535–6548.
18. Kishor, K. "Cloud computing in blockchain." In: Kishor K., Saxena, N., and Pandey, D. (eds), *Cloud-based Intelligent Informative Engineering for Society 5.0*, 1st edition. Chapman and Hall/CRC, New York, 2023, pp. 79–105. ISBN: 9781003213895. https://doi.org/10.1201/9781003213895-5.
19. Khaled, A., K. Mishchenko, and P. Richtárik. "Tighter theory for local SGD on identical and heterogeneous data." In: *Proceedings of International Conference on Artificial Intelligence and Statistics*, 2020 Aug 26–28; online, 2020.
20. Yang, H.H., Z. Liu, T.Q.S. Quek, and H.V. Poor. "Scheduling policies for federated learning in wireless networks." *IEEE Transactions on Communications* 68, no. 1 (2020): 317–333.
21. Wei, K., J. Li, M. Ding, C. Ma, H.H. Yang, F. Farokhi, S. Jin, T. Q. S. Quek, and H. V. Poor. "Federated learning with differential privacy: Algorithms and performance analysis." *IEEE Transactions on Information Forensics and Security* 15 (2020): 3454–3469.

22. Kishor, K., P. Nand, and P. Agarwal. "Secure and efficient subnet routing protocol for MANET." *Indian Journal of Public Health*, 9, no. 12 (2018): 200. https://doi.org/10.5958/0976-5506.2018.01830.2.
23. Cheng, Y., Y. Liu, T. Chen, and Q. Yang. "Federated learning for privacy-preserving AI." *Communications of the ACM* 63, no. 12 (2020): 33–36.
24. Chen, B., J. Wan, A. Celesti, D. Li, H. Abbas, and Q. Zhang. "Edge computing in IoT-based manufacturing." *IEEE Communications Magazine* 56, no. 9 (2018): 103–109.
25. Anshari, M., M. N. Almunawar, S. A. Lim, and A. Al-Mudimigh. "Customer relationship management and big data enabled: Personalization & customization of services." *Applied Computing and Informatics* 15, no. 2 (2019): 94–101.
26. Nedić, A., A. Olshevsky, and M. G. Rabbat. "Network topology and communication-computation tradeoffs in decentralized optimization." *Proceedings of the IEEE* 106, no. 5 (2018): 953–976.
27. Kishor, K., P. Nand, and P. Agarwal. Notice of retraction design adaptive subnetting hybrid gateway MANET protocol on the basis of dynamic TTL value adjustment. *Aptikom Journal on Computer Science and Information Technologies* 3, no.2 (2018): 59–65. https://doi.org/10.11591/APTIKOM.J.CSIT.115.
28. Karie, N. M., N. M. Sahri, and P. Haskell-Dowland. "IoT threat detection advances, challenges and future directions." In: *2020 Workshop on Emerging Technologies for Security in IoT (ETSecIoT)*, IEEE, Sydney, 2020, pp. 22–29.
29. Mothukuri, V., R. M. Parizi, S. Pouriyeh, Y. Huang, A. Dehghantanha, and G. Srivastava. "A survey on security and privacy of federated learning." *Future Generation Computer Systems* 115 (2021): 619–640.
30. Biswas, S., K. Singh, O. Taghizadeh, and T. Ratnarajah. "Design and analysis of FD MIMO cellular systems in coexistence with MIMO radar." *IEEE Transactions on Wireless Communications* 19, no. 7 (2020): 4727–4743.
31. Higgins, S. E., E. Mercier, E. Burd, and A. Hatch. "Multi-touch tables and the relationship with collaborative classroom pedagogies: A synthetic review." *International Journal of Computer-Supported Collaborative Learning* 6 (2011): 515–538.
32. Zhu, H., and Y. Jin. "Multi-objective evolutionary federated learning." *IEEE Transactions on Neural Networks and Learning Systems* 31, no. 4 (2019): 1310–1322.
33. Abreha, H. G., M. Hayajneh, and M. A. Serhani. "Federated learning in edge computing: A systematic survey." *Sensors* 22, no. 2 (2022): 450.
34. Singh, P., M. K. Singh, R. Singh, and N. Singh. "Federated learning: Challenges, methods, and future directions." In: Yadav, S. P., Bhati, B. S., Mahato, D. P., and Kumar, S. (eds), *Federated Learning for IoT Applications.* Springer International Publishing, Cham. 2022, pp. 199–214.
35. Nguyen, J., K. Malik, H. Zhan, A. Yousefpour, M. Rabbat, M. Malek, and D. Huba. "Federated learning with buffered asynchronous aggregation." In: *International Conference on Artificial Intelligence and Statistics.* PMLR, Virtual Conference, 2022, pp. 3581–3607.
36. Kishor, K. "Impact of cloud computing on entrepreneurship, cost, and security." In: Kishor K., Saxena, N., Pandey, D. (eds), Cloud-based Intelligent Informative Engineering for Society 5.0, 1st edition. CRC Press, New York, 2023, pp. 171–191. ISBN: 9781003213895. https://doi.org/10.1201/9781003213895-10.
37. Parra-Ullauri, J. M., X. Zhang, A. Bravalheri, S. Moazzeni, Y. Wu, R. Nejabati, and D. Simeonidou. "Federated analytics for 6G networks: Applications, challenges, and opportunities." *IEEE Network* 38, no. 2 (2024): 9–17.
38. Nguyen, D. C., Q.-V. Pham, P. N. Pathirana, M. Ding, A. Seneviratne, Z. Lin, O. Dobre, and W.-J. Hwang. "Federated learning for smart healthcare: A survey." *ACM Computing Surveys (CSUR)* 55, no. 3 (2022): 1–37.

39. Zhu, H., J. Xu, S. Liu, and Y. Jin. "Federated learning on non-IID data: A survey." *Neurocomputing* 465 (2021): 371–390.
40. Kishor, K., N. Saxena, and D. Pandey (eds) *Cloud-Based Intelligent Informative Engineering for Society 5.0*, 1st edition. Chapman and Hall/CRC, New York, 2023, pp. 1–234. ISBN: 9781003213895. https://doi.org/10.1201/9781003213895.
41. Bassoli, R., F. H. Fitzek, and E. C. Strinati. "Why do we need 6G?" *ITU Journal on Future and Evolving Technologies* 2, no. 6 (2021): 1–31.
42. Kishor, K. Using a half cheetah habitat for random augmentation computing. *Multimedia Tools and Applications* (2024). https://doi.org/10.1007/s11042-024-19084-0

3 Unleash Federated Machine Learning and Internet of Medical Things (IoMT) for Disease Screening and Enhancement of Smart Healthcare

Bhupinder Singh and Christian Kaunert

3.1 INTRODUCTION

Federated machine learning (FML) and the Internet of Medical Things (IoMT) have emerged as a transformative force in the evolving landscape of healthcare technology. At the forefront of this revolution is FML, a decentralized approach that allows machine learning (ML) models to be trained collaboratively across multiple local datasets without the need to centralize sensitive information. In the context of healthcare, where patient privacy is paramount, FML emerges as a game changer. Preserving data within individual healthcare nodes, it ensures that sensitive patient information remains secure while enabling the collective learning of powerful predictive models. The complementing of FML is the IoMT, a network of interconnected medical devices, and wearable sensors that continuously collect and transmit health data. This real-time stream of information offers unprecedented opportunities for healthcare providers to monitor patients remotely, detect anomalies early on, and provide timely interventions. This chapter explores the synergy between FML and IoMT, presenting a comprehensive overview of their integration for disease screening and the enhancement of smart healthcare systems and emphasizing the pivotal role in reshaping the future of healthcare. IoMT transforms traditional healthcare delivery by bringing healthcare beyond the confines of hospitals and clinics, into the daily lives of individuals [1]. Traditional screening methods often involve periodic and sometimes invasive tests, leading to delayed diagnoses and suboptimal patient outcomes [2]. The integration of FML and IoMT allows for continuous, real-time analysis of health data, enabling the early detection of abnormalities and

DOI: 10.1201/9781003489368-3

the implementation of personalized treatment plans [3]. The advantages of using the FML and IoMT for disease screening are more as they not only enhance the accuracy of disease screening but also facilitate personalized treatment plans tailored to individual health profiles [4,5]. However, challenges such as data security, interoperability, and the standardization of protocols must be addressed to ensure the seamless integration and effectiveness of FML and IoMT in healthcare settings [6].

The IoMT expedites the processes of data collection, transmission, and analysis while also increasing the amount of health data that caregivers can access and diversifying its sources [7,8]. Both patients and healthcare professionals can make better decisions as a result of the enhanced data exchange [9]. The networks and equipment required for telemedicine and virtual treatment are provided by IoMT. The use of remote healthcare capabilities became more commonplace as a way to reduce patient visits to medical facilities and lessen the burden on overburdened hospitals and other healthcare facilities [10]. For those who live farther out from major cities, telemedicine makes it easier to get medical treatment. It also makes it easier for folks all around the world who may otherwise find it difficult to visit professionals [11].

3.1.1 Background of Study

The Internet of Things (IoT) and other communication technologies have advanced recently, dramatically changing a wide range of application fields [12]. The IoMT has evolved as a result of IoT's current integration into a wide range of medical apparatus and gadgets. As a result, several IoMT-based healthcare apps are currently being deployed and used in regular situations [13]. Because of the growing worries about data security and privacy, classic ML models often rely on centralized data aggregation and learning. This presents problems for real-world healthcare systems [14].

Federated learning (FL) has become a dynamic distributed collective paradigm that is especially well-suited to modern healthcare settings [15]. Through the involvement of several stakeholders, such as hospitals, laboratories, and patients, FL makes it possible to train models without actually exchanging sensitive medical data [16,17]. With an emphasis on the security features of FL-based IoMT applications inside smart healthcare frameworks, this chapter offers an extensive review. It starts with an introduction to IoMT devices, going over their many kinds, uses, datasets, and security architecture [18]. The topic of FL is then explored, along with its application areas and several tools used in FL application development. This showcases FL-based IoMT applications, patents, practical healthcare initiatives using FL, and pertinent datasets to demonstrate the critical role FL plays in building safe IoMT systems [19].

3.1.2 Objectives of This Chapter

The objectives of this chapter are as to:

- explore the IoMT framework and related security threats;
- clarify FL function in protecting IoMT applications in healthcare;

- examination of IoMT devices and equipment, FL applications, and how FL helps design secure IoMT applications;
- express the FL-based IoMT applications, patents filed or published in the field, real-world initiatives, and publicly accessible healthcare datasets; and
- utilizing FL and diving into the potential of FL approaches in a variety of IoMT applications with security features of FL-based IoMT are also contrasted with those of other medical technologies (Figure 3.1).

3.1.3 Structure of this Chapter

This chapter comprehensively explores the various dimensions of FML and IoMT for disease screening concerning the enhancement of smart healthcare. Section 3.2 elaborates on the role of IoMT in collecting and transmitting healthcare data. Section 3.3 expresses the FML in healthcare. Section 3.4 discusses the concept of IoMT in smart healthcare. Section 3.5 specifies the integration of FML and IoMT for disease screening. Section 3.6 highlights the challenges and concerns: of implementing FML

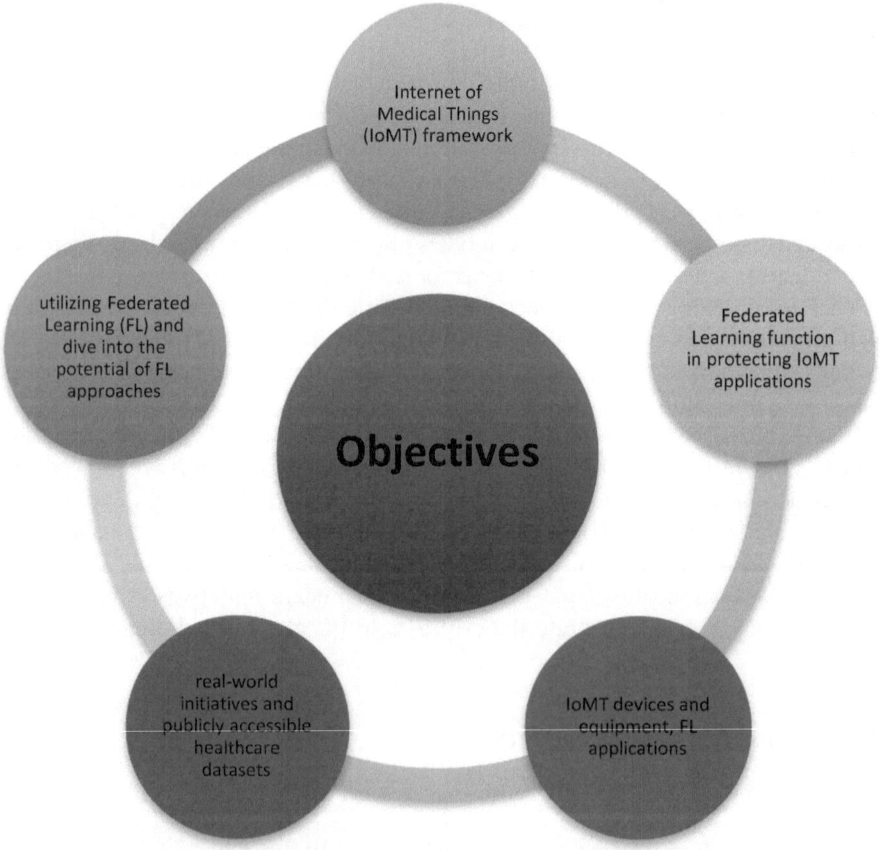

FIGURE 3.1 Objectives of the chapter.

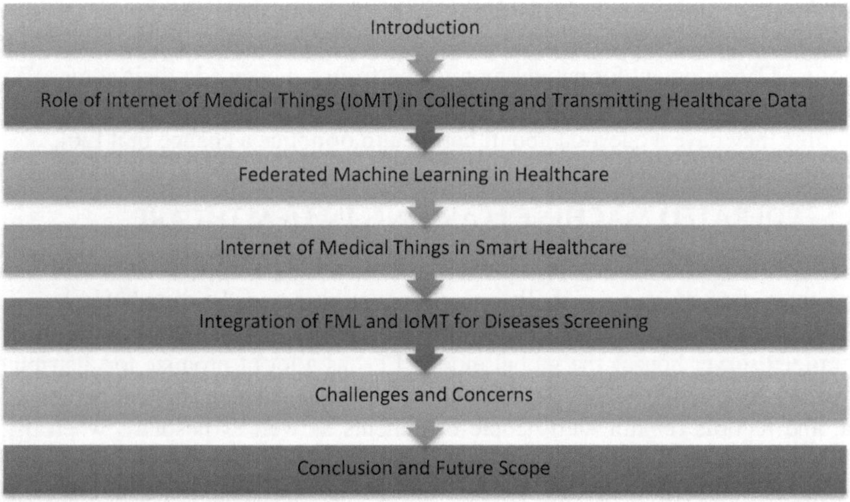

FIGURE 3.2 Structure of the chapter.

and IoMT in healthcare. And, finally, Section 3.7 concludes the chapter with future scope (Figure 3.2).

3.2 ROLE OF IOMT IN COLLECTING AND TRANSMITTING HEALTHCARE DATA

The term "Internet of Medical Things" (IoMT) describes a collection of apps and gadgets for use in healthcare that link to online computer networks and healthcare IT systems [20]. Machine-to-machine (M2M) communication is made easier by Wi-Fi-enabled medical equipment, which serves as the backbone of IoMT [21,22]. Because of the complicated architecture of IoMT inside healthcare systems, the enormous amounts of healthcare data it generates are frequently confusing and fragmented. The various parties may have access to vital patient medical information, such as hospitals, labs, and physicians. Secure access and quick communication are crucial since this health information is sensitive [23]. Strict laws like the Health Insurance Portability and Accountability Act are put in place to control how these data are interpreted and used [24]. This presents problems for sophisticated data processing methods like deep learning (DL), ML, and so forth, which need large amounts of training data. The examples of IoMT deployment include the following:

- using remote patient monitoring (RPM) for people with long-term illnesses and chronic disorders.
- keeping track of prescription orders for patients.
- following the locations of individuals who are hospitalized.
- collecting information from patients' wearable mobile health devices.
- establishing communication between ambulances traveling to hospitals and medical personnel.

IoMT devices connect to cloud platforms so that the collected data may be saved and processed. Another name for IoMT is the IoT [25]. Telemedicine also refers to the use of IoMT equipment for patient monitoring from a distance in their homes. With this method, patients may get care without having to visit a hospital or doctor's office each time they have a question about their health or notice a change in it [26].

3.3 FEDERATED MACHINE LEARNING IN HEALTHCARE

FL is becoming a well-recognized paradigm that enables learning from sensitive and fragmented data. It uses a globally shared model based on the distributed learning concept, keeping data localized at the various locations or institutions of origin. The centralized server houses the global model. FL has a lot of promise for distributed healthcare data analysis [27]. So, depending on the particular application, medical data, and records remain with people or patients as well as hospitals when using the FL model guaranteeing privacy protection [28]. With the vast volumes of medical data that modern healthcare systems amass, the application of data-driven ML has emerged as a promising technique for building accurate and robust statistical models [29]. Because current data are frequently contained behind data silos and privacy concerns restrict access to it, ML's potential in the medical industry remains unrealized [30]. However, the lack of sufficient data prevents ML from reaching its full potential and ultimately impedes its advancement from this stage to real-world application in clinical settings [31].

The successful implementation of FL could have considerable potential in facilitating large-scale precision medicine [32]. This could lead to the development of models that produce unbiased decisions, accurately reflect an individual's physiology, and demonstrate sensitivity to rare diseases, all while addressing governance and privacy concerns [33]. The application of FL requires meticulous technical considerations to ensure optimal algorithm performance without compromising safety or patient privacy. Despite these challenges, FL holds the promise of overcoming the limitations associated with approaches that rely on a single centralized pool of data [34,35].

3.3.1 Federated Machine Learning Enables Collaborative Model Training across Multiple Decentralized Healthcare Nodes

FL is a paradigm for collaborative algorithm training that eliminates the need to communicate real data, therefore addressing concerns about data privacy and governance [36]. FL was first created for a variety of industries including mobile and edge device applications but it has since become more well-known in the healthcare industry [37]. It makes it possible to develop cooperative insights like a consensus model without transferring patient data outside of the institutional firewalls where it is kept. So, only the model's parameters and gradients which are particular to each participating institution are shared throughout this process of ML [38,39].

The participating institutions in a FL process with an aggregate server can even stay anonymous to one another. It has been shown that models are capable of memorizing data in certain situations [40]. As covered in the "Technical considerations" approaches like differential privacy and learning from encrypted data have been proposed to improve privacy in the FL scenario. All things considered, the community is interested in FL because of its potential for healthcare applications, and FL techniques are a rapidly expanding field of healthcare [41].

3.3.2 Advantages of Federated Machine Learning in Preserving Privacy and Security of Healthcare Data

FL has simple primary advantages of enabling ML from non-co-located data, it addresses privacy and data governance issues [42]. Within the FL framework, every data controller sets up their own privacy rules and governance procedures [43]. They also maintain control over data access, including the option to withdraw it throughout both the validation and training stages. FL opens up new avenues for research, such facilitating large-scale institutional validation or stimulating creative studies on uncommon illnesses with low occurrence rates and sparse institutional datasets [44].

The notable benefit of relocating the model to the data instead of centralizing the data is that high-dimensional, storage-intensive medical data does not need to be duplicated for each user doing local model training or transferred from local institutions to a central pool [45]. With moving the concept to regional organizations, it becomes possible to naturally scale with a possibly growing global dataset without seeing a disproportionate increase in the amount of data storage needed. There are different topologies and computation strategies can be used to achieve FL operations [46]. For healthcare applications, the two most common ones are peer-to-peer methods and aggregation servers. Since the participants in FL only ever obtain aggregated model parameters rather than direct access to data from other organizations, FML always offers some degree of anonymity [47,48].

3.4 INTERNET OF MEDICAL THINGS IN SMART HEALTHCARE

The IoMT, which has entered almost every aspect of healthcare, created by the IoT being integrated into various medical equipment and gadgets [49]. The medical sensors are becoming a common feature of wearable technology and hospital and home equipment in smart healthcare systems [50]. Even the medical robots are capable of doing operations and other medical activities like remote monitoring and patient care. These sophisticated medical gadgets can also continually monitor patients' health in real-time [51]. Because of the large volume of data gathered by IoMT, there are now more prospects for the development and integration of sophisticated computational tools to store, process, analyze, and use the data to improve treatment and care quality [52]. In the context of telemedicine as a whole, the advantages of IoMT include the following:

Patient Monitoring Systems: IoMT makes it possible to continuously monitor patients who have long-term medical issues. This gives medical practitioners important information about the living circumstances of their patients, which helps them make decisions about their treatment [53].

Accessibility: IoMT improves patient access to medical care and educational opportunities. With telehealth applications, patients have additional service alternatives and may get medical attention when they need it [54].

Logistics: IoMT devices send out notifications for maintenance and other concerns while keeping an eye on the equipment in healthcare institutions. They also act as trackers, tracking patients and their prescriptions around hospital campuses to minimize confusion and mistakes [55].

Cost Control: Telehealth and RPM lower the price of in-person patient visits to medical institutions. In addition to saving providers money and time, accelerated processing of health data enables resources to be allocated to areas where they are most needed [56].

Better Patient Experience: IoMT brings in new technologies that let patients take care of themselves and reduce the need for in-person appointments. Consumer-grade wearables provide patients access to data, which were previously only available during medical appointments [57].

Accuracy: IoMT produces more data, giving medical practitioners more accurate information on the health problems of their patients. An IoMT-enabled blood pressure monitor, for example, can provide measurements for many days, which can result in a more accurate diagnosis than data from a single visit to the doctor [58] (Figure 3.3).

3.4.1 Components of IoMT—Wearable Devices, Sensors, and Health Monitoring Systems

IoMT devices are categorized into several sorts based on the situations in which they are used.

In-Home IoMT: Devices placed within or close to a patient's house are referred to as in-home IoMT. For example, personal emergency response systems use gadgets that are located at home to alert a hospital in case of an emergency. Devices known as RPM follow patients who have long-term medical issues while they are at home [59].

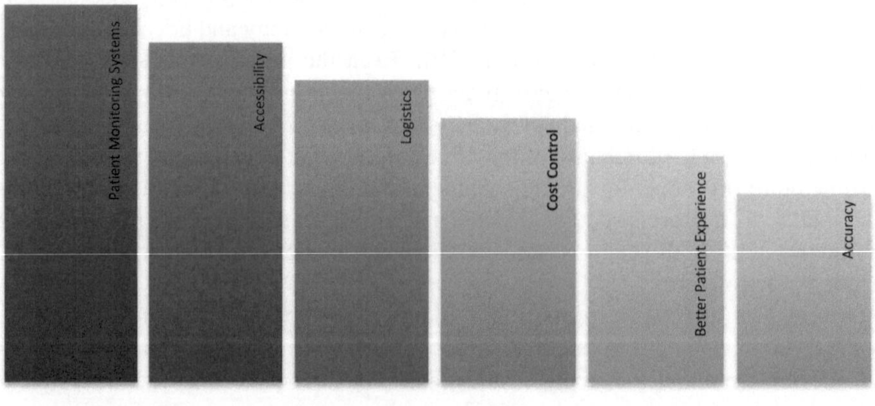

FIGURE 3.3 Advantages of IoMT.

Wearable IoMT: Also referred to as on-body IoMT, these gadgets are affixed to a person's body and collect health information. Wearables come in two varieties: consumer and medical grades. For example, consumer-grade smartwatches track personal well-being by recording vital signs like blood pressure and heart rate. Medical-grade on-body IoMT involves medically supervised equipment such as neuromodulation devices, which affect the neurological system to control pain. The other kind of wearable IoMT is smart pills, which are edible sensors that gather data from within the body [60].

Mobile IoMT: Near-field communication and RFID tags are used by many consumer mobile devices, such as smartphones, to enable information exchange with other IT systems. For instance, individuals may use their smartphones to control networked glucose monitors, and medical staff can get data via the network of their company [61].

Public IoMT: Also known as community IoMT, these devices are dispersed around a certain region. One example would be point-of-care kiosks, which link patients and healthcare practitioners while also providing medical supplies. They are essential in enabling people who live in isolated places without access to conventional medical facilities to get healthcare [62] (Figure 3.4).

3.4.2 IoMT Contributes to Real-time Data Collection and Patient Monitoring

The IoMT contributes to real-time data collection and patient monitoring and for this, it is essential that every user sends their learned model through the server in an efficient manner [63]. FML employs an iterative process on the server to aggregate local data into a global network which shows the training process [64]. So, every device has a local copy of the centralized ML program that users may access at any time, the model gains more intelligence as it gradually picks up new skills and trains itself using the data that users supply. The devices then send back to the central server the training data that they have obtained from their local versions of the ML algorithm [65].

Real-time data are actively collected in healthcare organizations from end devices, layered protocol stacks, networked or implanted sensors, and lightweight communication frameworks collectively known as the Internet of Medical Things or IoMT,

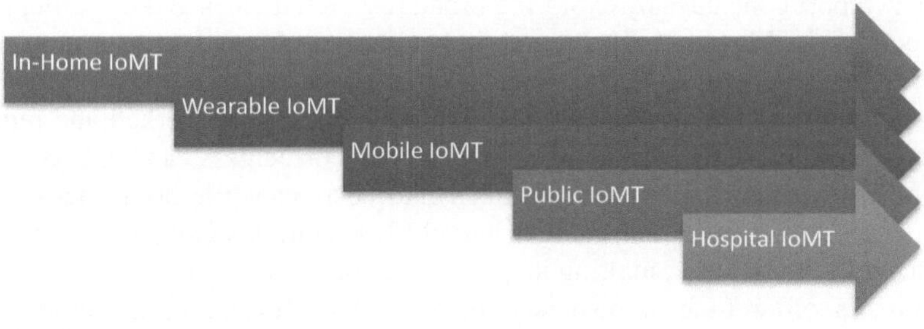

FIGURE 3.4 Components of IoMT.

ecosystems [66]. The extraction of significant data-driven insights is facilitated by the IoMT, which is a critical component in the advancement of healthcare analytics (HA) [67]. The privacy restrictions have raised concerns about data exchange across IoMT and stored electronic health records (EHRs), which might result in inaccurate analytics models with sparse data. This has led to a radical change in HA, moving away from centralized learning paradigms and toward edge- or distributed learning paradigms. FL is unique in distributed learning since it enables training on local data without requiring explicit data exchange [68,69]. FL has advantages, but it also has drawbacks, including fragmentation, different levels of data partitioning, and significant statistical variability in learning models, which might jeopardize accuracy throughout the learning and updating processes [70]. FL in healthcare has not fully addressed issues with sparsification, scalability, and large dispersed datasets. The possible FL integration in IoMT, FL aggregation strategies, reference architecture, and the usage of distributed learning models to facilitate FL in IoMT ecosystems in an effort to close this gap [71].

3.5 INTEGRATION OF FEDERATED MACHINE LEARNING AND IOMT FOR DISEASES SCREENING

In the field of smart healthcare, FL, IoMT, artificial intelligence (AI) and explainable AI (XAI) are the most well-known and intriguing technologies [72]. The healthcare systems function through centralized agents exchanging unprocessed data, which gives rise to notable weaknesses and difficulties in the system. But when AI is included, the system becomes a network of many agent collaborators that can effectively communicate with their individual hosts [73]. FML also offers a fascinating feature: it functions in a decentralized way, allowing communication inside the selected system based on models rather than raw data transfers. The integration of FL, IoMT, AI, and XAI methodologies has the ability to mitigate many constraints and difficulties within the healthcare system [74]. FL uses AI in intelligent medical applications as it starts by going over modern conceptions of developing technologies, such as FL, AI, and XAI, and how they interact with the healthcare system [75]. It investigates how FL–AI is integrated and categorized with healthcare technology across several areas. The discussion and concerns of security, privacy, stability, and dependability issue of patients that are currently plaguing the healthcare industry [76]. It also directs readers toward FL- and AI-based solutions for healthcare problems. The report concludes by exploring broad research themes and outlining potential directions for FL-based AI research in healthcare administration systems in the future [77,78].

FL, highlighting the gradient aggregation policy across networked and remotely connected hospitals [79]. System latency, model correctness, and the consensus mechanism's trustworthiness are all taken into account while doing performance analysis [80]. This shows that the distributed FL performs significantly better than centralized FL techniques, making it appropriate for use in actual IoMT prototypes. Therefore, in order to help dispersed networked healthcare organizations cover important solutions and investigate the potential of FL in IoMT [81,82].

3.5.1 Potential Synergies between Federated Machine Learning and IoMT in Healthcare

There have been notable developments of FML during the last years, with applications spanning industries including smart healthcare and IoT. These applications include patient health data recording, medical picture analysis, analysis of illnesses associated with cancer, and relevant economic data [83]. The conventional intelligent medical systems have often relied on cloud-based, centralized AI resources for data analytics and learning [84]. This centralized method, which is typified by raw data transfer, is not very scalable on the network and results in inefficient communication delay. The centralized AI system can be judged unsuitable for future healthcare systems, especially in e-healthcare where personal data are subject to restrictions [85].

3.6 CHALLENGES AND CONCERNS: IMPLEMENTING FEDERATED MACHINE LEARNING AND IOMT IN HEALTHCARE

The IoMT has significantly changed how medical institutions operate, improving the standard of care. IoMT devices are widely used to gather healthcare data since they can sense and communicate a person's health updates [86,87]. FL is then used to process which leads to a plethora of healthcare applications, such as illness prognosis and RPM [88]. DL techniques, like those used in biomedical image analysis, have shown promise in improving healthcare efficiency by efficiently managing enormous amounts of relevant data that may be used for the early diagnosis of acute diseases. In order to solve privacy problems related to datasets, this performed a thorough investigation that focused on blockchain-based FL, offering insights into the present state of this issue [89]. There are many obstacles to IoT technology integration in healthcare settings.

Ownership of Data: It is frequently unclear who is the true owner of data created by IoMT. Throughout its lifetime, a number of parties may create or interact with the data, including the patient, software provider, and other healthcare institutions. The extent of each party's data rights may vary depending on the circumstances [90].

Data Classification: Determining what types of data are medical gets more difficult as IoMT develops and gathers a wider variety of data. Emerging medical data, for instance, consists of health information deduced from non-health-related data using AI applications. It is possible to employ location monitoring, which usually only involves non-medical data, to monitor the spread of infectious illnesses [91].

Implementation: Ensuring that devices adhere to industry standards and resolving interoperability concerns hinder the deployment of IoMT infrastructure [92].

Costs Up Front: When deploying IoMT technology, there may be significant upfront costs. It may also take some time to see a favorable return on investment [93].

User Experience: It might be difficult to create user-friendly medical software and equipment. Devices with poor design may capture data insufficiently or inaccurately, particularly if patients remove the device or utilize it improperly [94].

FIGURE 3.5 Challenges and concerns.

Security: IoMT technology raises questions about secured health information in particular. To reduce risks like fraud and data breaches, compliance with laws like the Health Insurance Portability and Accountability Act (HIPAA) is crucial. The possibility the theft of credentials might be used by hackers to get prescription drugs or medical treatments without authorization, with serious consequences and fines for noncompliance [95] (Figure 3.5).

3.7 CONCLUSION AND FUTURE SCOPE

Healthcare providers will integrate IoMT equipment and networks with the latest developments in IoT technology. For example, it is expected that sensor technology will advance and provide more and better real-time data. Predictive healthcare is expected to improve concurrently with the deployment of AI-driven data analytics and ML. IoMT-generated healthcare data will not only make more thorough research possible but also create new opportunities for creative treatment approaches. IoMT devices have the ability to improve data exchange and collaboration among healthcare companies, expedite medical management, and save costs. The vision anticipates a federated and IoMT future for digital health. Through this perspective, it aims to convey the shared viewpoint which offers context and details.

REFERENCES

1. Rayan, R. A., Tsagkaris, C., & Iryna, R. B. (2021). The Internet of Things for healthcare: Applications, selected cases and challenges. In: Marques G., Bhoi A. K., de Albuquerque V. H. C., Hareesha K.S. (eds), *IoT in Healthcare and Ambient Assisted Living* (pp. 1–15). Springer, Singapore.
2. Attaran, M. (2017). The Internet of Things: Limitless opportunities for business and society. *Journal of Strategic Innovation and Sustainability* 12(1), 11.
3. Pujar, S. M., & Satyanarayana, K. V. (2015). Internet of Things and libraries. *Annals of Library and Information Studies (ALIS)*, 62(3), 186–190.
4. Habibzadeh, H., Dinesh, K., Shishvan, O. R., Boggio-Dandry, A., Sharma, G., & Soyata, T. (2019). A survey of Healthcare Internet of Things (HIoT): A clinical perspective. *IEEE Internet of Things Journal*, 7(1), 53–71.
5. Singh, B. (2024). Legal dynamics lensing metaverse crafted for videogame industry and e-sports: Phenomenological exploration catalyst complexity and future. *Journal of Intellectual Property Rights Law*, 7(1), 8–14.
6. Hassan, A., Prasad, D., Khurana, M., Lilhore, U. K., & Simaiya, S. (2021). Integration of Internet of Things (IoT) in health care industry: An overview of benefits, challenges, and applications. In: Taneja K., Taneja H., Kumar K., Selwal A., Ouh E. L. (eds), *Data Science and Innovations for Intelligent Systems* (pp. 165–180). Boca Raton.
7. Rghioui, A., & Oumnad, A. (2018). Challenges and opportunities of Internet of Things in healthcare. *International Journal of Electrical & Computer Engineering*, 8(5), 2753–2761. ISSN: 2088-8708.
8. Islam, S. R., Kwak, D., Kabir, M. H., Hossain, M., & Kwak, K. S. (2015). The Internet of Things for health care: A comprehensive survey. IEEE Access, 3, 678–708.
9. Singh, B. (2023). Blockchain technology in renovating healthcare: Legal and future perspectives. In: Kaushik K, Dahiya S, Aggarwal S, Dwivedi A. D. (eds), *Revolutionizing Healthcare through Artificial Intelligence and Internet of Things Applications* (pp. 177–186). IGI Global. IGI Global USA
10. Maksimović, M., & Vujović, V. (2017). Internet of Things based e-health systems: Ideas, expectations and concerns. In: Khan S. U., Zomaya A. Y., Abbas A. (eds),*Handbook of Large-Scale Distributed Computing in Smart Healthcare* (pp. 241–280). Springer, Cham.
11. Venu, D. N., ArunKumar, D. A., & Vaigandla, K. K. Investigation on Internet of Things (IoT): Technologies, challenges and applications in healthcare. *International Journal of Research*, 11, 143–153.
12. Singh, B. (2023). Federated learning for envision future trajectory smart transport system for climate preservation and smart green planet: Insights into global governance and SDG-9 (Industry, Innovation and Infrastructure). *National Journal of Environmental Law*, 6(2), 6–17.
13. Kadhim, K. T., Alsahlany, A. M., Wadi, S. M., & Kadhum, H. T. (2020). An overview of patient's health status monitoring system based on Internet of Things (IoT). *Wireless Personal Communications*, 114(3), 2235–2262.
14. Adhikary, T., Jana, A. D., Chakrabarty, A., & Jana, S. K. (2020). The Internet of Things (IoT) augmentation in healthcare: An application analytics. In: Gunjan V. K., Diaz V. G., Cardona M., Solanki V. K., Sunitha K. V. N. (eds), *ICICCT 2019-System Reliability, Quality Control, Safety, Maintenance and Management: Applications to Electrical, Electronics and Computer Science and Engineering* (pp. 576–583). Springer, Singapore
15. Paul, M., Maglaras, L., Ferrag, M. A., & AlMomani, I. (2023). Digitization of healthcare sector: A study on privacy and security concerns. *ICT Express*, 9(4), 571–588.
16. Sharma, A., Kaur, S., & Singh, M. (2021). A comprehensive review on blockchain and Internet of Things in healthcare. *Transactions on Emerging Telecommunications Technologies*, 32(10), e4333.

17. Ray, P. P., Chowhan, B., Kumar, N., & Almogren, A. (2021). BIoTHR: Electronic health record servicing scheme in IoT-blockchain ecosystem. *IEEE Internet of Things Journal*, 8(13), 10857–10872.
18. Egala, B. S., Pradhan, A. K., Badarla, V., & Mohanty, S. P. (2021). Fortified-chain: A blockchain-based framework for security and privacy-assured Internet of Medical Things with effective access control. *IEEE Internet of Things Journal*, 8(14), 11717–11731.
19. Hussien, H. M., Yasin, S. M., Udzir, N. I., Ninggal, M. I. H., & Salman, S. (2021). Blockchain technology in the healthcare industry: Trends and opportunities. *Journal of Industrial Information Integration*, 22, 100217.
20. Antonius, N., & Dachyar, M. (2020). The Internet of Things (IoT) design for cardiac remote patient monitoring using business process re-engineering. In: *2020 3rd International Conference on Applied Engineering (ICAE)* (pp. 1–7). IEEE. Batam, Indonesia.
21. Sharma, A., & Singh, B. (2022). Measuring impact of e-commerce on small scale business: A systematic review. *Journal of Corporate Governance and International Business Law*, 5(1), 34–38.
22. Tarikere, S., Donner, I., & Woods, D. (2021). Diagnosing a healthcare cybersecurity crisis: The impact of IoMT advancements and 5G. *Business Horizons*, 64(6), 799–807.
23. Farouk, A., Alahmadi, A., Ghose, S., & Mashatan, A. (2020). Blockchain platform for industrial healthcare: Vision and future opportunities. *Computer Communications*, 154, 223–235.
24. Gupta, S., Sharma, H. K., & Kapoor, M. (2022). Integration of IoMT and blockchain in smart healthcare system. In: Gupta S., Sharma H. K., Kapoor M. (eds) *Blockchain for Secure Healthcare Using Internet of Medical Things (IoMT)* (pp. 79–91). Springer International Publishing, Cham.
25. Singh, B. (2022). Understanding legal frameworks concerning transgender healthcare in the age of dynamism. *Electronic Journal of Social and Strategic Studies*, 3, 56–65.
26. Kagita, M. K., Thilakarathne, N., Gadekallu, T. R., & Maddikunta, P. K. R. (2022). A review on security and privacy of Internet of Medical Things. In: Ghosh U., Chakraborty C., Garg L., Srivastava G. (eds), *Intelligent Internet of Things for Healthcare and Industry* (pp. 171–187). Springer, Cham.
27. Shah, V., & Khang, A. (2023). Internet of Medical Things (IoMT) driving the digital transformation of the healthcare sector. In: Khang A., Rana G., Tailor R. K., Abdullayev V. (eds), *Data-Centric AI Solutions and Emerging Technologies in the Healthcare Ecosystem* (pp. 15–26). CRC Press. Boca Raton.
28. Singh, B. (2022). Relevance of agriculture-nutrition linkage for human healthcare: A conceptual legal framework of implication and pathways. *Justice and Law Bulletin*, 1(1), 44–49.
29. Saikia, A., & Jaiswal, S. (2023). Internet of Things in Healthcare: Changing the landscape. In: Sharma A., Sarma H. K. D., Biradar S. R. (eds), *IoT and Cloud Computing-Based Healthcare Information Systems* (pp. 247–260). Apple Academic Press, Palm Bay, FL.
30. Repass, C. (2022). The Applications of the Internet of things in the Medical Field. Theses and Dissertations.
31. Singh, B. (2022). COVID-19 pandemic and public healthcare: Endless downward spiral or solution via rapid legal and health services implementation with patient monitoring program. *Justice and Law Bulletin*, 1(1), 1–7.
32. Gorrepati, R. R., Jonnala, P., Guntur, S. R., & Kim, D. H. (2023). Semantic web of things for healthcare interoperability using IoMT technologies. In: Patel A., Debnath N. C. (eds), *Semantic Technologies for Intelligent Industry 4.0 Applications* (pp. 49–82). River Publishers, New York.

33. Ramzan, S., Aqdus, A., Ravi, V., Koundal, D., Amin, R., & Al Ghamdi, M. A. (2022). Healthcare applications using blockchain technology: Motivations and challenges. *IEEE Transactions on Engineering Management*, 70(8), 2874–2890.
34. Singh, B. (2020). Global science and jurisprudential approach concerning healthcare and illness. *Indian Journal of Health and Medical Law*, 3(1), 7–13.
35. Polu, S. K., & Polu, S. K. (2019). IoMT-based smart health care monitoring system. *International Journal for Innovative Research in Science & Technology*, 5(11), 58–64.
36. Almusallam, N., Alabdulatif, A., & Alarfaj, F. (2021). Analysis of privacy-preserving edge computing and Internet of Things models in healthcare domain. *Computational and Mathematical Methods in Medicine*, 2021(21), 1–6.
37. Akhtar, N., Rahman, S., Sadia, H., & Perwej, Y. (2021). A holistic analysis of Medical Internet of Things (MIoT). *Journal of Information and Computational Science*, 11(4), 209–222.
38. Singh, B. (2019). Profiling public healthcare: A comparative analysis based on the multidimensional healthcare management and legal approach. *Indian Journal of Health and Medical Law*, 2(2), 1–5.
39. Alturki, R., Alharbi, A. I., AlQahtani, S. S., Khan, S., & Truong Hoang, V. (2021). A proposed application for controlling overweight and obesity within the framework of the Internet of Things. *Mobile Information Systems*, 2021(1), 1–9.
40. Kaushik, K., Dahiya, S., Aggarwal, S., & Dwivedi, A. D. (Eds.). (2023). *Revolutionizing Healthcare through Artificial Intelligence and Internet of Things Applications*. IGI Global, Hershey, Pennsylvania, USA.
41. Zaabar, B., Cheikhrouhou, O., Ammi, M., Awad, A. I., & Abid, M. (2021). Secure and privacy-aware blockchain-based remote patient monitoring system for internet of healthcare things. In: *2021 17th International Conference on Wireless and Mobile Computing, Networking and Communications (WiMob)* (pp. 200–205). IEEE, Bologna.
42. Malik, H., Anees, T., Faheem, M., Chaudhry, M. U., Ali, A., & Asghar, M. N. (2023). Blockchain and Internet of Things in smart cities and drug supply management: Open issues, opportunities, and future directions. *Internet of Things*, 23, 100860.
43. Kandasamy, K., Srinivas, S., Achuthan, K., & Rangan, V. P. (2020). IoT cyber risk: A holistic analysis of cyber risk assessment frameworks, risk vectors, and risk ranking process. *EURASIP Journal on Information Security*, 2020(1), 1–18.
44. Kishor K. (2022). Personalized federated learning. In: Yadav S. P., Bhati B. S., Mahato D. P., Kumar S. (eds) *Federated Learning for IoT Applications*. EAI/Springer Innovations in Communication and Computing (pp. 31–52). Springer, Cham. https://doi.org/10.1007/978-3-030-85559-8_3.
45. Kishor, K. (2024). Using a half cheetah habitat for random augmentation computing. *Multimedia Tools and Applications*. https://doi.org/10.1007/s11042-024-19084-0
46. Sharma, A., Jha, N., & Kishor, K. (2022). Predict COVID-19 with chest X-ray. In: Gupta D., Polkowski Z., Khanna A., Bhattacharyya S., Castillo O. (eds) *Proceedings of Data Analytics and Management*. Lecture Notes on Data Engineering and Communications Technologies, vol 90. Springer, Singapore. https://doi.org/10.1007/978-981-16-6289-8_16.
47. Rani, S., Kataria, A., Kumar, S., & Tiwari, P. (2023). Federated learning for secure IoMT-applications in smart healthcare systems: A comprehensive review. *Knowledge-Based Systems*, 274, 110658.
48. Gadekallu, T. R., Alazab, M., Hemanth, J., & Wang, W. (2023). Guest editorial federated learning for privacy preservation of healthcare data in Internet of Medical Things and patient monitoring. *IEEE Journal of Biomedical and Health Informatics*, 27(2), 648–651.
49. Ali, M., Naeem, F., Tariq, M., & Kaddoum, G. (2022). Federated learning for privacy preservation in smart healthcare systems: A comprehensive survey. *IEEE Journal of Biomedical and Health Informatics*, 27(2), 778–789.

50. Zheng, X., Shah, S. B. H., Ren, X., Li, F., Nawaf, L., Chakraborty, C., & Fayaz, M. (2021). Mobile edge computing enabled efficient communication based on federated learning in Internet of Medical Things. *Wireless Communications and Mobile Computing*, 2021, 1–10.
51. Sachin, D. N., Annappa, B., Hegde, S., Abhijit, C. S., & Ambesange, S. (2024). FedCure: A heterogeneity-aware personalized federated learning framework for intelligent healthcare applications in IoMT environments. *IEEE Access*, 12, 15867–15883.
52. Muazu, T., Yingchi, M., Muhammad, A. U., Ibrahim, M., Samuel, O., & Tiwari, P. (2023). IoMT: A medical resource management system using edge empowered blockchain federated learning. *IEEE Transactions on Network and Service Management*, 21(1), 517–534.
53. Nguyen, D. C., Pham, Q. V., Pathirana, P. N., Ding, M., Seneviratne, A., Lin, Z., Dobre, O., & Hwang, W. J. (2022). Federated learning for smart healthcare: A survey. *ACM Computing Surveys (CSUR)*, 55(3), 1–37.
54. Wang, X., Hu, J., Lin, H., Liu, W., Moon, H., & Piran, M. J. (2022). Federated learning-empowered disease diagnosis mechanism in the Internet of Medical Things: From the privacy-preservation perspective. *IEEE Transactions on Industrial Informatics*, 19(7), 7905–7913
55. Lakhan, A., Hamouda, H., Abdulkareem, K. H., Alyahya, S., & Mohammed, M. A. (2024). Digital healthcare framework for patients with disabilities based on deep federated learning schemes. *Computers in Biology and Medicine*, 169, 107845.
56. Messinis, S., Temenos, N., Protonotarios, N. E., Rallis, I., Kalogeras, D., & Doulamis, N. (2024). Enhancing Internet of Medical Things security with artificial intelligence: A comprehensive review. *Computers in Biology and Medicine*, 170, 108036.
57. Lian, Z., Wang, W., Han, Z., & Su, C. (2023). Blockchain-based personalized federated learning for Internet of Medical Things. *IEEE Transactions on Sustainable Computing*, 8(4), 694–702.
58. Rehman, A., Abbas, S., Khan, M. A., Ghazal, T. M., Adnan, K. M., & Mosavi, A. (2022). A secure healthcare 5.0 system based on blockchain technology entangled with federated learning technique. *Computers in Biology and Medicine*, 150, 106019.
59. Bhattacharya, P., Verma, A., & Tanwar, S. (Eds.). (2023). *Federated Learning for Internet of Medical Things: Concepts, Paradigms, and Solutions*. CRC Press, Boca Raton, FL.
60. Zhou, X., Ye, X., Kevin, I., Wang, K., Liang, W., Nair, N. K. C., Shimizu, S., Yan, Z., & Jin, Q. (2023). Hierarchical federated learning with social context clustering-based participant selection for Internet of Medical Things applications. *IEEE Transactions on Computational Social Systems*, 10(4), 1742–1751.
61. Prasad, V. K., Bhattacharya, P., Maru, D., Tanwar, S., Verma, A., Singh, A., Tiwari, A. K., Sharma, R., Alkhayyat, A., Ţurcanu, F. E., & Raboaca, M. S. (2022). Federated learning for the Internet-of-Medical-Things: A survey. *Mathematics*, 11(1), 151.
62. Abbas, S., Issa, G. F., Fatima, A., Abbas, T., Ghazal, T. M., Ahmad, M., Yeun, C. Y., & Khan, M. A. (2023). Fused weighted federated deep extreme machine learning based on intelligent lung cancer disease prediction model for healthcare 5.0. *International Journal of Intelligent Systems*, 2023(1), 2599161.
63. Gupta, S., Tyagi, S., & Kishor, K. (2022). Study and development of self sanitizing smart elevator. In: Gupta D., Polkowski Z., Khanna A., Bhattacharyya S., Castillo O. (eds) *Proceedings of Data Analytics and Management*. Lecture Notes on Data Engineering and Communications Technologies, vol 90. Springer, Singapore. https://doi.org/10.1007/978-981-16-6289-8_15.

64. Srivastava, G., Yenduri, K. D. R., Hegde, P., Gadekallu, T. R., Maddikunta, P. K. R., & Bhattacharya, S. (2023). Federated learning enabled edge computing security for Internet of Medical Things: Concepts, challenges and open issues. In: Srivastava G., Ghosh U., Chun-Wei Lin J. (eds), *Security and Risk Analysis for Intelligent Edge Computing* (pp. 67–89). Springer International Publishing, Cham.
65. Mathkor, D. M., Mathkor, N., Bassfar, Z., Bantun, F., Slama, P., Ahmad, F., & Haque, S. (2024). Multirole of the Internet of Medical Things (IoMT) in biomedical systems for managing smart healthcare systems: An overview of current and future innovative trends. *Journal of Infection and Public Health*, 17(4), 559–572.
66. Rahman, M. A., & Hossain, M. S. (2021). An Internet-of-Medical-Things-enabled edge computing framework for tackling COVID-19. *IEEE Internet of Things Journal*, 8(21), 15847–15854.
67. Gupta, D., Kayode, O., Bhatt, S., Gupta, M., & Tosun, A. S. (2021). Hierarchical federated learning based anomaly detection using digital twins for smart healthcare. In: *2021 IEEE 7th International Conference on Collaboration and Internet Computing (CIC)* (pp. 16–25). IEEE, Atlanta, GA.
68. Tiwari, P., Lakhan, A., Jhaveri, R. H., & Gronli, T. M. (2023). Consumer-centric Internet of Medical Things for cyborg applications based on federated reinforcement learning. *IEEE Transactions on Consumer Electronics*, 69(4), 756–764.
69. Tamilarasi, K., Maheswari, K., Ramesh, S., Isaac, S., & Rajaram, A. (2023). A decentralized smart healthcare monitoring system using deep federated learning technique for IoMT. https://doi.org/10.21203/rs.3.rs-3339998/v1.
70. Wagan, S. A., Koo, J., Siddiqui, I. F., Attique, M., Shin, D. R., & Qureshi, N. M. F. (2022). Internet of Medical Things and trending converged technologies: A comprehensive review on real-time applications. *Journal of King Saud University-Computer and Information Sciences*, 34(10), 9228–9251.
71. Alamleh, A., Albahri, O. S., Zaidan, A. A., Albahri, A. S., Alamoodi, A. H., Zaidan, B. B., Qahtan, S., Alsatar, H. A., Al-Samarraay, M. S., & Jasim, A. N. (2022). Federated learning for IoMT applications: A standardization and benchmarking framework of intrusion detection systems. *IEEE Journal of Biomedical and Health Informatics*, 27(2), 878–887.
72. Khan, A. T., Fatima, A., Shahzad, T., Alissa, K., Ghazal, T. M., Al-Sakhnini, M. M., Abbas, S., Khan M. A., & Ahmed, A. (2023). Secure IoMT for disease prediction empowered with transfer learning in healthcare 5.0, the concept and case study. *IEEE Access*, 11, 39418–39430.
73. Ahmed, S. F., Alam, M. S. B., Afrin, S., Rafa, S. J., Rafa, N., & Gandomi, A. H. (2024). Insights into Internet of Medical Things (IoMT): Data fusion, security issues and potential solutions. *Information Fusion*, 102, 102060.
74. Ghosh, S., & Ghosh, S. K. (2023). FEEL: FEderated LEarning framework for ELderly healthcare using edge-IoMT. *IEEE Transactions on Computational Social Systems*, 10(4), 1800–1809.
75. Aminizadeh, S., Heidari, A., Toumaj, S., Darbandi, M., Navimipour, N. J., Rezaei, M., Talebi, S., Azad, P., & Unal, M. (2023). The applications of machine learning techniques in medical data processing based on distributed computing and the Internet of Things. *Computer Methods and Programs in Biomedicine*, 241, 107745.
76. Hermawan, D., Putri, N. M. D. K., & Kartanto, L. (2022). Cyber physical system based smart healthcare system with federated deep learning architectures with data analytics. *International Journal of Communication Networks and Information Security*, 14(2), 222–233.

77. Sindhusaranya, B., Yamini, R., Manimekalai Dr, M. A. P., & Geetha Dr, K. (2023). Federated learning and blockchain-enabled privacy-preserving healthcare 5.0 system: A Comprehensive approach to fraud prevention and security in IoMT. *Journal of Internet Services and Information Security*, 13(4), 199–209.
78. Ding, W., Abdel-Basset, M., Hawash, H., Abdel-Razek, S., & Liu, C. (2023). Fed-ESD: Federated learning for efficient epileptic seizure detection in the fog-assisted Internet of Medical Things. *Information Sciences*, 630, 403–419.
79. Kishor K. (2022). Communication-efficient federated learning. In: Yadav S. P., Bhati B. S., Mahato D. P., Kumar S. (eds) *Federated Learning for IoT Applications*. EAI/Springer Innovations in Communication and Computing. Springer, Cham. https://doi.org/10.1007/978-3-030-85559-8_9.
80. Eskandari, Z., & Rezaee, M. (2023). Blockchain-enabled federated learning to enhance security and privacy in Internet of Medical Things (IoMT). *International Journal of Web Research*, 6(1), 87–93.
81. Khan, M. F., Ghazal, T. M., Said, R. A., Fatima, A., Abbas, S., Khan, M. A., Issa, G. F., Ahmad, M., & Khan, M. A. (2021). An IoMT-enabled smart healthcare model to monitor elderly people using machine learning technique. *Computational Intelligence and Neuroscience*, 2021(1), 2487759.
82. Alahmadi, A., Khan, H. A., Shafiq, G., Ahmed, J., Ali, B., Javed, M. A., Khan, M.Z., Alsisi, R.H., & Alahmadi, A. H. (2023). A privacy-preserved IoMT-based mental stress detection framework with federated learning. *The Journal of Supercomputing*, 80, 10255–10274.
83. Yaqoob, M. M., Nazir, M., Khan, M. A., Qureshi, S., & Al-Rasheed, A. (2023). Hybrid classifier-based federated learning in health service providers for cardiovascular disease prediction. *Applied Sciences*, 13(3), 1911.
84. Kishor, K. (2023). Impact of cloud computing on entrepreneurship, cost, and security. In: Kishor K., Saxena, N., Pandey, D. (eds), *Cloud-based Intelligent Informative Engineering for Society 5.0*, 1st edition (pp. 171–191). CRC Press, New York. ISBN: 9781003213895. https://doi.org/10.1201/9781003213895-10.
85. Rauniyar, A., Hagos, D. H., Jha, D., Håkegård, J. E., Bagci, U., Rawat, D. B., & Vlassov, V. (2023). Federated learning for medical applications: A taxonomy, current trends, challenges, and future research directions. *IEEE Internet of Things Journal*, 11(5), 7374–7398.
86. Rahman, A., Hossain, M. S., Muhammad, G., Kundu, D., Debnath, T., Rahman, M., Khan, M. S. I., Tiwari, P., & Band, S. S. (2023). Federated learning-based AI approaches in smart healthcare: Concepts, taxonomies, challenges and open issues. *Cluster Computing*, 26(4), 2271–2311.
87. Siddiqui, S., Khan, A. A., Dev, K., & Dey, I. (2021). Integrating federated learning with IoMT for managing obesity in smart city. In: *Proceedings* of the 1st Workshop on Artificial Intelligence and Blockchain Technologies for Smart Cities with 6G (pp. 7–12). Association for Computing Machinery, New York.
88. Aouedi, O., Sacco, A., Piamrat, K., & Marchetto, G. (2022). Handling privacy-sensitive medical data with federated learning: Challenges and future directions. *IEEE Journal of Biomedical and Health Informatics*, 27(2), 790–803.
89. Kishor, K. (2023). Cloud computing in blockchain. In: Kishor K., Saxena, N., and Pandey, D. (eds), *Cloud-based Intelligent Informative Engineering for Society 5.0*, 1st edition, (pp. 79–105). Chapman and Hall/CRC, New York. ISBN: 9781003213895. https://doi.org/10.1201/9781003213895-5.
90. Singh, P., Gaba, G. S., Kaur, A., Hedabou, M., & Gurtov, A. (2022). Dew-cloud-based hierarchical federated learning for intrusion detection in IoMT. *IEEE Journal of Biomedical and Health Informatics*, 27(2), 722–731.

91. Namratha, M., Anusree, M. K., Niha, Pooja, S., & Arpana, M. R. (2023). Anomaly detection in medical IoT devices using federated learning. In: Senjyu T., So–In C., Joshi A. (eds), *International Conference on Smart Trends in Computing and Communications* (pp. 259–270). Springer Nature Singapore, Singapore.
92. Vishwakarma, S., Goswami, R. S., Nayudu, P. P., Sekhar, K. R., Arnepalli, P. R. R., Thatikonda, R., & Abdel-Rehim, W. M. (2023). Secure federated learning architecture for fuzzy classifier in healthcare environment. *Soft Computing*, 11, 1–12.
93. Aggarwal, K., Vimal, V. R., Patil, U., Senthilkumar, C., Ingle, A., & Kumar, T. S. (2023). Federated machine learning for cardiac disease detection using Internet of Medical Things. In: *2023 International Conference on Self Sustainable Artificial Intelligence Systems (ICSSAS)* (pp. 1326–1331). IEEE, India.
94. Kishor, K. (2023). Chapter 12 Review and significance of cryptography and machine learning in quantum computing". In: Yadav S. P., Singh R., Yadav V., Al-Turjman F., Kumar S. A. (eds) *Quantum-Safe Cryptography Algorithms and Approaches: Impacts of Quantum Computing on Cybersecurity* (pp. 159–176). De Gruyter, Berlin, Boston, MA. https://doi.org/10.1515/9783110798159-012.
95. Putra, K. T., Arrayyan, A. Z., Hayati, N., Damarjati, C., Bakar, A., & Chen, H. C. (2024). A Review on the application of Internet of Medical Things in wearable personal health monitoring: A cloud-edge artificial intelligence approach. *IEEE Access*, 12, 21437–21452.

4 Federated Machine Learning in Medical Science

A Perspective Investigation

Vivek Tomar, Swati Sharma, Sangeeta Arora, and Aditya Singhal

4.1 INTRODUCTION

Machine learning leads to a new era with possibilities in the healthcare industry providing an accurate diagnosis, personalized treatments, and better care of patients. The medical data are varied that is sensitive and distributed with different organizations due to traditional machine learning techniques. Federated machine learning transforms by providing a decentralized solution that addresses the involvement of healthcare-related data, and balances the vital collaboration, privacy, security, and efficiency.

Nowadays, federated learning (FL), artificial intelligence (AI), and explainable AI are the most demanding technologies in the healthcare industry. The healthcare industry previously exchanged unprocessed data with the help of centralized agents. FL is the most prominent feature, a decentralized operation that does model-based communication without exchanging the raw data.

The data on healthcare are highly sensitive and comes under strict laws. The healthcare data include images and e-health records with genomic information. The healthcare data are complicated in nature that require collaborative research and model training across the institutions, which balance data sharing and privacy protection. The evolution of federated machine learning excludes the need for raw data and to do the model training without sharing the raw data from dispersed data sources. The objective of the model is to enhance the model performance while maintaining the localized data. This technique benefits the field of medical science by benefiting in various aspects.

In FL, it will not be fetching the data from the local devices, such as Internet of Things (IoT) devices, servers, and so on. The devices will be communicating the information regarding the updated model. It protects the specific patient data residing in the actual device to maintain the protection law of data and manage patient confidence. The privacy preservation feature of FL is playing a key role regarding healthcare data. The patients are assured that their data will not be misused by

DOI: 10.1201/9781003489368-4

researchers for sharing. The patients have faith that their data are used only for the benefit of the healthcare industry, which helps in the treatment of patients in the future and provides better facilities.

The healthcare industry data are spread across hospitals, pathology labs, clinics, and various research facilities. All these places of data work jointly for the development of the model without the involvement of data sources. It trains the data with its local data and all data sources help to train the model. The model will use patient demographic data and medical history due to diverse data sources and collaborating ways to train the robust model. For critical diseases, the data may be considered by focusing on areas of collaborative learning.

Machine learning models take more time to train models on the large data sets instead of FL which makes it easy to train models from diverse data resources. The time used for the model training can be reduced with parallel training which provides a faster and more accurate model of the healthcare industry.

The decentralized nature of using data in FL solves the security issues of medical data. It reduces the chances of data theft during transmission during transmission among the devices. It is very common for theft of data in the healthcare industry to compromise patient confidence and legal issues that need attention. FL also reduces the threat of attacking the hacker at the time of data sharing. The framework used for FL improves security when various organizations work together without having any central authority to look into the data flow.

Data privacy cannot be compromised in the healthcare industry and must be adhere to rules for conducting the research and development of the new models. FL fulfills all the legal requirements and keeps the patient data confidential while using it. The healthcare data have multiple legalities for using it for research purposes. With FL, organizations can easily adapt by following legal and ethical issues for sharing the data for research purposes.

Federating learning also faces many challenges in the field of medical science with promising solutions. Researchers and medical practitioners are continuously addressing the challenges in the field of medical science. The FL approach enhances the outcomes of research effectively and is also applied to various healthcare issues. The decentralized approach to data management made it easy to implement with various tools, frameworks, and rules.

This chapter focuses on the various complex issues in medical data in the time of AI and machine learning. It provides the solution with the use of federated machine learning to solve security issues with assurance. In this chapter, privacy and security issues are highlighted using federated machine learning. The risks of providing solutions with federated machine learning in the field of medical science are also highlighted.

Machine learning is the subfield of computer science and AI that focuses on data and algorithms for simulation of human learning process and progressively increase their accuracy [1,2]. This is an essential element of data science called machine learning. The algorithms are used to train the models, which further produce the prediction and classification to give useful insights through various techniques, such as data mining, and so on. These insights are also useful for business and application decisions which is important to influence the growth. Machine learning is not limited, nowadays, it is converted into specialized learning which is helpful in various sectors, such as in the field of medical science.

Healthcare organizations can use machine learning to discover healthcare trends, collect and manage patient data, and other tasks. Hospitals and other healthcare organizations are starting to realize how machine learning may enhance decision-making and lower risk in the medical industry. Rapid insight-gathering is made possible by the rapidly developing field of machine learning in healthcare. Healthcare may benefit from machine learning by using it to develop new solutions, personalize care, and increase diagnosis accuracy. Machine learning is the process of programming computers to uncover patterns, forecast outcomes, and extract important information from vast amount of data that medical professionals cannot access and that directly affects patient health.

The main objective of machine learning in medical science is to enhance patient outcomes by generating previously unattainable medical insights. Using prediction algorithms, machine learning verifies the justifications and judgments made by medical professionals. Using predictive algorithms, it offers a means of verifying the logic and judgment of medical professionals. Let's say a patient is prescribed a particular drug by a doctor. If so, machine learning can discover a patient who had comparable health issues and responded well to the same treatment, validating this treatment approach. With the use of a computer model, machine learning enables users to adjust to changing conditions and identify and infer patterns by using learning from unprocessed data. It is possible to declare a model to be learning if it increases task performance [3].

Additionally, machine learning is being used for data analysis, which includes interpreting continuous data utilized in the intensive care unit, for detection of the patterns in the provided data to handle effectively defective data and alarm the patient which further leads to efficient treatment and monitoring. It is believed that the effective use of machine learning techniques can support the integration of computer-based systems in the healthcare setting, offering chances to support and improve the work of medical professionals and, eventually, raise the standard and efficiency of medical care [1].

Various algorithms use different functions that convert the input to the expected output in supervised machine learning. A common way to formulate a supervised learning assignment is a classification issue, where the learner must look at multiple input–output examples of the function to learn a function that maps a vector into one of several classes [2]. The classes used in the supervised algorithms are predefined, and these data are labeled with these classifications. These predefined classes are formed through a limited set method which is determined manually. The machine learning algorithm is used to build the mathematical models for the identification of trends. These models predict the data by evaluating the variance measurements [4].

The unsupervised learning is a type of inductive learning that involves working without the assistance of an outside teacher on projects like theory formation, taxonomic hierarchical classification, and discovery systems. More inference must be done by the learner in this type of unsupervised learning than in any other method that has been covered thus far. The learner is neither given access to an oracle that can categorize internally generated examples as positive or negative instances of any given concept, nor is it given a set of instances of a certain concept. Furthermore, a serious focus-of-attention issue is introduced since the observations may cover multiple ideas that need to be learned rather than concentrating on one at a time [5].

The field of healthcare management can be difficult and complex at times. These obstacles are addressed, and patient care is improved with the aid of the e-healthcare management system. The processes can be streamlined, costs can be decreased, and efficiency can be increased by utilizing patient portals, telemedicine, and electronic health records (EHRs), all while providing high-quality treatment to those who need it most.

Healthcare management also faces the issue of cost increases in healthcare services. In the case of critical disease, it is a great challenge to provide healthcare solutions to all patients due to higher demand. For example, during COVID-19, patients faced issues in hospital admission and proper treatment due to the lack of availability of experts. Now healthcare systems are keeping the records in digital form to address the issues to reduce the wait time, in case of packed hospitals and business of the experts. It offers telemedicine, remote patient monitoring, and other digital health solutions that ease the strain on medical facilities while enhancing patient access to care. It enhances interoperability across different healthcare systems even more. To manage the records, these systems must employ compatible systems. By offering a single platform for managing patient data, e-healthcare management systems can assist in overcoming this difficulty by making it simple for healthcare providers to access and share data across various platforms.

A range of gadgets are employed to collect personal health data for monitoring purposes. Critical patterns or behaviors are found using machine learning techniques. These devices are programmed to notify the user if they see something unusual. A promising tool that could reduce the cost of medical equipment and improve our understanding of the doctor–patient relationship is machine learning. Large datasets and machine learning have shown to be beneficial in the healthcare sector, helping with everything from patient follow-up and reminders to individualized treatment plans and prescriptions. The healthcare sector generates and stores a variety of medical images and signal data online, including electrocardiograms, CT scans, MRIs, X-rays, and other scans [6].

4.2 FUNDAMENTALS OF FEDERATED MACHINE LEARNING

FL offers a pathway to leverage data for the development of innovative AI applications while ensuring the security of sensitive data. By enabling AI model training without direct access to the underlying data, FL emerges as a pioneering solution, aligning well with evolving privacy standards. FL uses raw data from various sources, such as IoT healthcare devices, machines used, smart devices, and so on. [7].

FL benefited professionals and patients with the advantage of prescribing precise medicine in critical conditions. It provides unbiased decisions by training the models accurately and used for patients who focus on the exceptional disease carefully. It also ensures that the algorithm used provides a solution without a breach of patient data. This ensures that the solution in the benefit of the patient and helps to reduce stress [8].

FL handles all the issues related to privacy and data in the field of machine learning. In FL, various data have different issues regarding governance and privacy. These issues can provide and stop data access both which applies to validation and training processes. Moreover, by decentralizing the model to the data sources, the necessity to replicate high-dimensional, storage-intensive medical data from local institutions into a central repository is eliminated. This alleviates the burden on storage resources and enables users to perform local model training without inflating data storage requirements unnecessarily. Consequently, the model can expand organically as it is deployed across local institutions, leveraging the growing global dataset without the need for extensive additional storage.

Federated machine learning focuses on decentralized processing by distributing the data across multiple computing devices and servers rather than bringing all the data to one central location. The algorithm works collectively on the data stored on each device/server without sharing the data. After applying an algorithm to all the subsets of the data, each device sends the updated data to the head server that aggregates all the data to create an improved model. This method is also known as federated optimization.

FL prevents privacy breaches by enabling training to take place locally on the edge device. The data security is ensured through the updates of the encrypted model which are communicated with the central server. In the next step, the results which are aggregated, are decrypted using secure aggregation approaches like the secure aggregation principle. Access to data dispersed across numerous devices, regions, and organizations is ensured by FL.

At the central server, there exists a standard base model. Client devices receive the replicas of this model, utilizing their locally generated data for model training. Subsequently, secure aggregation techniques come into play, facilitating the sharing of updates derived from the locally trained models with the central server's primary model. This process involves creating new insights by amalgamating and averaging diverse inputs. The incorporation of data from multiple sources enhances the model's potential for widespread applicability. In the subsequent iteration, the central model, now retrained with updated parameters, is once again distributed to client devices. Each cycle allows the models to gather diverse data and refine themselves continuously, all while safeguarding users' privacy [9] (Figure 4.1).

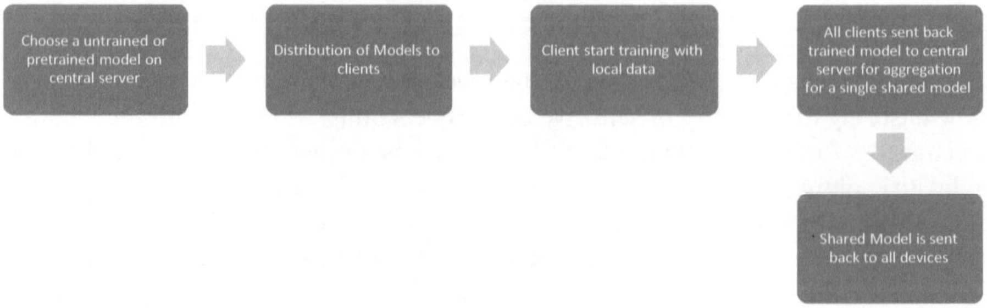

FIGURE 4.1 Steps of federated machine learning.

Various types of FL can be used, such as centralized, decentralized, and heterogeneous. In centralized FL, a central server plays a crucial role. It collects model updates during training and coordinates the initial selection of client devices. Communication occurs solely between individual edge devices and the central server. While this approach is straightforward and yields accurate models, network interruptions can halt the entire process due to the central server acting as a bottleneck. In decentralized FL, a central server is not required to coordinate learning. In this learning, model updates are exchanged on a regular basis with edge devices in the network. The final model produced by the edge device is further aggregated with the local updates from all connecting edge devices [10].

At the time of decentralized FL, which diminishes the risk of a single-point failure, the accuracy of the model hinges entirely on the network topology of the edge devices. Heterogeneous FL necessitates a diverse array of clients, including desktops, smartphones, and IoT devices. These devices vary in terms of hardware, software, computational power, and data types. Conventional FL techniques, if the properties of local models reflect those of the central model, have spurred the development of Heterogeneous FL (Hetero FL). However, such scenarios are relatively uncommon in the real-world settings. Hetero FL entails training over various local models, ultimately providing a single global model for inference.

Data analytics is used in the medical industry to analyze the patient's data. Most of the research is going on to analyze the medical data. The information encompasses details concerning patients, healthcare practitioners, affiliated entities handling insurance or financial transactions, as well as prescription and supply data. However, healthcare industry data are widely dispersed and comes in varying formats due to its origin within the medical sector. Of all the data types, information from the insurance sector is particularly sensitive, as it cannot be seamlessly transferred across sectors. A significant challenge arises in data analysis due to the inability of standard learning and training methodologies to process data without undergoing transfers [11].

4.3 E-HEALTHCARE SYSTEM

The term "e-healthcare system" describes how information technology, electronic processes, and communication tools are incorporated into the healthcare industry to enhance the provision of medical care. e-Healthcare can be defined as offering patients round-the-clock medical help via technological instruments and methods such wireless facilities, medical support, cellular technology, and information and communication technology [12].

The e-health systems can revolutionize healthcare services including health records in the electronic form, related scientific journals and consultation records either audio or video, and online consultations, feedback, medical records sharing, and more.

These systems leverage various technological tools to provide constant healthcare assistance to patients, streamline processes, enhance communication between healthcare providers and patients, and ultimately improve the overall efficiency and

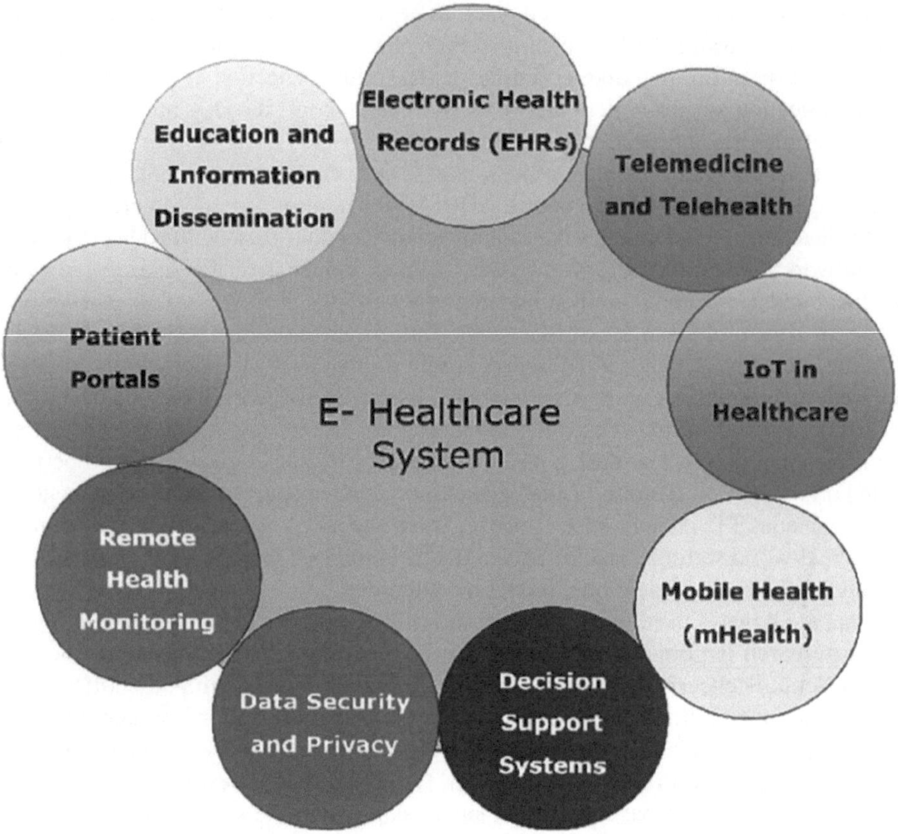

FIGURE 4.2 Key components of e-healthcare system.

effectiveness of healthcare services. Some of the key components and characteristics of e-healthcare systems are discussed as follows (Figure 4.2).

4.3.1 Electronic Health Records

It is the precise collection of population and patient-specific electronic data related to health in a digital format, which is used by many healthcare systems [13,14]. Various methods are used to share the records such as network connected systems, information network, or any other exchanges. E-Health records comprise extensive information including demographics, immunization, medication and allergy history, pathology lab results, images, personnel statistics related to body and billing information.

EHRs have been hailed as the key to improving care standards for a few years now [15]. Patient record data are currently being used by medical professionals to improve quality through care management programmes, in addition to being used for patient charts [14]. The integration of clinical data types from many clinics facilitates the identification and categorization of individuals with chronic illnesses.

Hospitalization rates are decreased thanks to these analytics and data, which may also enhance healthcare systems.

e-Health record systems are used to digitally store and routinely record correct patient data. The digital availability of all data saves clinicians' time when reviewing patient's medical histories. In addition, it permits direct connection between the patient and the healthcare professional and offers "privacy and security." Because there is only one editable file, there is less chance of data replication, less chance of papers being misplaced, and more cost-effectiveness. Also, there's a greater chance the file is up to date. Using e-health and medical records makes it simple to analyze them.

4.3.2 Telemedicine and Telehealth

This involves using telecommunications technology to provide therapeutic services remotely. Telemedicine enhances access to healthcare services by enabling virtual consultations, remote patient monitoring, and the exchange of medical information over great distances, especially in underserved or remote places. Physicians can perform more efficiently thanks to the use of telecommunications technologies for different medical diagnostics and treatments. It is simple to assist patients who are unable to visit and to perform diagnostic testing on a patient's recovery without hesitation [16].

Like telemedicine, telehealth provides more medical services remotely, away from the traditional doctor–patient relationship. It usually calls for the support of medical experts, such as pharmacists, nurses, and physicians, who assist with medication compliance, patient health improvement initiatives, and counseling patients regarding their medical concerns.

4.3.3 IoT in E-Healthcare

Using IoT-enabled devices, the healthcare industry can simply monitor patients thanks to innovations in the IoT. It is now able to freely monitor patients and give them outstanding therapy thanks to the IoT-based e-healthcare system. Patients feel more comfortable now that they can communicate with doctors more easily. This reduces hospital admissions as well because specialized care can be provided remotely [17].

IoT reduced healthcare costs while also having an impact on treatment outcomes. Using IoT devices to enable the remote delivery of healthcare solutions is revolutionizing the healthcare sector. Doctors, patients, hospitals, and the insurance sector all gain from IoT applications in the healthcare sector.

IoT devices collect and give medical professionals access to real-time health data. Wearables and monitoring devices are two types of these gadgets. These data, which may include vital signs, physical activity, and other relevant health metrics, enable proactive healthcare management.

4.3.4 Mobile Health

An important part of e-healthcare is mobile devices and software. Mobile apps allow patients to contact healthcare providers, get health information, remind themselves to take their medications, and arrange appointments. In underprivileged communities with high population densities and high rates of mobile phone usage, mobile health (mHealth) is gaining traction as a viable option. The non-profitable organizations are promoting mHealth in developing countries, for example, mHealth Alliance [18].

4.3.5 Decision Support Systems

Healthcare professionals may use the technologies related to decision support systems for the decision-making toward patient care. The decision support system handles volumes of medical data which is used by the various algorithms for the data analysis. Decision support systems provide better insight into the diverse, data such as imaging scans, lab results, patient history, and so on, collected from multiple resources [19].

The decision support system objective is the disease diagnosis. The professional uses the decision support system for the diagnosis of disease by analyzing the patient's symptoms with the medical history and test results. All elements are considered by the decision support system for the recommendation of the treatment. The decision support system improves the experience and efficient treatment strategies for the better treatment of the patient.

4.3.6 Data Security and Privacy

Data security and privacy are important factors for the patients while using the e-healthcare systems. The is a need to take strong security measures to protect the data from unauthorized access from hackers [20].

The medical data may be encrypted at the time of storage and transmission from one place to another for safety purposes. Encryption made it easy to personalize the data from hackers by transforming the data with the application of an algorithm that converts data into an unreadable form. Access control may be applied as privileges to provide access rights to various types of users in the healthcare industry. The authentication roles may be defined based on various factors for different types of users.

4.3.7 Remote Health Monitoring

The health monitoring previously was very difficult. With the help of remote health monitoring, a revolution in the healthcare industry came which changed the process of monitoring nowadays. Remote health monitoring made it easy to monitor the patient's health in real-time while professionals are not available in the hospitals. Various remote devices are available having sensors to collect and transmit the data to healthcare professionals to monitor patients remotely.

The patients are also advised to use fitness devices for the tracking of their health conditions. The health monitoring for diseases like diabetes, blood pressure, heart disease, and so on.

The Bluetooth sensor may be placed to transmit the data to a designated professional. Wireless internet devices are also helpful in gathering data from healthcare professionals. These data may be used by the healthcare professional to provide better solutions. Healthcare professionals are monitoring the data in regular intervals for their patients to assist them [21].

4.3.8 Patient Portals

The specialized patient portals are helpful for patients to keep their health records, test results, and appointment purpose for healthcare management. These portals help provide security to medical data and easy access for the healthcare professional. Healthcare professionals also ease toward providing better assistance to the patients [22]. Appointment scheduling is also effective in providing better streaming between healthcare professionals and patients. Patients also see the prescription or clarification through the data available on the portals. Healthcare professionals may track their patient's health progress and guide them. These types of platforms empower patients and healthcare professionals both for overall healthcare industry using technology.

Appointment scheduling features streamline administrative processes, improving efficiency for both patients and healthcare staff. Patients can easily request prescription refills, seek clarification on health information, and receive personalized health education resources. Through patient portals, individuals can track their health progress, set goals, and stay proactive about their wellness journey. These platforms contribute to patient empowerment, satisfaction, and overall healthcare quality by leveraging technology for seamless interaction and collaboration.

4.3.9 Education and Information Dissemination

Information availability plays a key role in the e-healthcare systems. The education regarding this may be provided with the help of videos or manuals. This helps patients understand how to share resources with medical professionals, diagnosis, treatment, and feeling of autonomy. The patients also feel confident about data security and privacy. This education helps patients and medical professionals with decision-making, which leads to the better monitoring of health. e-Healthcare portals provide comfort to patients by sitting in their homes without tension and getting personalized monitoring all the time [23].

Implementing and optimizing e-healthcare systems can lead to improved healthcare accessibility, efficiency, and patient outcomes. However, it also requires addressing challenges such as data security, regulatory compliance, and ensuring that technology complements and enhances the human aspect of healthcare delivery.

e-Healthcare is an all-encompassing strategy that combines technology and healthcare services. It is defined by the ongoing delivery of medical treatment to patients using a range of technological tools and approaches. This entails making use of wireless infrastructure, medical support, cellular technology, and information and communication technologies [12]. Practically speaking, this means EHRs, online scientific publications, audio and video consultations with medical specialists, and web-based procedures for tasks like sending test findings and giving comments to physicians. The way healthcare is delivered could be completely changed by these cutting-edge e-health solutions.

e-Healthcare has a wide range of applications, including the management of citizen health data and more specialized support like management assistance and healthcare service delivery [24]. The COVID-19 pandemic has hastened the deployment of e-Health systems, especially with the appearance of new viral strains [25]. A greater dependence on remote health management services enabled by computers and related equipment has resulted from the difficulties faced by medical staff, in addition to limited resources and growing healthcare expenses [11,24,26].

The IoT and smart gadgets play a crucial role in delivering automated remote healthcare services. These gadgets make it easier for patient health data to be sent to cloud servers, providing access to vital medical data [27]. Three essential components make up the use of IoT in healthcare management services: locating patients or medical personnel, recognizing and authenticating individuals, and automating the gathering of data from sensory devices.

EHRs, computer-assisted therapy sessions, online consultations, telemedicine, and decision-based support systems are all essential parts of e-Health. There are two types of ethical factors in e-Health: technical and professional. Similar to traditional healthcare settings, EHRs have fiduciary responsibility regarding how patient data are handled.

Researchers are looking at cutting-edge approaches for e-healthcare systems, and AI has become a crucial component of the healthcare industry. AI has the potential to be used in diagnosis and decision-making; for example, systems based on AI have been created to distinguish between pneumonia in patients infected with COVID-19 and common pneumonia features [21,28]. With its many enticing advantages, federated learning with artificial intelligence (FL–AI) emerges as a potential paradigm for smart healthcare.

By guaranteeing that only local updates are necessary for AI training in the smart healthcare system, FL–AI allays worries about data privacy. By storing raw data locally, client security is improved and the chance of private user data leaking to outside sources is decreased. Moreover, FL–AI outperforms conventional centralized learning techniques by striking a fair balance between accuracy and utility. It sacrifices just nominal accuracy, maintaining the model's generalizability. The scalability of the smart healthcare system is further improved by FL–AI's distributed learning feature, which offers flexibility in a range of healthcare network scenarios [29].

Most importantly, FL–AI helps to lower the communication expenses related to the transfer of raw data. FL–AI prevents the data from moving into the centralized network by reducing the risk of theft of data. The movement toward efficient saving costs and resources used in the network is reduced.

The e-healthcare industry revolutionized with the use of technology and provides efficient solutions for quality and overall improvement of the healthcare system. This not only provides solutions to developing countries but also to developed countries toward improvement in global health crises during a pandemic [10].

4.4 FL–HEALTHCARE INTEGRATION

FL innovates the healthcare industry by adding machine learning approaches to manage the security and privacy of the sensitive data of the patient. The FL and healthcare integration manages the challenges to ensure the machine learning models are trained efficiently without revealing the patient data.

FL uses patient-oriented approaches to ensure the control of patient's health information by sharing with the healthcare industry. The patient-oriented approach gives autonomy to the patient for getting professional advice at anytime from anywhere. The FL also integrates with edge computing for the enhancement of privacy and efficiency of the model training. Edge devices help to perform updates in local models by reducing the data transmission to centralized servers. Edge AI provides real-time training of the model. The patients are now able to get personalized timely support decision-making regarding health issues.

FL may also analyze longitudinal data of patients spread across multiple healthcare organizations. This helps to understand the disease and treatment on time and identify the patterns regularly. FL changes the traditional healthcare domains with the help of experts by using various technologies related to data science with AI. The integration of an interdisciplinary approach provides innovative solutions in the healthcare industry. The model trained for prediction will impact all the decisions for the patient. Here, FL models are using the various types of data because of their impact on patient health. FL models focus on transparency to improve the understanding of healthcare professionals to recommend solutions.

The organizations working toward creating frameworks to use the FL application because of its popularity in the healthcare industry. These organizations also address the standards for the validation and performance of the models using these frameworks. Collaboration in FL is also helpful in rare diseases by considering the historical data. The model training is done with various dataset integrations from different organizations to increase the precision of information.

Security and privacy using FL are improved using data analysis and prediction for the medical industry [25,30,31]. The use of FL enables patient treatment in the early stage by the prediction of models [32]. The application of the FL-based model is to predict the similarity among the patients [33]. The data can be examined for privacy including the treatment help for various diseases and causes [31].

To analyze the data in the FL-based models, it is necessary to manage the hospital's critical care units effectively, maintaining data privacy and predicting patient mortality, length of stay, and admission details [13,34].

FL–healthcare integration is a comprehensive strategy that extends beyond the technical components of training models. It includes working together across different areas, patient empowerment, regulatory compliance, and ethical issues. The fields of medical research, diagnostics, and personalized patient care are poised to undergo significant changes due to the convergence of FL and healthcare.

Below mentioned are some of the principles of FL–healthcare integration:

4.4.1 Privacy-Preserving Learning

For the integration of a FL–healthcare system, the maintenance of privacy is one of the main factors. In the traditional machine learning models, it is a challenge to maintain the security and privacy of medical data due to the centralized method of data training. The training of models locally in FL solves the problem of machine learning. It ensures that the data of the patient is safe in the healthcare organization due to decentralization which reduces data breaches. At the time of model updates, it is communicated to the centralized server in the FL framework. The local model is further combined with the global model for updation and to keep the patient data in the raw form. The changes in the model training must stick to the data protection requirements in all the fields, such as Health Insurance, and patient data while protecting the privacy of data.

4.4.2 Decentralized Healthcare Data

FL focuses on the data localization process which allows healthcare organizations to store their patient data on the local machines or server. Healthcare data follow the various regulatory frameworks and strictly control its storage. FL also supports the regulatory requirements by keeping data in the local devices or servers and maintaining the patient's confidence in the security and privacy of the data.

4.4.3 Collaborative Learning across Institutions

Learning in collaboration through FL improves the faith of healthcare organizations. The healthcare organization is collaborating in the process of model training without sharing patient data and contributing to the improvement of machine learning models. The collaborative contribution of the organizations supports robust models to address the challenges of the healthcare industry globally.

4.4.4 Improved Model Generalization

FL uses decentralized training for the development of models across various healthcare organizations. Traditional models are training the model on centralized datasets which is difficult to adapt to global healthcare settings. FL provides better insight into data due to the capability of accessing various datasets across the healthcare organization which further results in the model as versatile, and applicable to a wide range of patients.

4.4.5 Reduced Communication Overhead

In machine learning, the communication overhead to manage the volume of data in the healthcare sector. FL decreases the communication overhead by updating the model on time. The transmission of data is reduced in federated machine learning, which optimizes the use of the network but also reduces the risk of network congestion.

4.4.6 Customized Healthcare Solutions

Healthcare organizations are helped by providing flexibility in model training and analyzing data for specific types through FL. The customization may be done in the machine learning models to ensure the specific healthcare factors. For example, in a pediatric hospital, a model training of the data with consideration of any other data in other medical organizations.

4.4.7 Security and Regulatory Compliance

Data security in the healthcare industry is important due to the sensitive nature of patient data. FL takes care of security measures by using encryption, and secure aggregation to protect the student data during the training of the model. FL is integrated with healthcare to align with a standard framework for following healthcare regulations for data protection.

4.4.8 Optimizing Network Bandwidth

It is necessary to use the network resources efficiently for real-time data access in the healthcare industry. FL uses the network resources only at the time of model updation. This optimizes the wide healthcare networks to solve the network resources challenges with the machine learning.

4.4.9 Real-Time Learning and Adaptation

FL based on the continuous learning which uses the new data to update the model. The healthcare industry has the challenge of updating with environmental conditions from time to time to have the latest knowledge. This helps to deliver high-quality solutions and better treatment to patients (Figure 4.3).

4.5 CHALLENGES

The integration of FL and healthcare has many advantages and as well challenges that need attention. Heterogenous data used from various healthcare organizations are the key challenge. All organizations have different data formats, characteristics, and standards, used during the model training. There is a need to select an appropriate model for the aggregation of heterogeneous data. The selection of an efficient aggregation model helps to update the global model. Lots of research is going on in the direction of optimized aggregation methods that maintain privacy, accuracy, and efficiency. Federated machine learning enables the collaborative analysis of distributed healthcare data without breaching patient privacy in the development of a global model. Despite so many advantages, federated machine learning faces several challenges in the field of medical industry. There is a need to understand and address the challenges in the field of federated machine learning (Figure 4.4).

4.5.1 Data Heterogeneity

The healthcare industry deals with heterogeneous data with various factors such as formats, quality standards, and so on from the various healthcare industries from different geographical locations. It is a big challenge to deal with diversity in models of federated machine learning at the time of training. In this case, sometimes data

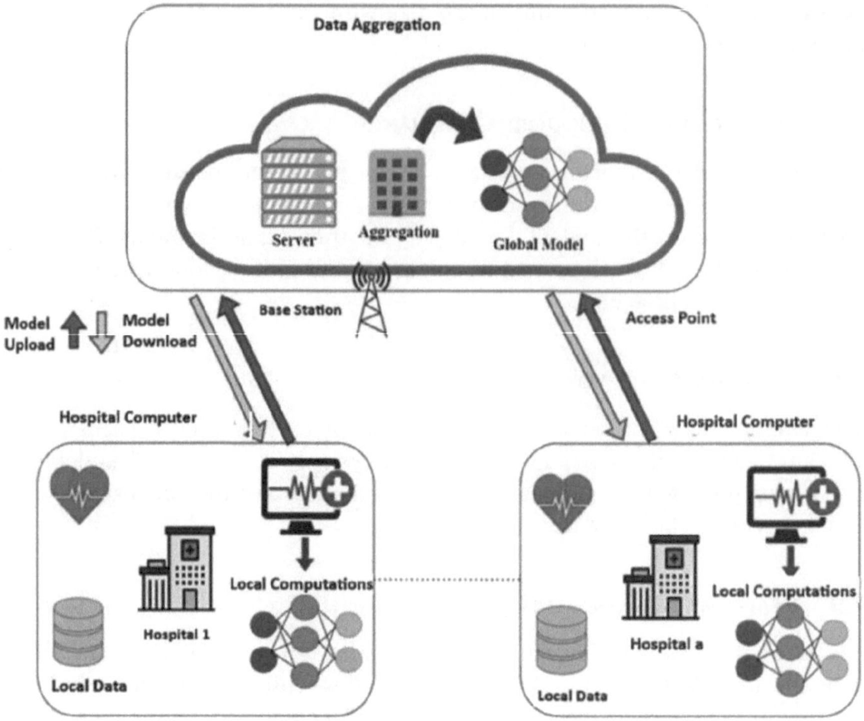

FIGURE 4.3 Federated learning with healthcare [30].

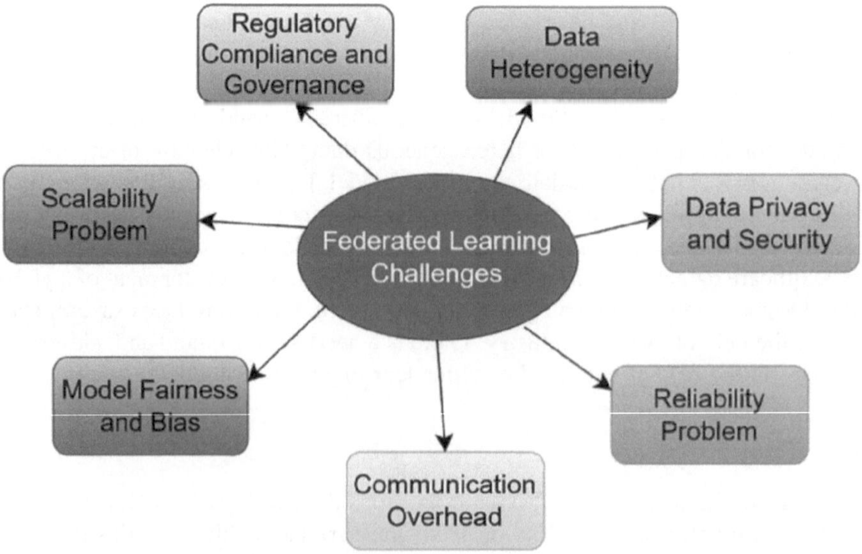

FIGURE 4.4 Federated learning challenges.

does not possess the quality. There is a need to give attention to heterogeneity for the development of robust models using FL algorithms with the added capability of different data types to ensure fairness and effectiveness.

4.5.2 Data Privacy and Security

The healthcare industry follows the regulatory requirements due to the sensitive nature of patient data. Federated machine learning keeps the data decentralized and local at their location by reducing the data breaching risk and unauthorized access. Federated machine learning uses various encryption techniques, secure communication protocols, etc. The risks are reduced if there is no risk of inference attacks.

4.5.3 Reliability Problem

The dependability of the current healthcare system has also come under scrutiny in the wake of security concerns. An attacker can take control of both the session and the data being exchanged in it by adopting session hijacking in the current smart healthcare system, which could have serious consequences for the system. A side script can then be used by the hacker to access the IoT network. The sensitive data on the system may be compromised if malicious SQL code is introduced into the database. When a user attempts to authenticate or log in, packet interception may potentially result in the loss of identity-related information. A ransomware attack can be used by an attacker to encrypt a healthcare system user's personal information. Therefore, the reliability of these attack possibilities is called into question. For example, when malicious actors can encrypt sensitive data from a system, it raises concerns about how trustworthy and secure the system is for data transfers [35].

4.5.4 Communication Overhead

Federated machine learning involves communication between participating nodes (e.g., healthcare institutions or devices) for exchanging model updates and aggregating gradients. This communication overhead can be significant, especially in large-scale FL settings with numerous participants and complex models. Minimizing communication overhead while maintaining model accuracy and convergence is a key challenge in FML. Techniques such as model compression, differential privacy, and decentralized optimization algorithms can help mitigate communication costs in FL.

4.5.5 Model Fairness and Bias

Biases inherent in healthcare data, such as disparities in patient demographics, disease prevalence, and healthcare access, can introduce biases into federated machine learning models. These biases can lead to unfair or discriminatory outcomes, undermining the reliability and trustworthiness of

Federated Machine Learning (FML) systems. Addressing model fairness and bias in FL requires careful consideration of data representativeness, algorithmic fairness metrics, and mitigation strategies such as bias correction and fairness-aware optimization techniques.

4.5.6 Scalability Problem

A massive number of incoming and outgoing networks need to be controlled in today's healthcare systems. Consequently, it is difficult to support numerous high-speed wireless systems without resulting in a decrease in system speed [26]. However, it is getting more and harder to guarantee speed, quality, and low cost at the same time as the number of networked devices increases. The network's bandwidth is once again being taxed by the increase in data traffic [36]. In addition, the architecture of the network presents scalability issues because it is hybrid and comprises devices from multiple suppliers. It is a big problem to manage the devices remotely in various terms such as time, administration, and so on.

4.5.7 Regulatory Compliance and Governance

Federated machine learning must comply with various regulatory frameworks such as HIPAA, GDPR, and the Health Information Technology for Economic and Clinical Health (HITECH) Act is essential for deploying federated machine learning in healthcare settings. It is mandatory to ensure compliance with legal and ethical considerations related to data privacy, consent management, etc. Establishing robust governance mechanisms, data-sharing agreements, and accountability frameworks is crucial for ensuring regulatory compliance and fostering trust among stakeholders in FL initiatives.

4.6 CONCLUSION

Integration of FL and healthcare signifies a change in machine learning which is applied to healthcare problems. FL handles issues of centralized methods by managing privacy and decentralized data by encouraging cooperation. It provides a better model with less communication overhead and adapts the different healthcare environments. FL provides the solution which expands the possibilities of machine learning. It also ensures the ethical and secure use of patient data for improved healthcare outcomes.

In FL, improving healthcare results and safeguarding patient privacy are two objectives that healthcare integration shows promise in tackling. FL emerges as a critical facilitator for collaborative and privacy-preserving developments in healthcare as technology advances and healthcare institutions increasingly employ data-driven approaches.

Federated machine learning faces several obstacles that need to be overcome in order to reach its full potential, even if it has enormous potential to advance medical research and healthcare delivery. Researchers, clinicians, legislators, and industry stakeholders must work together interdisciplinary to find creative solutions that strike a balance between the advantages of FML and the requirements to safeguard patient privacy, maintain data security, and advance equity and openness in healthcare AI applications.

The research community and FL are interested in incorporating AI into the intended networks due to the current advancements in healthcare. This survey thoroughly mines the vast amount of data on novel topics: FL, AI, XAI, and e-Healthcare. Furthermore, we go into great depth about the most recent advancements in AI and FL for smart healthcare applications. The integration of FL–AI, FL–Healthcare, and AI–Healthcare has been substantial. Moreover, we use the terms we've studied to handle a wide range of challenges, including security, privacy, dependability, scalability, and secrecy. Healthcare is not restricted to these methods, even though we have examined FL and AI–XAI tactics for the healthcare system and classed them into several types of solutions.

There are still certain problems and challenges that need to be fixed. This study looks at security advances, innovative dialogue, benefits of integration, taxonomies, and unresolved issues. Additionally, we have offered some suggestions for additional study in this field.

4.7 FUTURE SCOPE

4.7.1 Function of FL in Data Science

Every industry generates enormous amounts of data as IoT-based techniques proliferate within the current framework. The data scientist's present task is this. Data security is among the most crucial, and the sensors connected to the IoT mechanism have a lower memory capacity. Nevertheless, they produce enormous amounts of data every nanosecond. Therefore, it is both vital and challenging to store or convey these data quickly. The usage of FL in this scenario has the advantages of not requiring a data storage system and distributing the data across several clients; additionally, the security of individual data is guaranteed because the data is trained locally on each client [37].

4.7.2 Blockchain Technology with Artificial Intelligence in Security

Among all agents, AI is the most powerful. AI occasionally functions as a link and mediator for practical applications. However, blockchain offers a distributed ledger to protect against such weaknesses [38]. Blockchain (BC) additionally provides a provisional ledger to satisfy security and confidentiality requirements. For AI applications to function efficiently in application contexts, a secure platform is required. BC interacts with AI with ease and performs better than anticipated for the intended system. In the BC environment, AI agents are only connected while they are communicating with one another. When combined, they create a highly secure setting that can benefit both present and future customers [19,39].

4.7.3 Using FL–AI for Robot Control

Robots are utilized in numerous industries today to relieve people of a variety of hazardous activities in the context of modern technology. Once again, AI-based agents can handle a large variety of data exchanges and have a very high processing capacity. Furthermore, data and information can be efficiently stored in the cloud system with a remarkably big memory capacity thanks to state-of-the-art cloud storage technologies. The capacity to quickly obtain data from various sources has made continuing education and training possible. In the modern Industry 4.0 era, autonomous agents have more ramifications. Numerous businesses, including healthcare, education, the military, and even certain hotels, use robots. Safe data collection is essential, though, because training an agent requires a substantial amount of data from several sources. FL is therefore a wise decision since it avoids the most difficult task—data transfer—while protecting data privacy and ensuring the autonomous system's effectiveness [40].

4.7.4 Big Data Analytics in FL

The requirement for a huge volume, quick speed, and diverse range of data is the core idea of big data. Gathering information from a variety of sources and compiling it into a dataset containing a diversity of information from those sources is the main goal of comprehensive data analysis. This has become easier with the introduction of wireless networking technologies like Industry 4.0, IoT, and even 5G or 6G. These advancements are all helpful for data collection and work well in real-time settings. Big data are generated and gathered, among other places, by the industrial sector. This strategy is challenging because critical user privacy could be compromised or altered during data transit from its source to its destination. Machine learning models can be trained locally on system data thanks to a new type of AI technology called FL. Both the source data and the important data criterion are used to ensure continuous training. As a result, data privacy is guaranteed and there is no need to transfer sensitive customer data between locations [41].

4.7.5 FL in Blockchain-Based Industry 5.0 Applications

The main objective of the upcoming IR 5.0 is to establish a link between intelligent technology and the general people. IoT devices can be used to collect data even in remote locations, and once the data is acquired, connectivity can be created without any additional work on the part of the owner. They get the data, which has been collected from IoT devices. But the information that is collected is unprocessed and might not be in a format that a computer can comprehend. This is where the AI of the ML algorithm is used. Edge computing quickly completes the analysis after AI algorithms carry out the necessary calculations. However, it can be difficult to prevent an intruder from assaulting important data. As data are sent between devices, hackers may attack the hub's nodes. In this situation, blockchain technology and FL may be helpful. In this instance, each side gathers, trains, and processes input locally before sending the output to the server. The original configurations were then securely moved across the blockchain's blocks [42,43].

4.7.6 FL with Healthcare in Smart Industrial IoT

Many industries, including data research, education, healthcare, and so on, use IoT devices. This is a result of their growing usage. The smart IoT era has simplified the collection and transmission of data to a central server for further processing in the healthcare system. These data include MRI pictures, test samples for specific diseases, information about a patient staying in the hospital, and even data about hospital staff or management [44]. However, as these tools have developed, data collecting has emerged as the primary motivation for hackers. As a result, an attacker can easily read data while it is being transferred from the source to the destination and get access to the system. Given that FL maintains data confidentiality on the local system until processing activities are finished at each client, it might be a workable solution in this regard. After that, a secure scenario is used to send the output back to the central server for extra processing.

4.7.7 AI-FL in Intelligent Healthcare Powered by Big Data

Standardization in the procedures for gathering data, processing it, and producing results is highly desirable. Again, the healthcare sector would greatly benefit from a platform that could compile data from numerous sources, carry out a task, deal with pressing issues, or reach a decision without requiring human interaction. AI is critical to the healthcare industry because it increases computer awareness of human brain functions and provides access to a variety of powerful AI-based algorithms that simplify the administration of challenging tests for medical practitioners. Smart wearables with AI advantages and FL are used to aggregate data into a secure and intelligent system. The FL model is helping with the safe data collecting and clever AI model development. Predictive modeling of hospital stays duration and mortality using EHRs is also being done.

REFERENCES

1. Kishor, K., Singh, P., & Vashishta, R. (2023). "Develop model for malicious traffic detection using deep learning." In: Sharma, D. K., Peng, S. L., Sharma, R., Jeon, G. (eds) *Micro-Electronics and Telecommunication Engineering*. Lecture Notes in Networks and Systems, vol 617. Springer, Singapore. https://doi.org/10.1007/978-981-19-9512-5_8
2. Rieke, N., Hancox, J., Li, W., Milletari, F., Roth, H. R., Albarqouni, S., Bakas, S., Galtier, M. N., Landman, B. A., Maier-Hein, K., & Ourselin, S. (2020). "The future of digital health with federated learning." *NPJ Digital Medicine*, 3(1), 1–7.
3. Jayatilake, S. M. D. A. C., & Ganegoda, G. U. (2021). "Involvement of machine learning tools in healthcare decision making". *Journal of Healthcare Engineering*, 2021(1), 6679512.
4. Kishor, K., & Pandey, D. (2022). "Study and development of efficient air quality prediction system embedded with machine learning and IoT." In: Gupta, D., Khanna, A., Bhattacharyya, S., Hassanien, A. E., Anand, S., Jaiswal, A. (eds) *Proceeding International Conference on Innovative Computing and Communications*. Lecture Notes in Networks and Systems, vol. 471, Springer, Singapore, https://doi.org/10.1007/978-981-19-2535-1_24
5. Carbonell, J. G., Michalski, R. S., & Mitchell, T. M. (1983). "An overview of machine learning." In: Michalski R. S., Carbonell J. G., Mitchell T. M. (eds), *Machine Learning*. Elsevier Science Publishers, San Francisco, CA. https://doi.org/10.1016/B978-0-08-051054-5.50005-4

6. Javaid, M., Haleem, A., Singh, R. P., Suman, R., & Rab, S. (2022). "Significance of machine learning in healthcare: Features, pillars and applications." *International Journal of Intelligent Networks*, 3, 58–73. ISSN 2666-6030.
7. Wahab, O. A., Mourad, A., Otrok, H. & Taleb, T. (2021) "Federated machine learning: Survey, multi-level classification, desirable criteria and future directions in communication and networking systems." *IEEE Communications Surveys & Tutorials*, 23(2), 1342–1397.
8. Ge, L., Li, H., Wang, X., & Wang, Z. (2023). "A review of secure federated learning: Privacy leakage threats, protection technologies, challenges and future directions." *Neurocomputing*, 561, 126897. ISSN 0925-2312.
9. Kishor, K., & Nand, P. (2023). "Wireless networks based in the cloud that support 5G." In: Kishor K., Saxena, N., and Pandey, D. (eds), *Cloud-based Intelligent Informative Engineering for Society 5.0*, 1st edition, pp. 23–40. Chapman and Hall/CRC, New York. ISBN: 9781003213895. https://doi.org/10.1201/9781003213895-2
10. Sharma, A., Jha, N., & Kishor, K. (2022). "Predict COVID-19 with chest X-ray." In: Gupta, D., Polkowski, Z., Khanna, A., Bhattacharyya, S., Castillo, O. (eds) *Proceedings of Data Analytics and Management.* Lecture Notes on Data Engineering and Communications Technologies, vol. 90. Springer, Singapore. https://doi.org/10.1007/978-981-16-6289-8_16
11. Moulahi, W., Jdey, I., Moulahi, T., Alawida, M., & Alabdulatif, A. (2023). "A blockchain-based federated learning mechanism for privacy preservation of healthcare IoT data." *Computers in Biology and Medicine*, 167, 107630.
12. Hoque, M. R., Mazmum, M. F. A., & Bao, Y. (2014). "E-health in Bangladesh: Current status, challenges, and future direction." *International Technology Management Review*, 4(2), 87–96.
13. Chen, M., Yang, Z., Saad, W., Yin, C., Poor, H. V., & Cui, S. (2019). "Performance optimization of federated learning over wireless networks." In: *2019 IEEE Global Communications Conference (GLOBECOM)* (pp. 1–6). IEEE, Waikoloa, HI. DOI: 10.1109/GLOBECOM38437.2019.9013160
14. Cowie, M. R., Blomster, J. I., Curtis, L. H., Duclaux, S., Ford, I., Fritz, F., Goldman, S., Janmohamed, S., Kreuzer, J., Leenay, M., & Michel, A. (2017). "Electronic health records to facilitate clinical research." *Clinical Research in Cardiology*, 106(1), 1–9. https://doi.org/10.1007/s00392-016-1025-6.
15. Kishor, K. (2022). "Communication-efficient federated learning." In: Yadav, S. P., Bhati, B. S., Mahato, D. P., Kumar, S. (eds) *Federated Learning for IoT Applications.* EAI/Springer Innovations in Communication and Computing. Springer, Cham. https://doi.org/10.1007/978-3-030-85559-8_9
16. Payán, D. D., Frehn, J. L., Garcia, L., Tierney, A. A., & Rodriguez. H. P. (2022). "Telemedicine implementation and use in community health centers during COVID-19: Clinic personnel and patient perspectives." *SSM-Qualitative Research in Health*, 2, 100054.
17. Kumar, M., Kumar, A., Verma, S., Bhattacharya, P., Ghimire, D., Kim, S.H., & Hosen, A. S. (2023). "Healthcare Internet of Things (H-IoT): Current trends, future prospects, applications, challenges, and security issues." *Electronics*, 12(9), 2050.
18. Istepanian, R. S. H. (2022). "Mobile health (m-Health) in retrospect: The known unknowns." *International Journal of Environmental Research and Public Health*, 19(7), 3747.
19. Latif, S. A., Wen, F. B. X., Iwendi, C., Li-li, F. W., Mohsin, S. M., Han, Z., & Band, S. S. (2022). "AI-empowered, blockchain and SDN integrated security architecture for IoT network of cyber-physical systems." *Computer Communications*, 181, 274–283.
20. Abouelmehdi, K., Beni-Hssane, A., Khaloufi, H., & Saadi, M. (2017). "Big data security and privacy in healthcare: A Review." *Procedia Computer Science*, 113, 73–80.

21. Zhang, H., Jiang, S., & Xuan, S. (2024). "Decentralized federated learning based on blockchain: Concepts, framework, and challenges." *Computer Communications*, 216, 140–150.
22. Carini, E., Villani, L., Pezzullo, A. M., Gentili, A., Barbara, A., Ricciardi, W., & Boccia, S. (2021). "The impact of digital patient portals on health outcomes, system efficiency, and patient attitudes: Updated systematic literature review." *Journal of Medical Internet Research*, 23(9), e26189.
23. Kishor, K., Sharma, R., & Chhabra, M. (2022). "Student performance prediction using technology of machine learning. In: Sharma, D. K., Peng, S. L., Sharma, R., Zaitsev, D. A. (eds) *Micro-Electronics and Telecommunication Engineering.* Lecture Notes in Networks and Systems, vol. 373. Springer, Singapore. https://doi.org/10.1007/978-981-16-8721-1_53
24. Muhammad, G., Rahman, S. K. M. M., Alelaiwi, A., & Alamri, A. (2017). "Smart health solution integrating IoT and cloud: A case study of voice pathology monitoring." *IEEE Communications Magazine*, 55(1), 69–73.
25. Rahman, A., Chakraborty, C., Anwar, A., Karim, M., Islam, M., Kundu, D., Rahman, Z., & Band, S. S. (2021). "SDN-IoT empowered intelligent framework for industry 4.0 applications during COVID-19 pandemic." *Cluster Computing*, 25, 2351–236
26. Muhammad, G., Hossain, M. S., & Kumar, N. (2021). "EEG-based pathology detection for home health monitoring." *IEEE Journal of Selected Areas in Communications*, 39(2), 603–610.
27. Masud, M., Gaba, G. S., Alqahtani, S., Muhammad, G., Gupta, B., Kumar, P., & Ghoneim, A. (2021). "A lightweight and robust secure key establishment protocol for Internet of Medical Things in COVID-19 patients care." *IEEE Internet of Things Journal*, 8(21), 15694–15703.
28. Kishor, K. (2023). "Chapter 10 Study of quantum computing for data analytics of predictive and prescriptive analytics models". In: Yadav, S. P., Singh, R., Yadav, V., Al-Turjman, F., Kumar, S. A. (eds) *Quantum-Safe Cryptography Algorithms and Approaches: Impacts of Quantum Computing on Cybersecurity* (pp. 121–146). De Gruyter, Berlin, Boston. https://doi.org/10.1515/9783110798159-010
29. Khan, L. U., Pandey, S. R., Tran, N. H., Saad, W., Han, Z., Nguyen, M. N., & Hong, C. S. (2020). "Federated learning for edge networks: Resource optimization and incentive mechanism." *IEEE Communications Magazine*, 58(10), 88–93.
30. Kishor K. (2022). "Personalized federated learning". In: Yadav, S. P., Bhati, B. S., Mahato, D. P., Kumar, S. (eds) *Federated Learning for IoT Applications.* EAI/Springer Innovations in Communication and Computing, pp. 31–52. Springer, Cham. https://doi.org/10.1007/978-3-030-85559-8_3
31. Liu, Y., Yuan, X., Xiong, Z., Kang, J., Wang, X., & Niyato, D. (2020). "Federated learning for 6G communications: Challenges, methods, and future directions." *China Communications*, 17(9), 105–118.
32. Caldas, S., Duddu, S. M. K., Wu, P., Li, T., Konečný, J., McMahan, H. B., Smith, V., & Talwalkar, A. (2018). "Leaf: A benchmark for federated settings." arXiv preprint arXiv:1812.01097.
33. Bakopoulou, E., Tillman, B., & Markopoulou, A. (2019). "A federated learning approach for mobile packet classification." arXiv preprint arXiv:1907.13113.
34. Liu, B., Wang, L., & Liu, M. (2019). "Lifelong federated reinforcement learning: A learning architecture for navigation in cloud robotic systems." *IEEE Robotics Automation Letters*, 4(4), 4555–4562.
35. Rahman, A., Hossain, M. S., Muhammad, G., Kundu, D., Debnath, T., Rahman, M., Khan, M. S. I., Tiwari, P., & Band, S. S. (2022). "Federated learning-based AI approaches in smart healthcare: Concepts, taxonomies, challenges and open issues." *Cluster Computing,* 26, 2271–2311.

36. Hussain, F., Abbas, S. G., Shah, G. A., Pires, I. M., Fayyaz, U. U., Shahzad, F., Garcia, N. M., & Zdravevski, E. (2021). "A framework for malicious traffic detection in IoT healthcare environment". *Sensors*, 21(9), 3025.
37. Ahmed, U., Srivastava, G., & Lin, J. C. W. (2022). "Reliable customer analysis using federated learning and exploring deep-attention edge intelligence." *Future Generation Computer Systems*, 127, 70–79.
38. Gulati, P., Sharma, A., Bhasin, K., & Azad, C. (2020). "Approaches of blockchain with AI: Challenges & future direction." In: *Proceedings of the International Conference on Innovative Computing & Communications (ICICC)*, pp. 1–7. SSRN Elsevier, New Delhi.
39. Kishor, K. (2023). "Cloud computing in blockchain." In: Kishor K., Saxena, N., and Pandey, D. (eds), *Cloud-based Intelligent Informative Engineering for Society 5.0* (pp. 79–105). Chapman and Hall/CRC, New York. ISBN: 9781003213895. https://doi.org/10.1201/9781003213895-5
40. Xianjia, Y., Queralta, J.P., Heikkonen, J., & Westerlund, T. (2021). "Federated learning in robotic and autonomous systems." arXiv preprint arXiv:2104.10141.
41. Xu, W., Fang, W., Ding, Y., Zou, M., & Xiong, N. (2021). "Accelerating federated learning for IoT in big data analytics with pruning, quantization and selective updating." *IEEE Access*, 9, 38457–38466.
42. Xu, L., Zhou, X., Tao, Y., Yu, X., Yu, M., & Khan, F. (2021). "AF relaying secrecy performance prediction for 6G mobile communication networks in industry 5.0." *IEEE Transactions on Industrial Informatics*, 18(8), 5485–5493.
43. Chander, B., Pal, S., De, D., & Buyya, R. (2022). "Artificial intelligence-based internet of things for industry 5.0." In: Pal S., De D, Buyya R (eds), *Artificial Intelligence-Based Internet of Things Systems* (pp. 3–45). Springer.
44. Kishor, K. (2023). "Impact of cloud computing on entrepreneurship, cost, and security." In: Kishor K., Saxena, N., and Pandey, D. (eds), *Cloud-based Intelligent Informative Engineering for Society 5.0*, 1st edition (pp. 171–191). CRC Press, New York. ISBN: 9781003213895. https://doi.org/10.1201/9781003213895-10

5 Artificial Intelligence Techniques Based on Federated Learning in Smart Healthcare

Kanchan Naithani, Y. P. Raiwani, Shrikant Tiwari, and Alok Singh Chauhan

5.1 INTRODUCTION

The integration of artificial intelligence (AI) in smart healthcare systems has emerged as a transformative paradigm, promising enhanced patient care, efficient resource utilization, and improved diagnostic accuracy. Among various AI methodologies, federated learning has gained prominence as a robust and privacy-preserving approach for decentralized data processing in healthcare settings. This chapter explores the intersection of AI techniques and federated learning in the context of smart healthcare applications. Major state-of-the-art AI techniques employed in federated learning for healthcare applications are presented in this book chapter. Insights into ongoing research trends and future directions highlight the evolving landscape of this interdisciplinary field. This book chapter provides a holistic perspective on the integration of AI techniques with federated learning in smart healthcare, emphasizing the potential benefits, challenges, and ethical considerations. Real-world case studies and examples illustrate the successful implementation of federated learning in diverse healthcare scenarios, ranging from disease prediction and personalized treatment recommendations to remote patient monitoring. Furthermore, the chapter discusses the ethical considerations and regulatory aspects surrounding the deployment of AI techniques based on federated learning in smart healthcare. It serves as a valuable resource for researchers, practitioners, and policymakers interested in harnessing the power of AI to advance healthcare while prioritizing data privacy and security.

The healthcare industry is undergoing a profound transformation, propelled by advancements in technology, data analytics, and AI. The emergence of smart healthcare, a revolutionary paradigm, promises to enhance the quality, efficiency, and accessibility of healthcare services [1]. This transformation leverages cutting-edge technologies to create an interconnected ecosystem that not only diagnoses and treats medical conditions but also empowers patients, healthcare providers, and administrators with real-time insights.

DOI: 10.1201/9781003489368-5

Smart healthcare encompasses a spectrum of innovative technologies, that comprise wearable devices, Internet of Things (IoT) sensors, telemedicine platforms, and AI-driven applications [2]. These technologies converge to create an intelligent and responsive healthcare infrastructure capable of addressing the complex challenges facing the industry today.

The key components and objectives of smart healthcare, emphasize how it could transform patient care, streamline healthcare operations, and contribute to the overall well-being of individuals and communities. At the core of smart healthcare is the integration of technologies for information and communication, which enable the seamless exchange of data among different stakeholders in the healthcare ecosystem [3].

The integration of AI in smart healthcare is particularly transformative, as it introduces capabilities such as predictive analytics, personalized treatment plans, and data-driven decision-making. AI systems are able to examine enormous volumes of medical data, ranging from electronic health records (EHRs) to genomic information, to derive valuable insights that can inform diagnosis and treatment strategies. Additionally, AI makes it possible to create intelligent systems that can learn from data, adjust to new information, and enhance their performance over time [4].

The rise of smart healthcare is also marked by a shift toward patient-centric care. Empowered by wearable devices and mobile applications, individuals can actively participate in monitoring their health, receiving timely alerts, and engaging in preventive measures. Telemedicine facilitates remote consultations, making healthcare services more accessible, especially in underserved or remote areas [5].

With the evolution of smart healthcare, issues with data privacy, interoperability, and ethical considerations must be properly addressed. This introduction sets the stage for a comprehensive exploration of the various facets of smart healthcare, emphasizing its potential to redefine the healthcare landscape and contribute to the vision of a healthier, more connected world. Through the integration of technology and human-centric design, smart healthcare aims to not only treat illnesses but also proactively promote well-being and preventive healthcare practices.

5.2 OVERVIEW OF SMART HEALTHCARE SYSTEMS

Smart healthcare systems represent a paradigm shift in the way healthcare is delivered, leveraging advanced technologies to improve patient outcomes, streamline operations, and enhance overall efficiency [6]. These systems integrate a variety of cutting-edge technologies, including the IoT, AI, data analytics, and wearable devices as shown in Figure 5.1.

5.2.1 IoT in Healthcare

Connected medical devices and sensors enable real-time monitoring of patient vitals. Integration of IoT facilitates the seamless exchange of data between devices and healthcare platforms. Remote patient monitoring ensures timely intervention and personalized care.

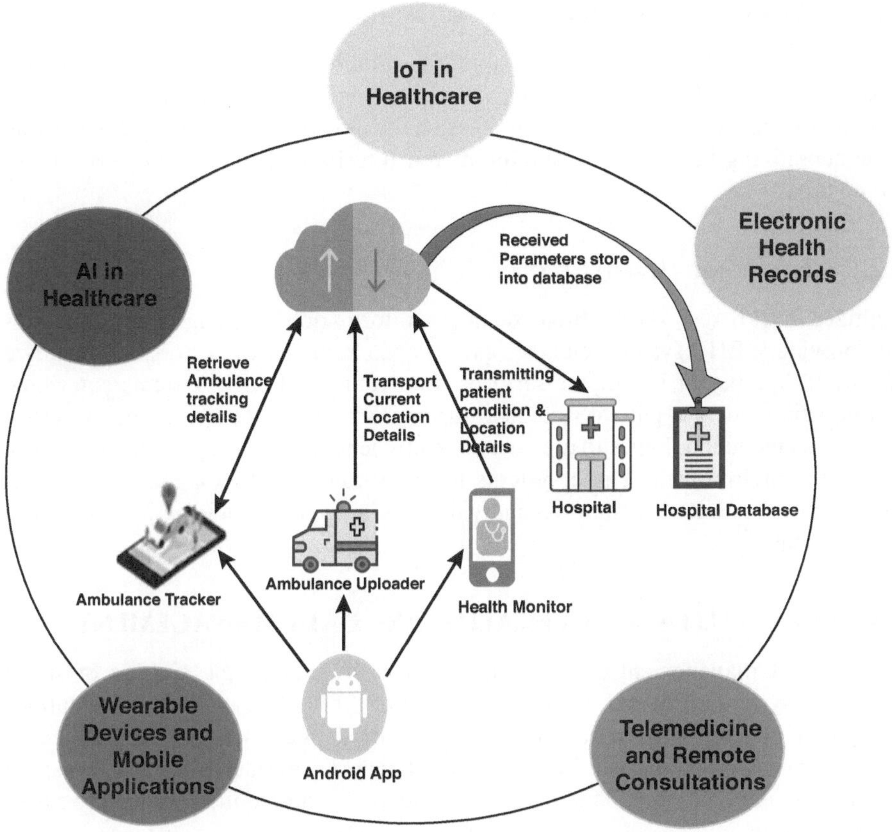

FIGURE 5.1 Smart healthcare systems.

5.2.2 Artificial Intelligence in Healthcare

AI algorithms analyze vast datasets, aiding in diagnostics, treatment planning, and drug discovery. Predictive analytics helps identify potential health risks, allowing for proactive and preventive measures. Chatbots and virtual health assistants improve patient involvement and offer on-demand information.

5.2.3 Wearable Devices and Mobile Applications

Wearables, like fitness trackers and smartwatches, facilitate ongoing health monitoring, while mobile applications empower users to actively oversee their health, access medical records, and receive personalized health recommendations. The integration of wearables and apps fosters a patient-centric approach to healthcare.

5.2.4 Telemedicine and Remote Consultations

Telemedicine platforms facilitate remote consultations, enhancing healthcare service accessibility. Video conferencing and virtual visits offer a convenient means for patients to engage with healthcare professionals. Telemedicine plays a crucial role in democratizing healthcare, extending its reach to individuals in remote or underserved areas.

5.2.5 Electronic Health Records

Digitized health records facilitate seamless information exchange among healthcare providers. EHR systems enhance care coordination, reduce errors, and improve the overall quality of healthcare services. Interconnected EHR systems contribute to a holistic view of a patient's medical history. Smart healthcare systems leverage interconnected technologies to create a more efficient, accessible, and patient-centric healthcare environment. These systems not only improve the diagnosis and treatment of diseases but also empower individuals to actively participate in their health management.

5.3 KEY CHALLENGES IN HEALTHCARE DATA MANAGEMENT

The efficient management of healthcare data is crucial for the success of smart healthcare systems; however, it has a unique set of difficulties and strategic solutions that should be carefully considered [7]. Healthcare data are sensitive and subject to stringent privacy regulations. Ensuring robust security measures to protect patient information is paramount. The increasing practice of interconnected devices raises concerns about data breaches and unauthorized access.

The integration of diverse systems and platforms often leads to interoperability challenges. Standardizing data formats and protocols is crucial for seamless data exchange between different healthcare entities. The volume, velocity, and variety of healthcare data pose challenges in storage, processing, and analysis. Implementing scalable and efficient big data solutions is essential for deriving meaningful insights from vast datasets. While taking ethical consideration into the picture, the use of healthcare data, especially in AI applications, requires careful consideration. Issues such as bias in algorithms, informed consent and responsible data handling must be addressed to maintain public trust.

As for regulatory compliance, healthcare data management must adhere to stringent regulatory frameworks, such as Health Insurance Portability and Accountability Act (HIPAA) in the United States or GDPR in Europe. Staying compliant with evolving regulations adds complexity to data management strategies. Apart from this, establishing effective data governance practices ensures data accuracy, reliability, and consistency. Poor data quality can lead to incorrect diagnoses, treatment errors, and compromised patient care.

Addressing these key challenges is essential to harness the full potential of smart healthcare systems. Strategic planning, collaboration, and the implementation of robust technologies, and policies are necessary to overcome these hurdles and pave the way for a more connected and efficient healthcare ecosystem.

5.4 FOUNDATIONS OF ARTIFICIAL INTELLIGENCE IN HEALTHCARE

AI has emerged as a transformative force in the field of healthcare, reshaping the foundations of how medical data are analyzed, diagnoses are made and patient care is administered [8]. The foundations of AI in healthcare rest on five key pillars Data Integration and Analysis, ML Algorithms, Predictive Analytics, and Robotics and Automation.

AI systems leverage vast datasets, including EHRs, medical imaging, and genomic information, to derive meaningful insights. The ability to integrate and analyze diverse data sources enables more accurate diagnostics and personalized treatment plans. Machine learning algorithms, a subset of AI, play a pivotal role in healthcare applications. These algorithms learn from data patterns, assisting in disease prediction, risk assessment, and identifying optimal treatment strategies.

Natural language processing (NLP) enables computers to understand and process human language, facilitating the extraction of valuable information from unstructured clinical notes, research papers, and patient narratives. This foundation enhances the efficiency of information retrieval and supports clinical decision-making processes. AI systems also use predictive analytics to forecast potential health issues and patient outcomes. Early identification of risks allows healthcare providers to intervene proactively, leading to improved patient care and outcomes.

AI-driven robotics enhance surgical precision, assist in repetitive tasks, and contribute to the automation of routine processes in healthcare settings. Robotics play a critical role in minimally invasive surgeries, reducing recovery times and improving patient experiences. The applications of AI in healthcare are diverse and continually expanding, showcasing the versatility of AI technologies in addressing complex challenges. Some notable applications are shown in Figure 5.2.

5.5 CURRENT TRENDS AND DEVELOPMENTS

The landscape of AI in healthcare is dynamic, with ongoing advancements shaping the future of medical practices [9]. Current trends and developments include:

- **Explainable AI**: There is a growing emphasis on developing AI systems that provide transparent and understandable explanations for their decisions, crucial for gaining trust in medical settings.
- **Federated Learning**: Federated learning is becoming increasingly popular as a privacy-conscious method, enabling the training of AI models using decentralized healthcare data sources while upholding patient privacy [10].
- **Continuous Monitoring with Wearables**: AI-enabled wearables offer continuous health monitoring, providing real-time data for early detection of health issues and personalized interventions.

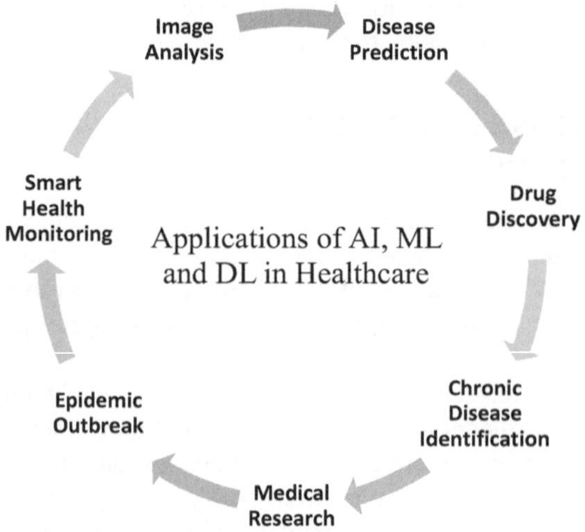

FIGURE 5.2 AL ML DL applications in health care.

- **Integration of AI with the Internet of Things**: The synergy between AI and the IoT is strengthening, enabling seamless data exchange and analysis for enhanced patient results and operational efficiency.
- **Ethical AI in Healthcare**: The ethical considerations of AI applications in healthcare are receiving increased attention, focusing on fairness, accountability, and transparency in algorithmic decision-making.

The foundations, applications, and current trends in AI showcase the transformative potential of these technologies in revolutionizing healthcare. As AI continues to evolve, its integration into healthcare practices promises to enhance precision, efficiency, and accessibility, ultimately improving patient outcomes and the overall healthcare experience.

5.6 FEDERATED LEARNING IN HEALTHCARE

Federated learning has come to light as an effective strategy to harness AI's potential in healthcare while resolving concerns related to data security and privacy [11]. This collaborative learning paradigm allows machine learning models to be trained across decentralized devices or servers without exchanging raw data, making it particularly suitable for the sensitive nature of healthcare information.

5.6.1 Understanding Federated Learning

Federated learning is a decentralized machine learning approach where models are trained collaboratively across multiple local devices or servers without sharing raw data. In this process, model updates, often in the form of weights or parameters, are exchanged and aggregated centrally [12]. This approach enables advancements in AI applications while maintaining privacy and security, making it particularly relevant in healthcare settings where sensitive patient data needs to be protected.

Federated learning operates on the principle of decentralized model training. Instead of centralizing data in a single location, models are trained locally on individual devices or servers. Raw data never leave the local device during the training process, ensuring that sensitive patient information remains secure. Only model updates, in the form of weights or parameters, are shared and aggregated centrally. The central server aggregates the locally trained models to create a global model that benefits from the knowledge learned across all decentralized sources. This iterative process continues, refining the global model without compromising individual data. Federated learning enables collaborative learning across multiple institutions, promoting knowledge sharing without compromising patient privacy. This approach is well-suited for scenarios where data are distributed across various healthcare entities [13].

The mathematical equations involved in federated learning can be quite complex and depend on the specific algorithm being used. A simplified representation of the general update equation in Federated Averaging, one of the common federated learning algorithms is shown in Equation 5.1.

Let w_i^k represent the model parameters on the kth client at iteration t. The global model w_t at iteration t is then updated by averaging the local models:

$$w_{t+1=} = \frac{1}{k}\sum_{k=1}^{K} w_i^k \tag{5.1}$$

Here,

- w_{t+1} is the updated global model at iteration $t+1$,
- w_t^k is the model parameters on the kth client at iteration t, and
- K is the total number of clients.

Along with the global model when federated learning is used for smart healthcare to preserve data privacy while still benefiting from collaborative model training. The mathematical equation for federated learning can be expressed as shown in Equation 5.2.

Let's denote:

- D as the global dataset containing N samples.
- $D_1, D_2,\ldots, D_m$ as m local datasets held by m different devices or servers.
- w was the global model parameters that we wanted to train collaboratively.
- $w_1, w_2,\ldots, w_m$ as local model parameters.

The federated learning optimization problem can be formulated as:

$$\min_w f(w) = \frac{1}{N}\sum_{i=1}^{N} f_i(w) \tag{5.2}$$

where $f_i(w)$ the local objective function is sample I and can be expressed as Equation 5.3.

$$f_i(w) = \frac{1}{m}\sum_{j=1}^{m}\frac{|D_j|}{|D|} f_{i,j}(w_j) \tag{5.3}$$

Here,

- $|D_j|$ is the size of the local dataset D_j,
- $|D|$ is the size of the global dataset D, and
- $f_{i,j}(w_j)$ is the local loss for the ith sample on the jth device using the local model parameters w_j.

The goal is to find the global model parameters w that minimize the average of the local losses across all devices while considering the contribution of each local dataset based on its size. The federated learning process typically involves multiple communication rounds, where local models are trained on local data, and then the global model is updated by aggregating these local models while preserving data privacy. The aggregation step might involve methods such as weighted averaging, where the weights are determined by the size of the local datasets.

5.6.2 Advantages and Challenges in Healthcare Context

Exploiting federated learning in healthcare provides benefits like safeguarding patient privacy, fostering collaborative research, and accepting varied datasets from different institutions. However, challenges include ensuring secure model aggregation, addressing communication overhead, and managing the heterogeneity of healthcare data sources [11]. Overcoming these difficulties can hold the key to federated learning's full potential, fostering a collaborative and privacy-conscious environment for advancing healthcare AI as shown in Figures 5.3 and 5.4.

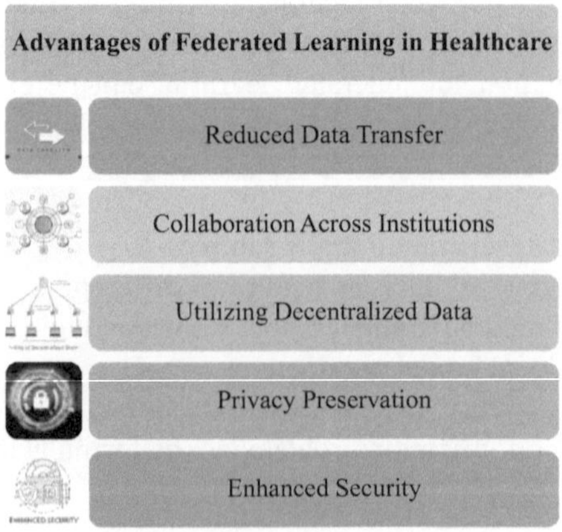

FIGURE 5.3 Advantages in healthcare federated learning.

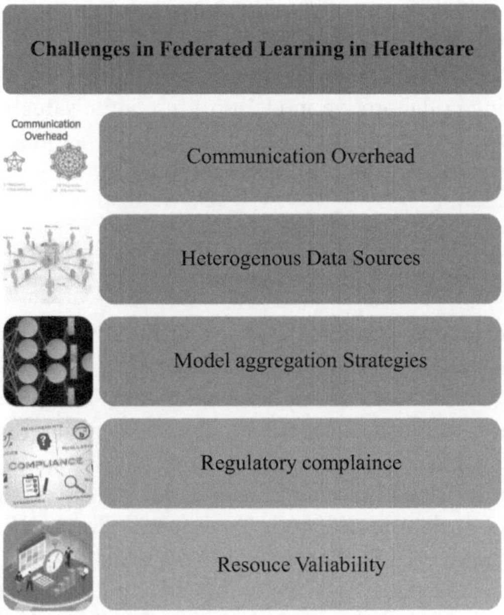

FIGURE 5.4 Challenges in healthcare federated learning.

5.6.3 Advantages of Federated Learning in Healthcare

By keeping patient data decentralized and lowering the possibility of data breaches or unauthorized access, federated learning solves privacy concerns. Healthcare data are often dispersed across different hospitals, clinics, and research institutions. Federated learning allows leveraging this distributed data without the need for centralized storage. It facilitates collaborative research and model training across multiple healthcare entities. The decentralized nature of federated learning inherently provides an additional layer of security, as compromising one node does not expose the entire dataset. Only model updates are transmitted, reducing the amount of data transfer compared to traditional centralized learning approaches. This is particularly advantageous in bandwidth-limited environments.

5.6.4 Challenges in Healthcare Federated Learning

Coordinating model updates and aggregating them centrally introduces communication overhead, requiring efficient protocols to manage this process. Healthcare institutions may use different data formats and standards. Harmonizing these diverse datasets for effective federated learning poses a challenge. Designing robust and efficient model aggregation strategies is crucial. Different federated learning approaches may be more suitable for specific healthcare applications. Ensuring compliance with healthcare regulations and standards, such as HIPAA, while implementing federated learning requires careful consideration and adherence. Variability in computing resources and capabilities across different healthcare institutions may impact the efficiency and speed of federated learning.

Federated learning holds immense potential for advancing AI applications in healthcare by allowing collaborative model training while safeguarding patient privacy. As research and development in this field progress, addressing the challenges associated with federated learning will be essential to unlock its full benefits for the healthcare industry.

5.7 PRIVACY CONCERNS IN HEALTHCARE DATA

The integration of advanced technologies, especially AI, in healthcare, comes with the imperative of safeguarding patient privacy. Privacy concerns are particularly paramount given the sensitive nature of healthcare data. To address these concerns, various privacy-preserving techniques are employed, especially in innovative paradigms like federated learning [12].

Protecting patient privacy is essential because healthcare data contain extremely sensitive information, including diagnoses, treatment plans, and medical histories. Strict laws, such as the HIPAA in the United States, which requires the safe processing and storage of patient data, must be complied with by healthcare institutions. Large datasets combined in healthcare settings increase the danger of data breaches, which could result in unauthorized access to and possible exploitation of patient health data. Ensuring privacy is vital for maintaining patient trust and encouraging active patient engagement, especially in scenarios where patients are required to share data for research or remote monitoring.

5.7.1 Techniques for Ensuring Patient Privacy in Federated Learning

Privacy-preserving techniques in federated learning play a pivotal role in safeguarding patient data, making it a trusted framework for collaborative healthcare AI without compromising confidentiality. Figure 5.5 illustrates how methods such as differential privacy, secure aggregation, and homomorphic encryption are integrated into the process.

Differential privacy, a technique introducing noise to individual data points, ensures that specific information about any individual remains challenging to infer.

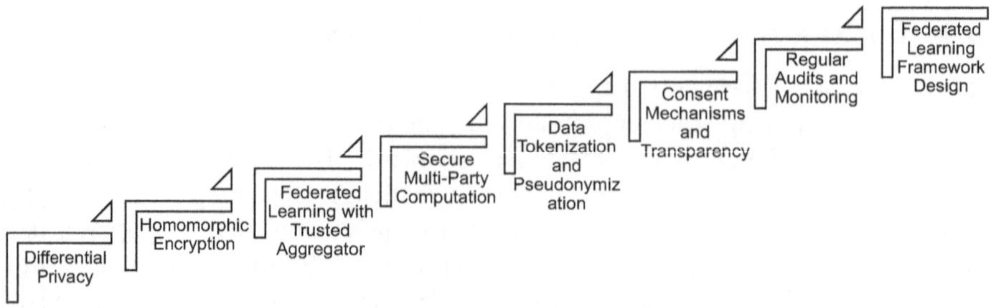

FIGURE 5.5 Techniques for ensuring patient privacy in federated learning.

This method prevents the undue influence of individual data on the overall outcome, contributing to the overall confidentiality of the process. Homomorphic encryption, allowing computations on encrypted data without decryption, adds an extra layer of protection during transmission and analysis in federated learning.

In the federated learning landscape, a trusted aggregator is a key player, aggregating model updates without accessing raw data, and preserving privacy. However, trust in the aggregator is paramount. Secure Multi-Party Computation (SMPC) enables joint computation of functions over private inputs, allowing model training without exposing individual data. Tokenization and pseudonymization further enhance patient privacy by replacing sensitive data with tokens and pseudonyms, respectively, making it difficult to directly associate information with specific individuals. Transparent data handling practices build trust among patients and stakeholders, while regular audits and monitoring of privacy-preserving techniques ensure compliance with regulations and identify potential vulnerabilities.

The federated learning framework's design itself incorporates privacy-preserving principles, such as minimizing data transmission and employing secure communication protocols. In the context of healthcare data, achieving patient privacy, especially in innovative approaches like federated learning, requires a combination of technical solutions, ethical considerations, and regulatory compliance. The implementation of these privacy-preserving techniques not only protects individuals but also fosters a trustworthy environment for the advancement of healthcare technologies.

5.8 DECENTRALIZED MODEL TRAINING

In the area of healthcare, where sensitive patient data are a primary concern, decentralized model training has emerged as a promising approach to control the power of AI while safeguarding confidentiality. This methodology, emphasizing collaboration over data centralization, has significant implications for the development of advanced healthcare applications.

5.8.1 Collaborative Model Training in Healthcare

Collaborative model training in healthcare involves the joint development of machine learning models by multiple institutions or entities without sharing raw patient data. Through decentralized processes, each participant trains models locally on their datasets, and only aggregated model updates are exchanged. This approach enhances privacy by keeping sensitive information at its source, while still allowing the collective intelligence of diverse datasets to contribute to the overall model's efficacy. The collaborative model training fosters cooperation, data diversity, and improved generalization across different populations, making it a pivotal methodology for advancing healthcare AI applications [13].

Decentralized model training involves training machine learning models on data that is distributed across various local devices or servers. Each local entity retains control over its data, ensuring that sensitive patient information remains on-premises. Healthcare institutions, research facilities, and clinics can collaborate on model training without sharing raw data. Model parameters are exchanged, aggregated, and refined collectively, enhancing the overall intelligence of the model. Healthcare data are diverse and often exhibits variations across different populations and regions. Decentralized model training allows models to be exposed to a broader range of data, improving their generalization capabilities. The decentralized nature of model training ensures that patient data remain at its source, addressing privacy concerns. Only model updates, typically in the form of weights or parameters, are shared and aggregated.

Decentralized model training aligns with the principles of edge computing, where computation occurs closer to the data source. This reduces the need for extensive data transfers and contributes to more efficient and localized processing.

5.8.2 Benefits of Decentralized Approaches

Decentralized approaches offer numerous benefits, particularly in healthcare, by distributing computational processes across local entities. These benefits include enhanced privacy and security, as sensitive data remains localized, collaborative model training without sharing raw data, and the ability to adapt to diverse datasets, improving generalization across different populations [14]. Furthermore, decentralized approaches optimize resource utilization, ensuring efficient and scalable AI development in healthcare systems.

Maintaining decentralization of sensitive healthcare data greatly lowers the danger of data breaches and unauthorized access. This enhances overall privacy and security, aligning with regulatory requirements. Decentralized model training enables collaborative efforts across healthcare institutions without the need to share raw data. This facilitates joint research and model development while maintaining data ownership.

Each local entity contributes to the training process using its resources, distributing the computational load, and optimizing resource utilization. This is particularly advantageous in resource-constrained environments. Exposure to diverse datasets from different geographic locations and demographic groups enhances the generalization capabilities of models. Decentralized training allows models to adapt to variations in healthcare data. Compared to centralized approaches, decentralized training reduces the need for extensive data transfers between local entities and the central server, minimizing communication overhead. Decentralized model training is inherently scalable, allowing healthcare systems to expand their collaborative efforts without significant infrastructural changes.

Decentralized model training in healthcare offers a transformative approach to AI development by fostering collaboration while prioritizing patient privacy and data security. As the healthcare industry continues to adopt advanced technologies, decentralized approaches provide a pathway to harness the collective intelligence of diverse datasets while respecting the principles of responsible data handling.

5.8.3 Innovative AI Techniques for Healthcare

In the rapidly evolving landscape of healthcare, innovative AI techniques are playing a pivotal role in transforming traditional practices, enhancing diagnostics, and improving patient outcomes. Here, we delve into advanced machine learning models and the integration of AI algorithms in the context of smart healthcare systems. Advanced Machine Learning Models in healthcare are shown in Figure 5.6.

Multiple-layer neural networks are used in deep learning, a subset of machine learning, to enable complicated feature extraction and representation learning. Applications include medical image analysis (e.g., MRI and CT scans), pathology recognition, and predictive analytics. Reinforcement learning models learn through trial and error, making them suitable for optimizing treatment plans and personalized medicine. These models are employed in decision-making processes, such as recommending treatment options based on patient responses.

Transfer learning leverages pre-trained models on large datasets and adapts them to new, smaller datasets in healthcare applications. This approach is valuable in scenarios where acquiring large labeled datasets is challenging. Ensemble learning combines predictions from multiple models to enhance overall performance and robustness. In healthcare, ensemble models contribute to improved accuracy in diagnostics and risk prediction. XAI focuses on making AI models interpretable and understandable for clinicians and patients. This is crucial in healthcare settings where transparent decision-making is essential, allowing stakeholders to trust and comprehend AI-driven insights.

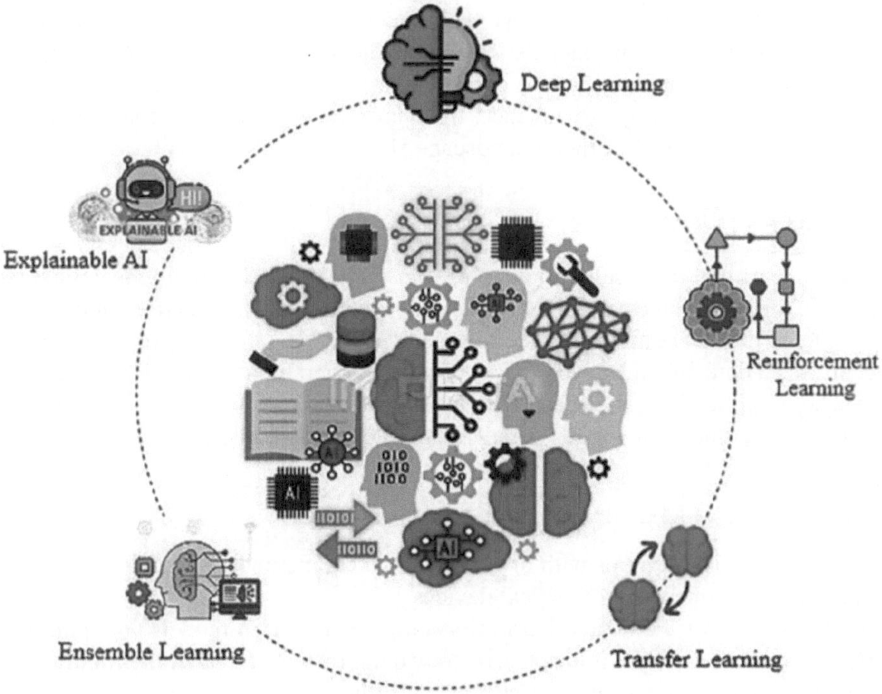

FIGURE 5.6 Innovative AI techniques for healthcare.

5.8.4 Integration of AI Algorithms in Smart Healthcare Systems

Smart healthcare systems incorporating AI algorithms, revolutionize patient care and diagnostics. These algorithms, ranging from deep learning for medical imaging to predictive analytics for disease management, enhance decision-making processes. Through applications like personalized treatment plans and telehealth support, the integration of AI algorithms improves healthcare accessibility, efficiency, and outcomes, ushering in a new era of intelligent healthcare solutions.

AI algorithms analyze patient data to predict the likelihood of diseases, enabling proactive preventive measures. This includes identifying risk factors and recommending personalized interventions for chronic diseases. Image recognition algorithms contribute to radiology, pathology, and dermatology diagnostics. AI-driven algorithms in telehealth platforms enable remote monitoring of patients with chronic conditions. These algorithms provide real-time insights, supporting timely interventions and reducing hospital readmissions. AI-powered virtual health assistants offer personalized health advice, medication reminders, and information on-demand. NLP enables these assistants to understand and respond to user queries.

AI algorithms contribute to fraud detection by analyzing billing patterns and identifying irregularities. Healthcare analytics powered by AI facilitates data-driven decision-making for administrators and policymakers. AI expedites the search for new drugs by evaluating large datasets, forecasting promising therapeutic candidates, and refining the plans for clinical trials. This speeds up the creation of cutting-edge remedies for a range of illnesses.

The integration of innovative AI techniques shown in Figure 5.7, including advanced machine learning models, into smart healthcare systems holds immense promise for revolutionizing patient care, diagnostics, and treatment strategies. As these technologies continue to advance, they pave the way for a more personalized, efficient, and accessible healthcare landscape [11].

5.9 CASE STUDIES

Federated learning finds real-world applications in healthcare by enabling collaborative model training across multiple institutions without sharing sensitive patient data. It facilitates predictive maintenance for medical devices, disease prediction using decentralized data, and the development of personalized treatment plans in oncology [15]. These applications showcase how federated learning contributes to improved healthcare outcomes, data privacy, and collaborative research efforts in the field.

5.9.1 Predictive Maintenance for Medical Devices

Scenario*:* Multiple hospitals with diverse medical equipment collaborate to enhance the predictive maintenance of critical devices.

Implementation*:* Federated learning is employed to train models on equipment performance data locally without sharing raw data. Insights on potential failures and maintenance requirements are aggregated centrally.

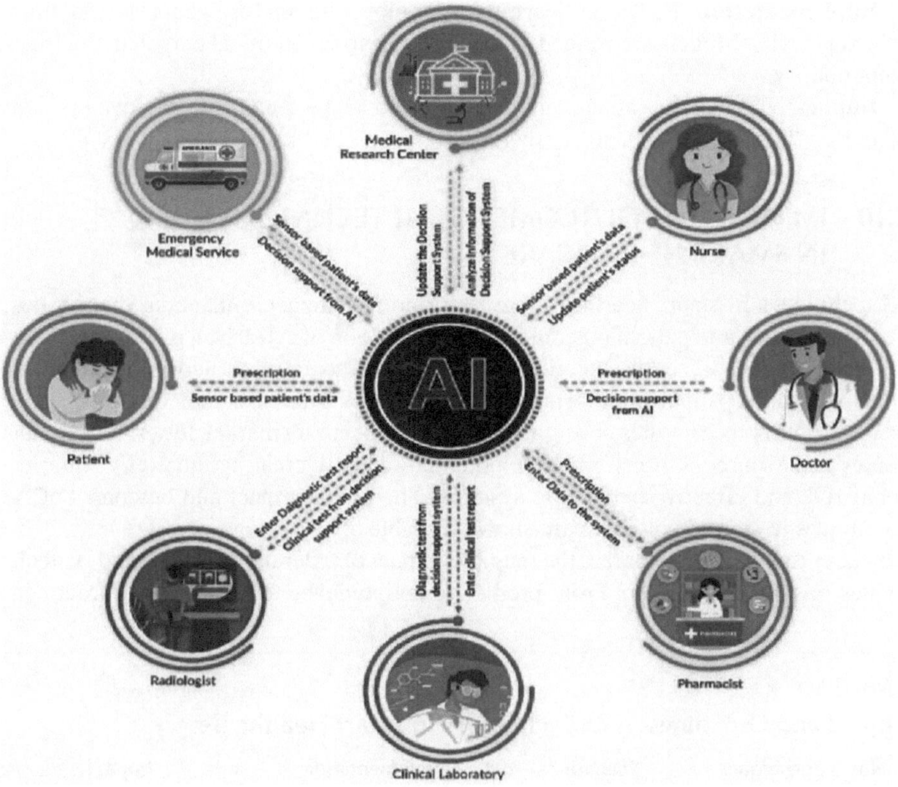

FIGURE 5.7 Integration of AI algorithms in smart healthcare systems.

Impact: Improved device reliability, reduced downtimes, and cost savings in equipment maintenance across participating healthcare facilities.

5.9.2 Disease Prediction Using Decentralized Data

Scenario: Healthcare institutions from different regions collaborate to predict the outbreak of infectious diseases without sharing patient-specific data.

Implementation: Federated learning allows each institution to train models on locally stored epidemiological data. Model updates are aggregated to predict disease trends collaboratively.

Impact: Early detection of outbreaks, efficient resource allocation, and timely public health interventions based on localized patterns [16].

5.9.3 Personalized Treatment Plans in Oncology

Scenario: Oncology centers collaborate to develop personalized treatment plans for cancer patients while preserving patient privacy.

Implementation: Federated learning is employed to analyze genomic and clinical data locally. Models are trained collaboratively to recommend personalized treatment options.

Impact: Enhanced treatment efficacy, reduced side effects, and improved patient outcomes through tailored cancer therapies.

5.10 IMPACT AND OUTCOMES OF AI TECHNIQUES IN SMART HEALTHCARE

AI techniques in smart healthcare have a profound impact, enhancing diagnostics, treatment plans, and patient outcomes. The integration of AI-driven diagnostic imaging leads to early disease detection, reducing healthcare costs and improving outcomes. Remote patient monitoring with AI enables proactive interventions, while drug discovery accelerates, bringing novel treatments to market faster. These outcomes underscore the transformative potential of AI in creating more efficient, personalized, and effective healthcare systems. The major impact and outcomes of AI techniques in smart healthcare are shown in Table 5.1.

These case studies illustrate the tangible impact of federated learning and AI techniques in smart healthcare. From predictive maintenance to personalized treatment

TABLE 5.1
Impact and Outcomes of AI Techniques in Smart Healthcare

S. No	Techniques	Scenario	Implementation	Impact
1	AI-driven Diagnostic Imaging	Integration of AI algorithms in medical imaging for the diagnosis of cardiovascular diseases	Deep learning models analyze cardiac images, identifying subtle abnormalities and predicting cardiovascular risks	Improved patient outcomes, lower healthcare costs, and more prompt interventions due to increased accuracy in early disease detection
2	Remote Patient Monitoring with Wearables	Implementation of AI in remote patient monitoring using wearable devices	Wearables collect continuous health data and AI algorithms analyze patterns to predict and alert healthcare providers to potential health issues	Proactive healthcare interventions, reduced hospital admissions, and improved patient engagement in managing chronic conditions
3	Drug Discovery Acceleration	AI-driven drug discovery for neurological disorders	Machine learning models analyze complex datasets to predict potential drug candidates, speeding up the drug discovery process	Faster development of novel treatments, reduced research costs, and increased efficiency in bringing therapies to market

(Continued)

TABLE 5.1 (*Continued*)
Impact and Outcomes of AI Techniques in Smart Healthcare

S. No	Techniques	Scenario	Implementation	Impact
4	Virtual Health Assistants for Chronic Disease Management	Integration of virtual health assistants powered by AI to support patients with chronic diseases	NLP enables virtual assistants to understand patient queries, provide information, and offer medication reminders	Enhanced patient education, improved medication adherence, and better management of chronic conditions through continuous support
5	Fraud Detection in Healthcare Billing	AI-based fraud detection in healthcare billing systems	Machine learning algorithms analyze billing patterns, identify anomalies, and flag potentially fraudulent activities	Reduced financial losses due to fraud, improved accuracy in billing processes, and streamlined revenue cycle management

plans and remote patient monitoring, these technologies contribute to improved patient care, enhanced diagnostics, and transformative outcomes in the healthcare ecosystem.

5.10.1 Security Measures in Federated Learning

Security measures in federated learning are crucial to protect sensitive information during collaborative model training. The privacy and confidentiality of individual data points are protected by methods including differential privacy mechanisms, secure aggregation protocols, and federated learning encryption. Authentication for trusted aggregators and secure model initialization further contribute to mitigating potential threats, making federated learning a secure framework for collaborative AI in healthcare and other sensitive domains [17].

5.10.2 Ensuring Data Security in Distributed Environments

In remote systems, ensuring data security requires putting strong encryption, access controls, and authentication procedures in place. By employing end-to-end encryption, delicate data are protected during transmission, reducing the risk of unauthorized access. Secure aggregation protocols and trusted authentication mechanisms further enhance the security of collaborative processes in distributed environments [18]. These measures collectively safeguard data integrity and confidentiality in complex, interconnected systems also Ensuring Data Security in Distributed Environments is shown in Table 5.2.

TABLE 5.2
Ensuring Data Security in Distributed Environments

S. No	Data Security Techniques	Description	Impact
1	Federated Learning Encryption	Using end-to-end encryption to protect the transfer of model changes throughout the federated learning process	Protects against potential eavesdropping and unauthorized access during data transmission, ensuring the confidentiality of information
2	Secure Aggregation Protocols	Employ secure aggregation techniques during the model aggregation phase to prevent malicious actors from gaining insights into individual model updates	Enhances the privacy of participants by safeguarding against potential attacks during the model aggregation process
3	Differential Privacy Mechanisms	Integrate differential privacy techniques to add noise to individual data points, making it harder to trace specific information back to individual contributors	Improvement of personal data protection while maintaining the ability to draw insightful conclusions from aggregated data
4	Trusted Aggregator Authentication	Implement authentication mechanisms for trusted aggregators, ensuring that only authorized entities can aggregate model updates	Mitigates the risk of model poisoning or attacks on the aggregation process, reinforcing the overall security of the federated learning system
5	Secure Model Initialization	Ensure secure model initialization at local devices, preventing the injection of compromised models into the federated learning process	Guards against adversarial attacks on the federated learning system by maintaining the integrity of the initial model parameters

5.10.3 Cybersecurity Measures for Healthcare Systems

Cybersecurity measures for healthcare systems are imperative to safeguard sensitive patient data, which is shown in Table 5.3. Strong access restrictions, frequent security audits, and encryption of health information while it's in transit and at rest are crucial elements. Healthcare systems are often more resilient to cyberattacks when endpoint security, personnel cybersecurity best practice training, and clearly defined incident response strategies are in place. To guarantee the availability, confidentiality, and integrity of medical records, a proactive, multifaceted cybersecurity plan is essential [19].

Implementing robust security measures in federated learning and healthcare systems is paramount to safeguarding sensitive data, maintaining the privacy of individuals, and ensuring the integrity of critical healthcare services. By combining encryption, authentication, cybersecurity best practices, and proactive measures, healthcare organizations can build resilient systems that withstand the evolving landscape of cyber threats.

TABLE 5.3
Cybersecurity Measures for Healthcare Systems

S. No	Techniques	Description	Impact
1	Access Controls and Authentication	Establishing strong authentication and access rules in place to prevent illegal access to private medical data	Prevents unauthorized users from gaining access to patient records, protecting against data breaches and unauthorized data manipulation
2	Regular Security Audits	Perform routine security audits to find weaknesses and evaluate healthcare systems' overall cybersecurity posture	Enables proactive identification and mitigation of potential security threats, reducing the risk of data breaches and system compromises
3	Encryption of Health Data	Protection against unwanted access and data interception by encrypting health data while it's in transport and at rest	Security of patient data from possible data breaches and guarantees the privacy of sensitive medical information
4	Endpoint Security	Implement robust endpoint security measures to protect against malware, ransomware, and other cyber threats targeting healthcare devices	Reduces the risk of malicious attacks on healthcare devices, ensuring the integrity and availability of critical healthcare systems
5	Employee Training and Awareness	Provide ongoing cybersecurity training to healthcare staff, raising awareness about potential threats and the importance of secure practices	Enhances the human factor in cybersecurity, reducing the likelihood of unintentional security breaches and promoting a culture of security awareness
6	Incident Response Plans	To guarantee a prompt and well-coordinated response to security incidents, create and update incident response plans regularly	Minimizes the impact of security incidents by enabling healthcare organizations to respond effectively, mitigate risks, and recover quickly
7	Vendor Security Assessments	To guarantee the security of outsourced healthcare services, thoroughly evaluate the security of third-party vendors and service providers	Reduces the dangers brought on by vulnerabilities in third parties and guarantees the general security of the healthcare ecosystem

5.11 INTEROPERABILITY AND STANDARDS

Interoperability in smart healthcare is achieved through common data standards, Health Information Exchanges (HIEs), and adherence to frameworks like Fast Healthcare Interoperability Resources (FHIR). These measures facilitate seamless data exchange and collaboration among diverse healthcare systems, ensuring efficient and secure communication. Standardization efforts for federated learning in healthcare involve the development of federated learning Application Programming Interfaces (FL APIs) and privacy-preserving standards, aligning with healthcare regulations to enhance interoperability while preserving patient privacy [20].

TABLE 5.4
Achieving Interoperability in Smart Healthcare

S. No	Techniques	Approach	Impact
1	Common Data Standards	Adopt and adhere to standardized data formats and protocols across healthcare systems, ensuring seamless data exchange	Enhances interoperability by facilitating the sharing of patient information, medical records, and other healthcare data among different entities
2	Health Information Exchanges (HIEs)	Establish Health Information Exchanges that serve as centralized platforms for sharing and accessing patient data	Promotes interoperability by providing a standardized infrastructure for healthcare providers to exchange patient information securely
3	Fast Healthcare Interoperability Resources (FHIR)	Embrace FHIR as a modern standard for healthcare data exchange, providing a robust framework for interoperability	FHIR facilitates the seamless integration of healthcare data across disparate systems, promoting interoperability and data sharing
4	Application Programming Interfaces (APIs)	Implement standardized APIs that adhere to industry specifications, allowing different healthcare applications to interact and share data	Enables interoperability by providing a common interface for applications, fostering collaboration and data exchange
5	Semantic Interoperability	Utilize standardized healthcare terminologies and ontologies to ensure a common understanding of medical concepts and terminology	Enhances semantic interoperability, enabling accurate interpretation and exchange of healthcare information among diverse systems

5.11.1 Achieving Interoperability in Smart Healthcare

Achieving interoperability in smart healthcare involves adopting common data standards, leveraging HIEs, and embracing modern frameworks like FHIR, as shown in Table 5.4. By implementing standardized Application Programming Interfaces (APIs) and ensuring semantic interoperability through shared healthcare terminologies, the seamless exchange of patient information becomes possible. These measures enhance collaboration among healthcare systems, promoting efficient data sharing and ultimately improving the quality and continuity of patient care across diverse healthcare settings.

5.11.2 Standardization Efforts for Federated Learning in Healthcare

Standardization efforts for federated learning in healthcare focus on developing FL APIs and privacy-preserving standards, as shown in Table 5.5. These initiatives aim to facilitate seamless collaboration and model sharing across diverse healthcare entities

TABLE 5.5
Standardization Efforts for Federated Learning in Healthcare

S. No	Techniques	Approach	Impact
1	Federated Learning Application Programming Interfaces (FL APIs)	Develop and adhere to standardized FL APIs that define how federated learning systems interact with each other	Facilitates interoperability among federated learning platforms, allowing seamless collaboration and model sharing across different healthcare entities
2	Privacy-Preserving Standards	Establish privacy-preserving standards for federated learning to ensure consistent implementation of techniques such as encryption and secure aggregation	Enhances interoperability by providing a common framework for preserving privacy during the federated learning process
3	Federated Learning Model Exchange Formats	Define standardized formats for exchanging federated learning models between different systems	Promotes interoperability by ensuring that models trained on one system can be easily integrated and utilized in another, fostering collaborative research and development
4	Compliance with Healthcare Regulations	Align federated learning standards with existing healthcare regulations, such as HIPAA, to ensure compliance with privacy and security requirements	Offers a legal framework that upholds the highest standards of privacy and data protection in the healthcare industry while promoting interoperability
5	Cross-Domain Federated Learning Standards	Extend federated learning standards to accommodate collaborations across different healthcare domains, encouraging interdisciplinary research and data sharing	Enables federated learning models to be applied across diverse healthcare settings, improving the generalizability and effectiveness of AI applications

while ensuring compliance with healthcare regulations. By aligning with industry standards and privacy-preserving techniques, standardization efforts enhance interoperability and promote responsible data sharing practices in federated learning for healthcare.

Achieving interoperability in smart healthcare and federated learning requires the establishment and adherence to common standards. By adopting standardized data formats, APIs, and privacy-preserving techniques, the healthcare industry can build a cohesive and interconnected ecosystem.

5.12 REGULATORY AND ETHICAL CONSIDERATIONS

Navigating the deployment of AI and federated learning in healthcare requires strict adherence to healthcare regulations such as HIPAA and GDPR. Securing Institutional Review Board (IRB) approval for research involving human subjects is essential to ensure ethical research practices. Compliance with FDA regulations for AI-based medical devices and alignment with standards for interoperability further contribute to responsible and ethical deployment, fostering patient trust and ensuring regulatory compliance in the evolving landscape of healthcare technology.

5.12.1 Ethical Implications of AI and Federated Learning in Healthcare

The ethical repercussions of AI and federated learning in healthcare underscore the importance of obtaining informed consent, mitigating bias, ensuring explainability in AI decisions, and prioritizing patient privacy. Adherence to ethical principles promotes transparency, fairness, and accountability in the development and deployment of healthcare AI models. Continuous monitoring for biases and the responsible use of emerging technologies help uphold patient rights, build trust among stakeholders, and ensure equitable access to the benefits of AI in healthcare.

- **Informed Consent**

  Ethical Principle: Obtains informed consent from patients before using their data in AI or federated learning research, ensuring transparency about the purpose and potential risks.

  Impact: Respects individuals' autonomy and rights over their data, fostering trust in AI applications.
- **Bias and Fairness**

  Ethical Principle: Mitigates bias in AI models and federated learning by addressing issues related to fairness, transparency, and accountability.

  Impact: Ensures equitable treatment and outcomes for diverse patient populations, preventing discrimination.
- **Explainability in AI Decisions**

  Ethical Principle: Prioritizes explainability in AI models, especially in critical healthcare decision-making processes, to provide understandable justifications for outcomes.

  Impact: Fosters trust among healthcare professionals, enabling them to interpret and validate AI-driven decisions.
- **Privacy Preservation**

  Ethical Principle: Implements robust privacy-preserving techniques, especially in federated learning, to protect patient privacy during collaborative model training.

  Impact: Upholds the ethical duty to safeguard sensitive health information, ensuring the confidentiality and trustworthiness of healthcare systems.
- **Accountability and Transparency**

  Ethical Principle: Establishes accountability mechanisms and transparent practices in the development and deployment of AI in healthcare.

  Impact: Demonstrates a commitment to ethical conduct, allowing stakeholders to understand the processes and decisions made by AI systems.
- **Continual Monitoring and Evaluation**

  Ethical Principle: Conducts ongoing monitoring and evaluation of AI models to identify and rectify potential biases, errors, or unintended consequences.

Impact: Demonstrates a commitment to continuous improvement and ethical responsibility in deploying AI technologies in healthcare.

Navigating the regulatory landscape and addressing ethical considerations is imperative in the development and deployment of AI and federated learning in healthcare. Compliance with regulations ensures legal standards are met, while ethical principles safeguard patient rights, autonomy, and trust in the evolving landscape of healthcare technology.

5.13 FUTURE DIRECTIONS AND EMERGING TECHNOLOGIES

Future directions in smart healthcare involve the prospects of AI and federated learning, driving advancements in personalized medicine, predictive analytics, and decentralized clinical trials. Emerging technologies shaping the future include edge computing for real-time data analysis, blockchain for enhanced health data security, and the transformative impact of 5G technology on telehealth. Additionally, breakthroughs in genomic medicine, augmented reality (AR), virtual reality (VR), and the rise of explainable AI (XAI) are contributing to a paradigm shift in healthcare, promising more precise diagnostics, personalized treatments, and improved patient outcomes [21].

5.13.1 Prospects of AI and Federated Learning in Healthcare

The prospects of AI and federated learning in healthcare are promising, offering transformative opportunities. AI enables personalized medicine, disease prediction, and decentralized clinical trials, fostering collaboration while preserving patient privacy [22]. Federated learning empowers diverse healthcare institutions to collectively advance research and model development without compromising the security of sensitive patient data, paving the way for more efficient, patient-centric, and collaborative healthcare solutions [23].

AI, especially federated learning, is poised to play a pivotal role in advancing personalized medicine. By analyzing diverse patient data sources, AI can tailor treatments based on individual genetic, clinical, and lifestyle factors, optimizing therapeutic outcomes [24]. AI-driven predictive analytics, fueled by federated learning, holds the potential to revolutionize early disease detection. By analyzing comprehensive datasets from diverse healthcare institutions, these technologies can identify subtle patterns indicative of diseases before symptoms manifest. Federated learning can transform the landscape of clinical trials by enabling decentralized and patient-centric approaches. This allows for the aggregation of insights across multiple locations without centralizing sensitive patient data, expediting the drug development process. AI-powered real-time health monitoring using wearable devices, integrated with federated learning principles, can provide continuous and personalized health insights. This approach facilitates proactive intervention for individuals with chronic conditions, improving overall healthcare management. AI is expected to augment healthcare professionals by assisting in diagnostics, treatment planning, and routine tasks [25,26]. Federated learning ensures that models are trained on diverse datasets, enhancing their adaptability and generalization across various healthcare settings.

Advanced AI algorithms, integrated with federated learning, will continue to enhance diagnostic accuracy in radiology and medical imaging. The collaborative nature of federated learning allows models to learn from diverse imaging datasets, improving their robustness.

5.13.2 Emerging Technologies Shaping the Future of Smart Healthcare

Emerging technologies are shaping the future of smart healthcare with innovations such as edge computing for real-time data processing, blockchain for secure health data management, and 5G technology for enhanced telehealth capabilities. Advances in genomic medicine, coupled with CRISPR gene-editing technology, promise breakthroughs in precision medicine. Additionally, AR, VR, and the advent of XAI contribute to a technologically advanced healthcare landscape, offering improved diagnostics, patient engagement, and overall healthcare outcomes as shown in Figure 5.8.

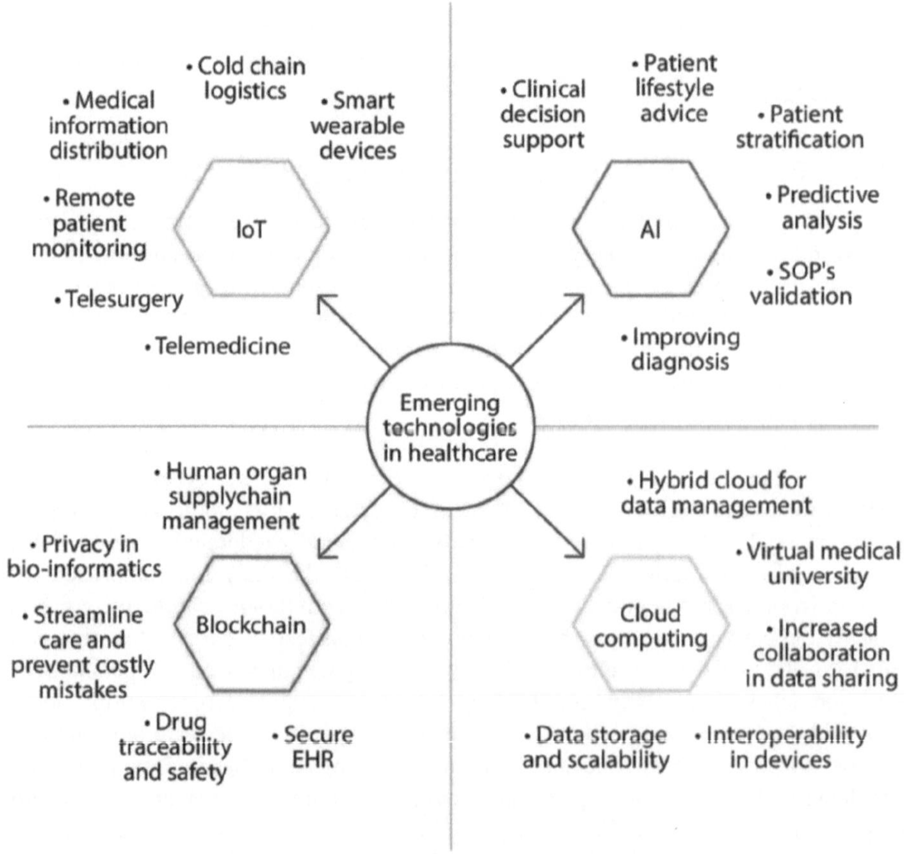

FIGURE 5.8 Emerging technologies and framework of smart healthcare systems with their essential characteristics.

The deployment of 5G technology is set to revolutionize telehealth by providing high-speed, low-latency connectivity. This enables seamless and real-time communication between healthcare providers and patients, especially in remote or underserved areas. Advances in genomic medicine, coupled with CRISPR gene-editing technology, are opening new frontiers in precision medicine. These technologies allow for targeted treatments based on an individual's genetic makeup. AR and VR are poised to transform healthcare by enhancing medical training, patient education, and surgical procedures. These technologies create immersive and interactive experiences, improving learning and patient outcomes. IoMT involves interconnected medical devices and sensors that collect and transmit health data. This technology facilitates remote patient monitoring, predictive analytics, and the seamless integration of healthcare devices [27]. The development of XAI aims to make machine learning models more interpretable and understandable. In healthcare, XAI is crucial for ensuring transparency in AI-driven decisions and building trust among healthcare professionals and patients.

The future of smart healthcare is shaped by the ongoing evolution of AI and federated learning, as well as the integration of emerging technologies [28,29]. From personalized medicine and decentralized clinical trials to the widespread adoption of edge computing and blockchain, these advancements promise to enhance patient care, improve diagnostics, and revolutionize the healthcare landscape [30,31].

5.14 CONCLUSION

Federated learning offers a decentralized paradigm for collaborative model training, preserving the privacy and security of sensitive healthcare data across multiple entities. Privacy concerns in healthcare data are addressed through techniques such as differential privacy, homomorphic encryption, and SMPC, ensuring the confidentiality of patient information. Decentralized approaches in model training enable collaborative efforts without sharing raw data, providing benefits such as enhanced privacy, resource utilization, and scalability. The integration of advanced AI techniques, including deep learning, reinforcement learning, and ensemble learning, holds immense promise for personalized medicine, disease prediction, and efficient healthcare analytics. Common data standards, HIEs, and adherence to frameworks like FHIR are crucial for achieving interoperability in smart healthcare.

Establish mechanisms for continual monitoring of AI models to identify and rectify biases, ensuring fair and equitable healthcare outcomes. Foster collaborative research efforts and decentralized clinical trials enabled by federated learning, promoting efficiency and inclusivity in healthcare research. Invest in education and training programs to equip healthcare professionals with the skills needed to navigate and leverage emerging technologies effectively. Align AI implementations with ethical guidelines, ensuring transparency, explainability, and accountability in decision-making processes. Stay informed about evolving healthcare regulations and update systems and practices accordingly to maintain compliance and uphold ethical standards. Embrace emerging technologies such as blockchain, 5G, and XAI responsibly, considering their potential impact on healthcare delivery, patient outcomes, and data security.

In conclusion, the integration of AI techniques, particularly through innovative approaches like federated learning, stands as a transformative force in smart healthcare. By implementing the recommendations and staying attuned to ethical and regulatory considerations, the healthcare industry can navigate this transformative journey responsibly, ensuring the well-being and trust of patients in the digital era of healthcare.

REFERENCES

1. Ahmad, K. A. B., Khujamatov, H., Akhmedov, N., Bajuri, M. Y., Ahmad, M. N., & Ahmadian, A. (2022). Emerging trends and evolutions for smart city healthcare systems. *Sustainable Cities and Society*, 80, 103695.
2. Mishra, P., & Singh, G. (2023). Internet of Medical Things healthcare for sustainable smart cities: Current status and future prospects. *Applied Sciences*, 13(15), 8869.
3. Moreira, M. W., Rodrigues, J. J., Korotaev, V., Al-Muhtadi, J., & Kumar, N. (2019). A comprehensive review on smart decision support systems for health care. *IEEE Systems Journal*, 13(3), 3536–3545.
4. Sahu, M., Gupta, R., Ambasta, R. K., & Kumar, P. (2022). Artificial intelligence and machine learning in precision medicine: A paradigm shift in big data analysis. *Progress in Molecular Biology and Translational Science*, 190(1), 57–100.
5. George, A. S., & George, A. H. (2023). Telemedicine: A new way to provide healthcare. *Partners Universal International Innovation Journal*, 1(3), 98–129.
6. Awotunde, J. B., Folorunso, S. O., Ajagbe, S. A., Garg, J., & Ajamu, G. J. (2022). AiIoMT: IoMT-based system-enabled artificial intelligence for enhanced smart healthcare systems. In: Al-Turjman F., Nayyar A. (eds) *Machine Learning for Critical Internet of Medical Things: Applications and Use Cases*, pp. 229–254. Springer, Cham.
7. Cai, Q., Wang, H., Li, Z., & Liu, X. (2019). A survey on multimodal data-driven smart healthcare systems: Approaches and applications. *IEEE Access*, 7, 133583–133599.
8. Husnain, A., Rasool, S., Saeed, A., Gill, A. Y., & Hussain, H. K. (2023). AI'S healing touch: Examining machine learning's transformative effects on healthcare. *Journal of World Science*, 2(10), 1681–1695.
9. Grassini, S. (2023). Shaping the future of education: Exploring the potential and consequences of AI and ChatGPT in educational settings. *Education Sciences*, 13(7), 692.
10. Kishor K. (2022). Personalized federated learning. In: Yadav S. P., Bhati B. S., Mahato D. P., Kumar S. (eds) *Federated Learning for IoT Applications*. EAI/Springer Innovations in Communication and Computing. Springer, Cham. https://doi.org/10.1007/978-3-030-85559-8_3
11. Rauniyar, A., Hagos, D. H., Jha, D., Håkegård, J. E., Bagci, U., Rawat, D. B., &Vlassov, V. (2023). Federated learning for medical applications: A taxonomy, current trends, challenges and future research directions. *IEEE Internet of Things Journal*, 11(5), 7374–7398. DOI: 10.1109/JIOT.2023.3329061
12. AbdulRahman, S., Tout, H., Ould-Slimane, H., Mourad, A., Talhi, C., & Guizani, M. (2020). A survey on federated learning: The journey from centralized to distributed on-site learning and beyond. *IEEE Internet of Things Journal*, 8(7), 5476–5497.
13. Kishor K. (2022). Communication-efficient federated learning. In: Yadav S. P., Bhati B. S., Mahato D. P., Kumar S. (eds) *Federated Learning for IoT Applications*. EAI/Springer Innovations in Communication and Computing. Springer, Cham. https://doi.org/10.1007/978-3-030-85559-8_9
14. Alanazi, A. (2023). Clinicians' views on using artificial intelligence in healthcare: Opportunities, Challenges and Beyond. *Cureus*, 15(9), e45255.

15. Haddad, A., Habaebi, M. H., Islam, M. R., Hasbullah, N. F., & Zabidi, S. A. (2022). Systematic review on AI-blockchain based e-healthcare records management systems. *IEEE Access*, 10, 94583–94615.
16. Kamruzzaman, M. (2020). Architecture of smart health care system using artificial intelligence. In: *2020 IEEE International Conference on Multimedia & Expo Workshops (ICMEW)*, pp. 1–6. IEEE, London. DOI: 10.1109/ICMEW46912.2020.9106026
17. Kishor, K. (2023). Cloud computing in blockchain. In: Kishor K., Saxena, N., and Pandey, D. (eds), *Cloud-based Intelligent Informative Engineering for Society 5.0*, 1st edition, pp. 79–105. Chapman and Hall/CRC, New York. ISBN: 9781003213895. https://doi.org/10.1201/9781003213895-5
18. Kishor, K. (2023). Impact of cloud computing on entrepreneurship, cost, and security. In: Kishor K., Saxena, N., and Pandey, D. (eds), *Cloud-based Intelligent Informative Engineering for Society 5.0*, 1st edition, pp. 171–191. CRC Press, New York. ISBN: 9781003213895. https://doi.org/10.1201/9781003213895-10
19. Kishor, K., Nand, P., & Agarwal, P. (2018). Secure and efficient subnet routing protocol for MANET. *Indian Journal of Public Health*, 9(12), 200. https://doi.org/10.5958/0976-5506.2018.01830.2
20. Gupta S., Tyagi S., & Kishor K. (2022) Study and development of self sanitizing smart elevator. In: Gupta D., Polkowski Z., Khanna A., Bhattacharyya S., Castillo O. (eds) *Proceedings of Data Analytics and Management*. Lecture Notes on Data Engineering and Communications Technologies, vol 90. Springer, Singapore. https://doi.org/10.1007/978-981-16-6289-8_15
21. Kishor, K., & Nand, P. (2023). Wireless networks based in the cloud that support 5G. In: Kishor K., Saxena, N., and Pandey, D. (eds), *Cloud-based Intelligent Informative Engineering for Society 5.0*, 1st edition, pp. 23–40. Chapman and Hall/CRC, New York. ISBN: 9781003213895. https://doi.org/10.1201/9781003213895-2
22. Alabdulatif, A., Khalil, I., & Rahman, M. S. (2022). Security of blockchain and AI-empowered smart healthcare: Application-based analysis. *Applied Sciences*, 12(21), 11039. https://doi.org/10.3390/app122111039
23. Singh, S. K., Tiwari, S., Abidi, A. I., & Singh A. (2017). Prediction of pain intensity using multimedia data. *Multimedia Tools and Applications*, 76(18), 19317–19342. ISSN:1380–7501. https://doi.org/10.1007/s11042-017-4718-6
24. Tiwari, K., & Singh, S. K. (2015). Does all newborn face lookalike to human and machine?. *Proceedings of the National Academy of Sciences, India Section A: Physical Sciences*, 5(17), 211–220. ISSN:0369-8203. https://doi.org/10.1007/s40010-014-0196-7
25. Naithani, K., & Raiwani, Y.P. (2023). Sentiment analysis on social media data: A survey. In: Saini H. S., Sayal R., Govardhan A., Buyya R. (eds) *Innovations in Computer Science and Engineering. ICICSE 2022*. Lecture Notes in Networks and Systems, vol 565. Springer, Singapore. https://doi.org/10.1007/978-981-19-7455-7_59
26. Tiwari, K., & Singh S. K. (2012). Face recognition for newborns. *IET Biometrics*, 1(4), 200–208. ISSN: 2047-4946. https://doi.org/10.1049/iet-bmt.2012.0040
27. Singh, R., Singh, R., Acharya, A., Tiwari, K., & Hari, O. (2021). Pose and illumination invariant hybrid feature extraction for newborn. *Source: Recent Advances in Computer Science and Communications (Formerly: Recent Patents on Computer Science)*, 14(2), 368–375. ISSN: 2666–2566. https://doi.org/10.2174/2213275912666190328201840
28. Naithani, K., & Tiwari, S. (2023). Chapter 8 Deep learning for the intersection of ethics and privacy in healthcare. In: Maruthi P. B., Prasad S., Tyagi A. K. (eds) *Machine Learning Algorithms Using Scikit and TensorFlow Environments*, pp. 154–191. IGI Global Publishing. ISBN: 9781668485316. https://doi.org/10.4018/978-1-6684-8531-6

29. Sharma, A., Jha, N., & Kishor, K. (2022) Predict COVID-19 with Chest X-ray. In: Gupta D., Polkowski Z., Khanna A., Bhattacharyya S., Castillo O. (eds) *Proceedings of Data Analytics and Management.* Lecture Notes on Data Engineering and Communications Technologies, vol 90. Springer, Singapore. https://doi.org/10.1007/978-981-16-6289-8_16.
30. Naithani, K., Raiwani, Y. P., Alam, I., & Aknan, M. (2023). Analyzing hybrid C4. 5 Algorithm for sentiment extraction over lexical and semantic interpretation. *Journal of Information Technology Management*, 15, 57–79.
31. Shrivastava, S., Kumar, A., Saxena, S., & Tiwari, S. (2020). Chapter 18 Internet of Things for control and prevention of infectious diseases. In: Singh S. K., Shankar R., Pandey A. K., Udmale S. S., Chaudhary A. (eds) *IoT Based Data Analytics for the Healthcare Industry*, pp. 277–284. Elsevier. ISBN: 9780128214725, Academic press. https://doi.org/10.1016/B978-0-12-821472-5.00012-0

6 Federated Machine Learning in Medical Science

A Prospective Investigation

Vrinda Sachdeva, Gaurav Agarwal, Arun Kumar, Shailja Varshney, and Dina Rajput

6.1 INTRODUCTION

Machine learning is applied in many fields of research and industry, including medicine. In medical science, machine learning plays a critical role by allowing researchers to examine large volumes of medical data and find trends, correlations, and predictions related to human health. Numerous facets of healthcare, like as illness diagnosis, treatment planning, drug development, and customized medicine, might be completely transformed by it. Clinical decision-making and new insights into disease causes can be gained by using machine algorithms to examine patient records, genetic data, medical imaging, and clinical trial results. Furthermore, machine learning models can assist healthcare practitioners in lowering expenses, enhancing patient experiences overall, and improving patient outcomes. Taking all factors into account, machine learning is gaining greater significance in the realm of medical science and is anticipated to maintain its pivotal role in the foreseeable future.

This chapter explores the application of machine learning and predictive analysis in the field of medical informatics. The initiation involves an assessment of the current landscape of predictive modeling in diagnostic medicine, highlighting essential task descriptions and addressing ongoing research challenges. The conventional taxonomy of supervised, unsupervised, and reinforcement learning is employed in this context. Subsequently, we investigate recent advancements in different computing methods, including semi-supervised learning, deep learning, and transfer learning. This chapter's primary focus is on circumstances involving self-supervised learning and the application of deep neural networks in medical science. For image processing, convolutional neural networks are employed, and for anomaly identification and differential diagnosis, generative adversarial models are utilized. Next, we evaluate the most significant links between machine learning research and healthcare analytics, concentrating on the main areas of study for predictive analytics in diagnostic imaging. In conclusion, we establish a connection between the previously discussed unsolved issue of utilizing machine learning for predictive medical informatics and the effect limits of present research, such as patient privacy and security.

DOI: 10.1201/9781003489368-6

This chapter aims to provide a comprehensive exploration of federated machine learning (FML) in the context of medical science, highlighting its potential applications, addressing challenges, and suggesting future directions for research and implementation in healthcare settings.

6.1.1 Overview of Machine Learning in Medical Science

In recent times, something exciting has been happening in the world of the medical field. The integration of the machine learning field into the realm of medical science has ushered in a transformative era, revolutionizing the landscape of healthcare delivery, diagnostics, and personalized treatment. In an era characterized by an unprecedented influx of health-related data, machine learning algorithms offer unparalleled potential to derive meaningful insights, and predict disease patterns. As the strength of medical information continues to grow exponentially, machine learning emerges as a catalyst for unlocking patterns, correlations, and hidden knowledge that can significantly enhance clinical decision-making.

Historically, medical science has relied on traditional methodologies for diagnosis and treatment, often constrained by the limitations of human cognitive capacity and the huge volume of available data. However, the advent of machine learning technologies has introduced a paradigm shift, empowering healthcare professionals with the ability to tackle vast datasets for precise diagnostics, prognostics, and treatment planning. From analyzing medical images and genomic data to optimizing hospital operations and predicting disease outbreaks, machine learning's applications in medical science are diverse and far-reaching.

This introduction delves into the multifaceted impact of machine learning on medical science, exploring its applications across various domains of healthcare and addressing the challenges and ethical considerations that accompany this technological evolution. Through examining real-world case studies and success stories, our goal is to offer a thorough and inclusive overview of the subject and ongoing revolution in healthcare, driven by the fusion of cutting-edge machine learning approaches and the wealth of medical field data at our disposal. It becomes clear as we proceed through this investigation that machine learning is a transformational force rather than just a tool, reshaping the landscape of medical practice and offering unprecedented opportunities for precision, efficiency, and personalized care.

6.1.2 Challenges in Centralized Machine Learning for Healthcare Data

Centralized machine learning for healthcare data faces several challenges, primarily because of the easy availability of the data and the complex regulatory environment. Here are some key challenges.

6.1.2.1 Data Privacy and Security

Sensitive Information: Healthcare data contain highly available information about individuals. Maintaining the privacy of medical data is a paramount concern, and centralized models must ensure that access is restricted to authorized personnel.

Data Breach Risks: Centralized systems are more susceptible to large-scale data breaches. Unauthorized access might result from a security breach, jeopardizing patient privacy and public confidence in the healthcare sector.

6.1.2.2 Data Integration

Medical imaging, wearable technology, electronic health records (EHRs), and other various sources are common sources of healthcare data. Because these heterogeneous data sources have different data formats, and structures, integrating them into a centralized system can be difficult.

6.1.2.3 Scalability

Healthcare generates a massive amount of data, and centralized systems need to be maintained and handle the increasing volume of information. Scalability challenges can arise when dealing with large datasets, leading to performance issues and slower processing time.

6.1.2.4 Computational Resources

Training and running machine learning models on healthcare data require substantial computational resources. Centralized systems may struggle to provide the necessary computing power, leading to slower model development and deployment.

6.1.2.5 Data Transfer and Latency

Centralized models often require transferring large amounts of data to a central location, incurring data transfer costs and potential delays.

Real-Time Processing: In healthcare, real-time processing is crucial for tasks like patient monitoring and emergency response. Centralized systems may face challenges in providing low-latency solutions.

6.1.2.6 Interoperability

Centralized machine learning systems must be interoperable with existing healthcare infrastructure and technologies. Achieving seamless integration can be challenging due to the varying standards and protocols used in different healthcare settings. Addressing these challenges requires a multidisciplinary approach involving expertise in data science, healthcare, ethics, and regulatory compliance. Additionally, advancements in privacy-preserving technologies, federated learning (FL), and decentralized approaches are being explored to mitigate some of these challenges while maintaining the benefits of the machine learning approach in the healthcare field.

6.1.3 Motivation for Federated Machine Learning

FML is an approach that employs decentralized devices or servers holding local data samples for training machine learning models without transferring the actual samples. The impetus for adopting FML stems from several challenges and considerations present in conventional, centralized machine learning methods. Here are some primary motivations for embracing FML.

6.1.3.1 Privacy Preservation

In many applications, sensitive data are distributed across different devices or locations, making it challenging to centralize all data in one location. The FL enables the localized training of various models on each device without the need to share raw data. There are many models where updates or gradients are exchanged, ensuring the preservation of individual data privacy.

6.1.3.2 Data Decentralization

In situations where data are produced and stored at the edge, such as IoT devices or mobile phones, centralizing all data for training becomes impractical or inefficient. FL allows machine learning models to be trained on the technical devices, thereby minimizing the necessity to transmit data to a central server.

6.1.3.3 Reduced Communication Overhead

Transferring significant volumes of unstructured data to a centralized server can be resource-intensive and may lead to increased communication costs. FL reduces communication overhead by transmitting only model updates, gradients, or aggregated information, making it more efficient in terms of bandwidth and latency.

6.1.3.4 Real-Time Learning

Some applications require real-time model updates or adaptation to changing conditions, which may not be feasible with centralized learning approaches due to the time needed to collect, aggregate, and process data. FL enables continuous learning on local devices, allowing models to adapt quickly to changing patterns or conditions, as shown in Figure 6.1.

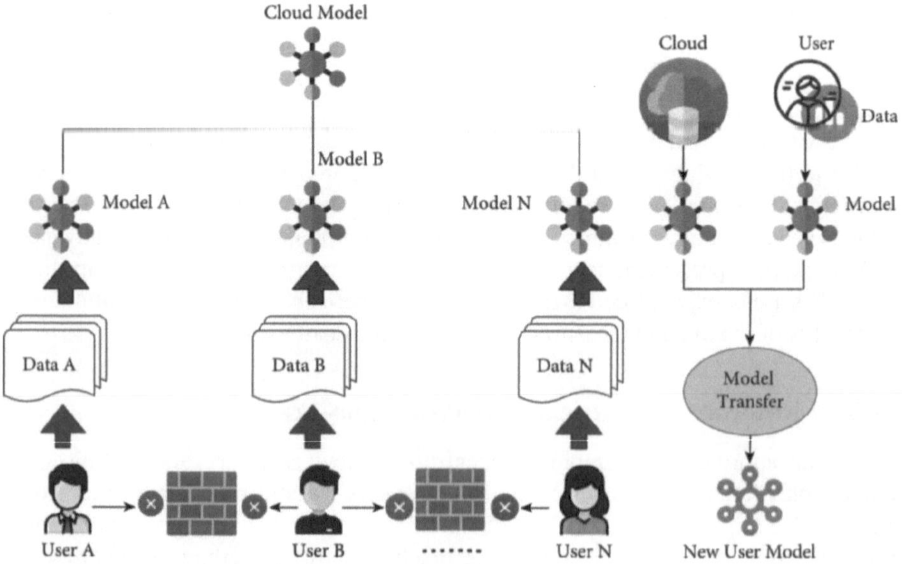

FIGURE 6.1 FML for machine learning models.

6.1.3.5 Edge Computing and Low Latency

FL is aptly suited for edge computing environments, where data processing takes place closer to the source, thereby diminishing latency and enhancing response times. Training models locally on edge devices align with the principles of edge computing, making it suitable for applications with stringent latency requirements.

6.1.3.6 Resource Efficiency

In cases where devices have limited computational resources or bandwidth, FL distributes the model training process, leveraging the capabilities of each device without overburdening any single node, as shown in Figure 6.1.

6.1.3.7 Security

Centralized models can be vulnerable to attacks on the central server, potentially exposing sensitive information. FL distributes the learning process, making it more resilient to certain types of attacks, as compromising one device doesn't compromise the entire dataset. FML addresses privacy concerns, enables decentralized data training, reduces communication overhead, supports real-time learning, aligns with edge computing principles, and enhances overall efficiency and security in machine learning applications across distributed environments.

6.2 FUNDAMENTALS OF FEDERATED MACHINE LEARNING

In today's artificial intelligence and machine learning scenario, FL has demonstrated remarkable achievement across various industries and scientific domains. FL can be described as a collaborative machine learning environment implemented on devices. It was first introduced by Google in 2015 [1].

6.2.1 Architecture of Federated Machine Learning

FL used as a global machine learning model hosted on a centralized service provider is distributed to the local edge devices within a pool of customers or users, as depicted in Figure 6.2. This model is used as the initial phase model, it is trained on the client's data directly on the local edge device, leaving behind a trained local model and client data [2]. In this process, the local model's training, and the selected and modified parameters of the client are transferred back to the centralized server, where they are amalgamated with the parameters from all other clients in the pool, updating the globalized model. After the globalized model is updated, a suitable round of training occurs by disseminating the revised globalized model to a different pool of particular customers. FL operates as a collaborative and iterative learning environment. Its modular design facilitates the creation of a secure, private setting for collaborative machine learning [3,4].

The fundamental principle of FML revolves around training machine learning models without exchanging datasets among decentralized devices used to store sample data of localized servers. This method differs from both approaches that presume uniform

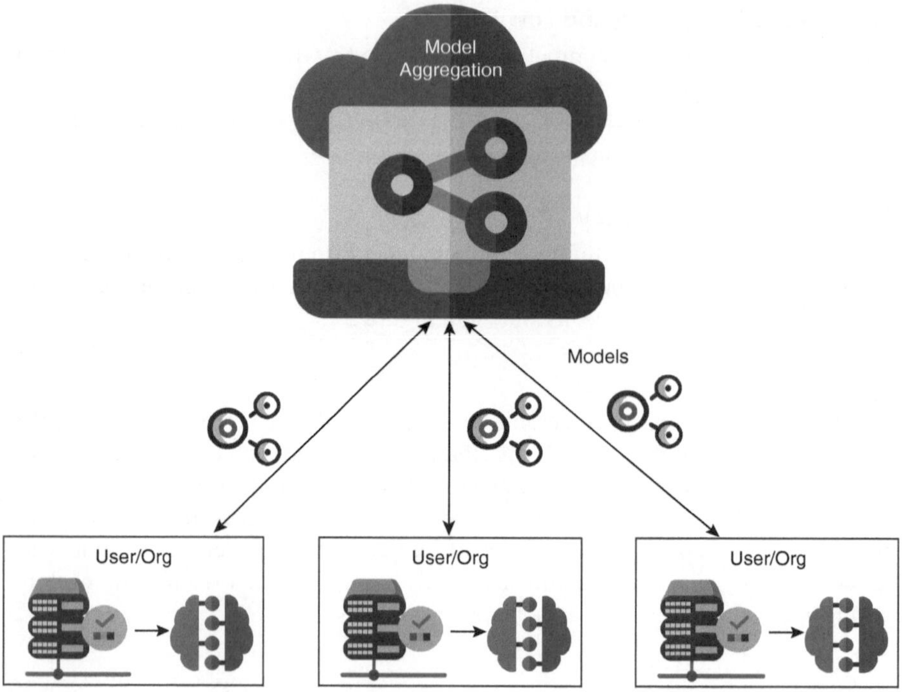

FIGURE 6.2 FL architecture.

distribution of localized data samples and standard centralized machine learning methods that combine can be used to store local datasets in a single training session.

6.2.2 Mathematical Formulation of Federated Learning

The following is the goal function for FL:

$$f(a_1,a_2,a_3,a_4,\ldots a_n)=\frac{1}{n}\sum_{i=1}^{n} fi(a_i)$$

where *f*(*i*) is the local objective function for node *i*, which specifies how the weights modeling a_i confirm to the local data set of node *i*, and node set *n*. The weights representing the node *i* in the modeling process are represented by a_i.

Training a shared model on all of the nodes' local datasets is the aim of FL or in other words:

- To optimize the goal function $f(a_1, a_2, a_3, a_4, \ldots, a_n)$.
- Achieving consensus on a_i. In other words, $a_1, a_2, a_3, a_4, \ldots, a_n$ converge to some common upon the conclusion of the training process.

The fundamentals of FML involve several key components and processes. Here are the fundamental concepts.

6.2.2.1 Decentralized Training

Conventional machine learning typically entails gathering all data in a central location for model training. In FL, the training process is decentralized, with individual devices or servers conducting local training on their respective datasets.

6.2.2.2 Local Model Training

Every participating device or server independently trains a model on its own data. This local model is then updated by leveraging traditional machine learning algorithms or deep learning methods, utilizing the specifics of the respective device's dataset [5,6].

6.2.2.3 Model Updates

FL focuses on transmitting model updates or gradients instead of sending raw data to a central server. After local training, each device calculates the changes made to its model and shares only these updates with the central server or other participating devices as shown in Figure 6.3.

6.2.2.4 Aggregation of Model Updates

The central server consolidates the modeling updates received from different devices to construct a globalized learning model. The aggregation process may incorporate techniques like weighted averaging or other aggregation methods, depending on the specific FL algorithm employed, as illustrated in Figure 6.3.

6.2.2.5 Communication Protocol

For FL to work, devices and the central server must be able to communicate model changes via a communication protocol. In order to protect data security and privacy throughout the communication process, this protocol needs to be created [7,8].

6.2.2.6 Privacy-Preserving Techniques

Ensuring privacy is a pivotal element in FL. Strategies like federated averaging, secure multi-party computing, and differential privacy are implemented to safeguard the confidentiality of individual data samples throughout the model training process [9,10].

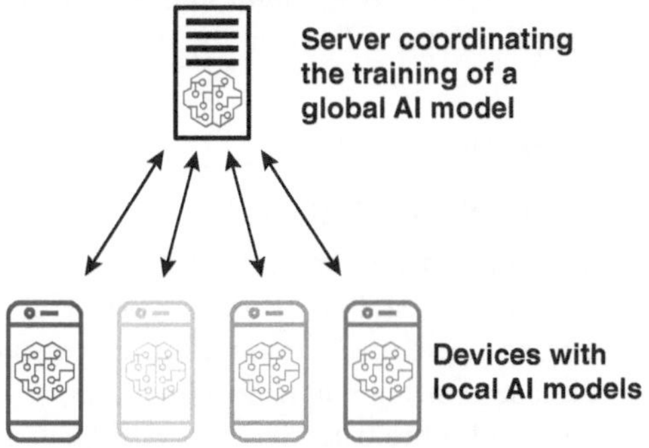

FIGURE 6.3 Illustrates the process of training a global AI model through cell phones using a federated learning protocol.

6.2.2.7 Initialization and Synchronization

FL models need to be initialized before the training process begins. Ensuring proper synchronization of model parameters across devices is crucial to maintaining a coherent and effective global model [11,12].

6.2.2.8 Hyperparameter Tuning

FL may involve tuning hyperparameters that govern the learning process. Techniques for adaptive learning rates and other hyperparameter adjustments are often used to enhance convergence and performance [13].

6.2.2.9 Heterogeneous Environments

FL is crafted to operate in diverse environments characterized by variations in hardware capabilities, data distributions, and processing capacities across devices. FL algorithms should be robust and adaptable to such variations [14,15].

6.2.2.10 Model Evaluation and Testing

Evaluating and testing the global model is an essential step to ensure its effectiveness. This process may involve sending a representative subset of data to the central server for evaluation or implementing other validation techniques.

6.2.2.11 Iterative Process

In FL, local training, model updates, and aggregation are often carried out iteratively to progressively enhance the global model. The iterative nature allows the modeling to adapt to various changes by distributing underlying data flow.

By combining these fundamental concepts, FML addresses privacy concerns, enables decentralized model training, and leverages the collective knowledge from distributed datasets to create a robust and adaptable global model [16,17].

6.3 APPLICATIONS OF MACHINE LEARNING IN THE MEDICAL FIELD USING FEDERATED APPROACHES

FML shows significant potential in the realm of medical science, presenting inventive solutions to tackle issues concerning privacy, data sharing, and model generalization. Below are some primary applications of FML in the field of medical field [18–20].

6.3.1 Disease Prediction and Diagnosis

6.3.1.1 Federated Models for Early Detection

Federated models can be employed for the early detection of diseases by training models across data from different healthcare institutions [6,21–23]. For instance, early detection of diseases like cancer or neurodegenerative disorders, heart disease, brain tumor, and skin disease can benefit from collaborative training on diverse patient populations as shown in Figure 6.4.

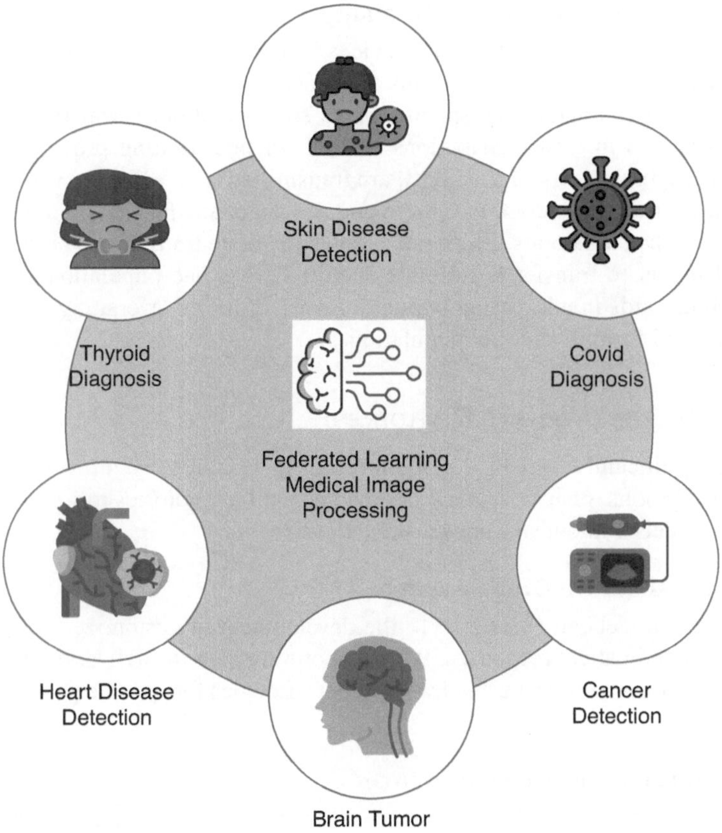

FIGURE 6.4 Application of ML in medical image processing.

6.3.1.2 Improved Accuracy with Decentralized Data

The creation of prognostic and diagnostic prediction models for diseases is made possible via FL [24,25]. Hospitals and clinics can collaborate to train models on diverse datasets, incorporating information from different demographics and regions. This can lead to more accurate and generalizable models for various medical conditions.

6.3.2 Treatment Personalization

6.3.2.1 Tailoring Treatments Based on Local Data

With FL, the modeling process can be trained on individual patient data to create personalized treatment plans [26,27]. This method takes into account an individual's distinct medical history, genetic makeup, and responses to treatments, resulting in more efficient and customized healthcare interventions.

6.3.2.2 Addressing Population Variability

Distribute the global model to local devices, servers, or institutions representing different populations. Each entity trains the model locally on its dataset, capturing the nuances and characteristics specific to its subpopulation. Local training helps address variations in data distributions [28,29]. In this training process, only the model updates, in the form of gradients, are transmitted to a central server for aggregation. The central server combines these updates to update the global model without revealing specific data points. Federated models facilitate transfer learning, enabling the global model to transfer knowledge acquired from one population to another. This capability aids in addressing population variability by leveraging insights and patterns learned from diverse subpopulations.

6.3.3 Drug Discovery and Development

FML can significantly benefit collaborative drug research by allowing different research institutions, pharmaceutical companies, and laboratories to jointly develop predictive models without sharing sensitive data.

6.3.3.1 Collaborative Drug Research

The federated model can contribute to the development of customized drugs based on individualized patient responses. By incorporating data from diverse sources, the model may assist in tailoring drug development strategies for specific patient populations [30,31].

6.3.3.2 Utilizing Diverse Data Sources

Collaborative federated training enhances the generalization capabilities of the model. By learning from diverse datasets, the model becomes more robust and adaptable, potentially improving its ability to predict drug responses across different populations or conditions.

FL supports continuous learning, allowing the model to adapt to new information and emerging trends in drug research [11,32,33]. As new data becomes available, the model can be updated without the need for centralized retraining.

6.3.4 Public Health Surveillance

6.3.4.1 Real-Time Monitoring

FML can be a valuable tool for real-time monitoring in public health surveillance [34,35]. This approach allows different health agencies, institutions, and regions to collaboratively analyze and learn from their data without the need to centralize sensitive information.

6.3.4.2 Rapid Response to Health Threats

FML can play a vital role in expediting responses to health threats by fostering collaboration among diverse healthcare institutions, public health agencies, and

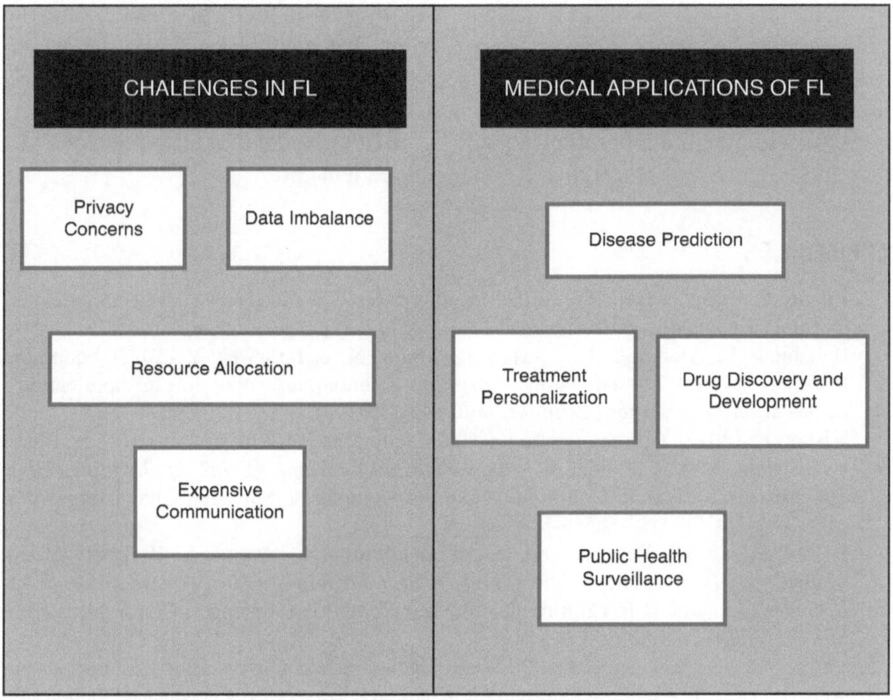

FIGURE 6.5 Challenges and medical applications of FL.

researchers [18,36,37]. By leveraging FL, rapid response to health threats becomes more feasible, efficient, and privacy-preserving. The decentralized nature of the approach allows for collective intelligence without compromising individual data privacy, ultimately contributing to more effective public health interventions [38,39] (Figure 6.5).

6.4 CONCLUSION

FL has acquired a significant interest in academic and industrial circles, emerging as a focal point for enhancing the efficacy and safeguarding the privacy of data-driven applications. In recent times, numerous applications and use cases have embraced FL to leverage its potential benefits [40–42].

The decentralized nature of FL, allowing model training across various institutions, the sharing of raw data is unnecessary and addresses crucial privacy concerns while fostering collaborative research [43]. The application of FL in medical science presents many opportunities, ranging from improved diagnostic accuracy to personalized treatment plans. The federated approach enables the utilization of diverse datasets across institutions, leading to robust and generalizable models. Moreover, the survey extensively explores the nuanced aspect of privacy in FL, emphasizing the importance of preserving patient confidentiality and data security.

In summary, FML stands as a beacon of innovation in medical science, offering a collaborative and privacy-preserving paradigm that holds the potential to revolutionize healthcare practices. As we navigate this evolving landscape, future efforts should be directed toward refining methodologies, addressing challenges, and fostering interdisciplinary collaborations to unlock the full potential for utilization of FML in progressing medical research and enhancing patient care.

REFERENCES

1. Zhang, C., Xie, Y., Bai, H., Yu, B., Li, W., & Gao, Y. (2021). A survey on federated learning. *Knowledge-Based Systems*, 216, 106775.
2. Banabilah, S., Aloqaily, M., Alsayed, E., Malik, N., & Jararweh, Y. (2022). Federated learning review: Fundamentals, enabling technologies, and future applications. *Information Processing & Management*, 59(6), 103061.
3. Kishor, K. (2022). Communication-efficient federated learning. In: Yadav, S. P., Bhati, B. S., Mahato, D. P., Kumar, S. (eds) *Federated Learning for IoT Applications*. EAI/Springer Innovations in Communication and Computing. Springer, Cham. https://doi.org/10.1007/978-3-030-85559-8_9
4. Kishor, K. (2022). Personalized federated learning. In: Yadav, S. P., Bhati, B. S., Mahato, D. P., Kumar, S. (eds) *Federated Learning for IoT Applications*. EAI/Springer Innovations in Communication and Computing. Springer, Cham. https://doi.org/10.1007/978-3-030-85559-8_3
5. Verma, R. K., & Kishor, K. (2024). Image processing applications in agriculture with the help of AI. In: Khan, M. A., Khan, R., Praveen, P., Verma, A., Panda, M. (eds) *Infrastructure Possibilities and Human-Centered Approaches With Industry 5.0* (pp. 162–181). IGI Global, Hershey, Pennsylvania, USA. https://doi.org/10.4018/979-8-3693-0782-3.ch010
6. Brisimi, T. S., Chen, R., Mela, T., Olshevsky, A., Paschalidis, I. C., & Shi, W. (2018). Federated learning of predictive models from federated electronic health records. *International Journal of Medical Informatics*, 112, 59–67.
7. Kishor, K., & Nand, P. (2023). "Wireless networks based in the cloud that support 5G." In: Kishor K., Saxena, N., and Pandey, D. (eds), *Cloud-based Intelligent Informative Engineering for Society 5.00*, 1st edition (pp. 23–40). Chapman and Hall/CRC, New York. ISBN: 9781003213895. https://doi.org/10.1201/9781003213895-2
8. Kishor, K. (2023). Cloud computing in blockchain. In: Kishor K., Saxena, N., and Pandey, D. (eds), *Cloud-based Intelligent Informative Engineering for Society 5.0*, 1st edition (pp. 79–105). Chapman and Hall/CRC, New York. ISBN: 9781003213895. https://doi.org/10.1201/9781003213895-5
9. Kishor, K. (2023). Impact of cloud computing on entrepreneurship, cost, and security. In: Kishor K., Saxena, N., and Pandey, D. (eds), *Cloud-based Intelligent Informative Engineering for Society 5.0*, 1st edition (pp. 171–191). CRC Press, New York. ISBN: 9781003213895. https://doi.org/10.1201/9781003213895-10
10. Kishor, K., Nand, P., & Agarwal, P. (2018). Secure and efficient subnet routing protocol for MANET. *Indian Journal of Public Health*, 9(12), 200. https://doi.org/10.5958/0976-5506.2018.01830.2
11. Li, L., Fan, Y., Tse, M., & Lin, K. Y. (2020). A review of applications in federated learning. *Computers & Industrial Engineering*, 149, 106854.
12. Aledhari, M., Razzak, R., Parizi, R. M., & Saeed, F. (2020). Federated learning: A survey on enabling technologies, protocols, and applications. *IEEE Access*, 8, 140699–140725.

13. Ouyang, F., & Jiao, P. (2021). Artificial intelligence in education: The three paradigms. *Computers and Education: Artificial Intelligence*, 2, 100020.
14. Alqahtani, T., Badreldin, H. A., Alrashed, M., Alshaya, A. I., Alghamdi, S. S., bin Saleh, K., & Albekairy, A. M. (2023). The emergent role of artificial intelligence, natural learning processing, and large language models in higher education and research. *Research in Social and Administrative Pharmacy*, 19(8), 1236–1242.
15. Laupichler, M. C., Aster, A., Schirch, J., & Raupach, T. (2022). Artificial intelligence literacy in higher and adult education: A scoping literature review. *Computers and Education: Artificial Intelligence*, 3, 100101.
16. Kishor, K., Nand, P., & Agarwal, P. (2018). Notice of retraction design adaptive subnetting hybrid gateway MANET protocol on the basis of dynamic TTL value adjustment. *Aptikom Journal on Computer Science and Information Technologies*, 3(2), 59–65. https://doi.org/10.11591/APTIKOM.J.CSIT.115
17. Gupta, S., Tyagi, S., & Kishor, K. (2022). Study and development of self sanitizing smart elevator. In: Gupta, D., Polkowski, Z., Khanna, A., Bhattacharyya, S., Castillo, O. (eds) *Proceedings of Data Analytics and Management*. Lecture Notes on Data Engineering and Communications Technologies, vol. 90. Springer, Singapore. https://doi.org/10.1007/978-981-16-6289-8_15
18. Kishor, K., & Pandey, D. (2022). Study and development of efficient air quality prediction system embedded with machine learning and IoT. In: Gupta, D., Khanna, A., Bhattacharyya, S., Hassanien, A.E., Anand, S., Jaiswal, A. (eds) *Proceeding International Conference on Innovative Computing and Communications*. Lecture Notes in Networks and Systems, vol. 471, Springer, Singapore. https://doi.org/10.1007/978-981-19-2535-1_24
19. Chu, Y. W., Hosseinalipour, S., Tenorio, E., Cruz, L., Douglas, K., Lan, A., & Brinton, C. (2022). Mitigating biases in student performance prediction via attention-based personalized federated learning. In: *Proceedings of the 31st ACM International Conference on Information & Knowledge Management, 31st ACM International Conference on Information and Knowledge Management* (pp. 3033–3042). CIKM, Atlanta.
20. Wen, J., Zhang, Z., Lan, Y., Cui, Z., Cai, J., & Zhang, W. (2023). A survey on federated learning: Challenges and applications. *International Journal of Machine Learning and Cybernetics*, 14(2), 513–535.
21. Chen, H., Li, H., Xu, G., Zhang, Y., & Luo, X. (2020). Achieving privacy-preserving federated learning with irrelevant updates over e-health applications. In: *ICC 2020-2020 IEEE International Conference on Communications (ICC)* (pp. 1–6). IEEE, Dublin. DOI: 10.1109/ICC40277.2020.9149385
22. Chen, Y., Luo, F., Li, T., Xiang, T., Liu, Z., & Li, J. (2020). A training-integrity privacy-preserving federated learning scheme with trusted execution environment. *Information Sciences*, 522, 69–79.
23. Chen, Y., Qin, X., Wang, J., Yu, C., & Gao, W. (2020). Fedhealth: A federated transfer learning framework for wearable healthcare. *IEEE Intelligent Systems*, 35(4), 83–93.
24. Kishor, K., Rani, R., Rai, A.K., & Sharma, V. (2023). 3D application development using unity real time platform. In: Swaroop, A., Kansal, V., Fortino, G., Hassanien, A. E. (eds) *Proceedings of Fourth Doctoral Symposium on Computational Intelligence. DoSCI 2023.* Lecture Notes in Networks and Systems, vol. 726. Springer, Singapore. https://doi.org/10.1007/978-981-99-3716-5_54
25. Kishor, K., Sangal, S., & Shahroz (2023). Real-time traffic signs and lane line detection. In: Swaroop, A., Kansal, V., Fortino, G., Hassanien, A. E. (eds) *Proceedings of Fourth Doctoral Symposium on Computational Intelligence. DoSCI 2023.* Lecture Notes in Networks and Systems, vol. 726. Springer, Singapore. https://doi.org/10.1007/978-981-99-3716-5_71

26. Kishor, K. & Verma, R. K. (2023). Cloud computing-based smart agriculture. In: Sharma, A., Chanderwal, N., Khan, R. (eds) *Convergence of Cloud Computing, AI, and Agricultural Science* (pp. 120–136). IGI Global, Hershey, Pennsylvania, USA. https://doi.org/10.4018/979-8-3693-0200-2.ch006
27. Kishor, K. (2023). Chapter 17 Application of quantum computing for digital forensic investigation. In: Yadav, S. P., Singh, R., Yadav, V., Al-Turjman, F., Kumar, S. A. (eds) *Quantum-Safe Cryptography Algorithms and Approaches: Impacts of Quantum Computing on Cybersecurity* (pp. 231–248). De Gruyter, Berlin, Boston. https://doi.org/10.1515/9783110798159-017
28. Kishor, K. (2023). Chapter 10: Study of quantum computing for data analytics of predictive and prescriptive analytics models. In: Yadav, S. P., Singh, R., Yadav, V., Al-Turjman, F., Kumar, S. A. (eds) *Quantum-Safe Cryptography Algorithms and Approaches: Impacts of Quantum Computing on Cybersecurity* (pp. 121–146). De Gruyter, Berlin, Boston, MA. https://doi.org/10.1515/9783110798159-010
29. Kishor, K. (2023). Chapter 12: Review and significance of cryptography and machine learning in quantum computing. In: Yadav, S. P., Singh, R., Yadav, V., Al-Turjman, F., Kumar, S. A. (eds) *Quantum-Safe Cryptography Algorithms and Approaches: Impacts of Quantum Computing on Cybersecurity* (pp. 159–176). De Gruyter, Berlin, Boston, MA. https://doi.org/10.1515/9783110798159-012
30. Fang, C., Guo, Y., Wang, N., & Ju, A. (2020). Highly efficient federated learning with strong privacy preservation in cloud computing. *Computers & Security*, 96, 101889.
31. Goecks, J., Jalili, V., Heiser, L. M., & Gray, J. W. (2020). How machine learning will transform biomedicine. *Cell*, 181(1), 92–101.
32. Huang, L., Shea, A. L., Qian, H., Masurkar, A., Deng, H., & Liu, D. (2019). Patient clustering improves efficiency of federated machine learning to predict mortality and hospital stay time using distributed electronic medical records. *Journal of Biomedical Informatics*, 99, 103291.
33. Kim, Y. J., & Hong, C. S. (2019). Blockchain-based node-aware dynamic weighting methods for improving federated learning performance. In: *2019 20th Asia-Pacific Network Operations and Management Symposium (APNOMS)* (pp. 1–4). IEEE, Matsue. DOI: 10.23919/APNOMS.2019.8893114
34. Balta, D., Sellami, M., Kuhn, P., Schöpp, U., Buchinger, M., Baracaldo, N., Anwar, A., Ludwig, H., Sinn, M., Purcell, M., & Altakrouri, B. (2021). Accountable federated machine learning in government: Engineering and management insights. In: *Electronic Participation: 13th IFIP WG 8.5 International Conference, ePart 2021, Granada, Spain, September 7–9, 2021, Proceedings 13* (pp. 125–138). Springer International Publishing, Granada.
35. Persson, J. A., & Olsson, C. M. (2021). A federated interactive learning IoT-based health monitoring platform. In: *New Trends in Database and Information Systems: ADBIS 2021 Short Papers, Doctoral Consortium and Workshops: DOING, SIMPDA, MADEISD, MegaData, CAoNS, August 24–26, 2021, Proceedings* (p. 235). Springer Nature, Tartu
36. Cai, L., Lin, D., Zhang, J., & Yu, S. (2020). Dynamic sample selection for federated learning with heterogeneous data in fog computing. In: *ICC 2020-2020 IEEE International Conference on Communications (ICC)* (pp. 1–6). IEEE, Dublin. DOI: 10.1109/ICC40277.2020.9148586
37. Doku, R., Rawat, D. B., & Liu, C. (2019, July). Towards federated learning approach to determine data relevance in big data. In: *2019 IEEE 20th International Conference on Information Reuse and Integration for Data Science (IRI)* (pp. 184–192). IEEE, Los Angeles, CA. DOI: 10.1109/IRI.2019.00039

38. Yang, Q., Liu, Y., Chen, T., & Tong, Y. (2019). Federated machine learning: Concept and applications. *ACM Transactions on Intelligent Systems and Technology (TIST)*, 10(2), 1–19.
39. Sharma, A., Jha, N., & Kishor, K. (2022). Predict COVID-19 with chest X-ray. In: Gupta, D., Polkowski, Z., Khanna, A., Bhattacharyya, S., Castillo, O. (eds) *Proceedings of Data Analytics and Management*. Lecture Notes on Data Engineering and Communications Technologies, vol. 90. Springer, Singapore. https://doi.org/10.1007/978-981-16-6289-8_16
40. Wahab, O. A., Mourad, A., Otrok, H., & Taleb, T. (2021). Federated machine learning: Survey, multi-level classification, desirable criteria and future directions in communication and networking systems. *IEEE Communications Surveys & Tutorials*, 23(2), 1342–1397.
41. Lim, W. Y. B., Xiong, Z., Miao, C., Niyato, D., Yang, Q., Leung, C., & Poor, H. V. (2020). Hierarchical incentive mechanism design for federated machine learning in mobile networks. *IEEE Internet of Things Journal*, 7(10), 9575–9588.
42. Kishor, K., Singh, P., & Vashishta, R. (2023). Develop model for malicious traffic detection using deep learning. In: Sharma, D. K., Peng, S. L., Sharma, R., Jeon, G. (eds) *Micro-Electronics and Telecommunication Engineering*. Lecture Notes in Networks and Systems, vol. 617. Springer, Singapore. https://doi.org/10.1007/978-981-19-9512-5_8
43. Hu, K., Li, Y., Xia, M., Wu, J., Lu, M., Zhang, S., & Weng, L. (2021). Federated learning: A distributed shared machine learning method. *Complexity*, 2021, 1–20.

7 Healthcare Informatics Security Issues and Solutions Using Federated Learning

Sachin A. Goswami, Saurabh Dave, and Kashyap C. Patel

7.1 INTRODUCTION

In recent years, the healthcare industry has seen a significant transition toward digitalization, which encompasses patient data, diagnostic tools, and treatment approaches. This transformation has been fueled by the incorporation of healthcare informatics technology. Even though these technological improvements bring about a number of benefits, including improved patient care and operational efficiency, they also bring about a number of serious concerns surrounding the privacy and security of data. Security concerns in healthcare informatics, which include data breaches, cyberattacks, and the complexity of regulatory compliance, pose significant threats to the security of patient information and the dependability of healthcare infrastructure [1–3].

It is vital that creative solutions be developed in order to properly address these security problems. These solutions must be able to effectively maintain the confidentiality of sensitive patient data while also promoting collaborative healthcare activities. A decentralized machine learning strategy that enables different institutions to cooperatively train models utilizing dispersed data without sharing raw data is referred to as federated learning. This is one of the developing solutions in the healthcare industry [4,5].

The purpose of this research is to investigate the current security concerns in contemporary healthcare systems that are related to informatics and to provide potential solutions that make use of federated learning capabilities. The purpose of this chapter is to contribute to the ongoing conversation about strengthening healthcare informatics systems by analyzing the landscape of healthcare data security, providing an explanation of the idea of federated learning and its benefits, and digging into its applications and challenges in healthcare settings [6,7].

The purpose of this chapter is to provide healthcare stakeholders with actionable insights that will strengthen data privacy, confidentiality, and security in healthcare informatics. This will be accomplished by conducting a comprehensive examination

 DOI: 10.1201/9781003489368-7

of security challenges, potential risks, and the effectiveness of federated learning in mitigating these risks. In the end, the goal is to cultivate a healthcare environment that is safe, collaborative, and places a priority on the safety of patients, while also instilling faith in digital healthcare technology [8,9].

7.1.1 Overview of Healthcare Informatics Security

To protect sensitive healthcare data, systems, and networks from unauthorized access, breaches, and cyber threats, healthcare informatics security entails the installation of various methods, protocols, and technologies. Because the healthcare sector is increasingly adopting digital technologies for the management of patient information, it is essential to ensure the security of healthcare informatics in order to preserve the privacy of patients, maintain their confidentiality, and maintain the integrity of their data as a whole (Figure 7.1)

The following are important aspects and factors that pertain to the security of healthcare informatics:

7.1.1.1 Data Protection

Protecting patient information requires the use of effective encryption, access restrictions, and authentication systems, which are all used by healthcare companies. On the other hand, access controls restrict system access based on user roles and permissions, while encryption guarantees that data is rendered unreadable to those who are not permitted to see it [10,11].

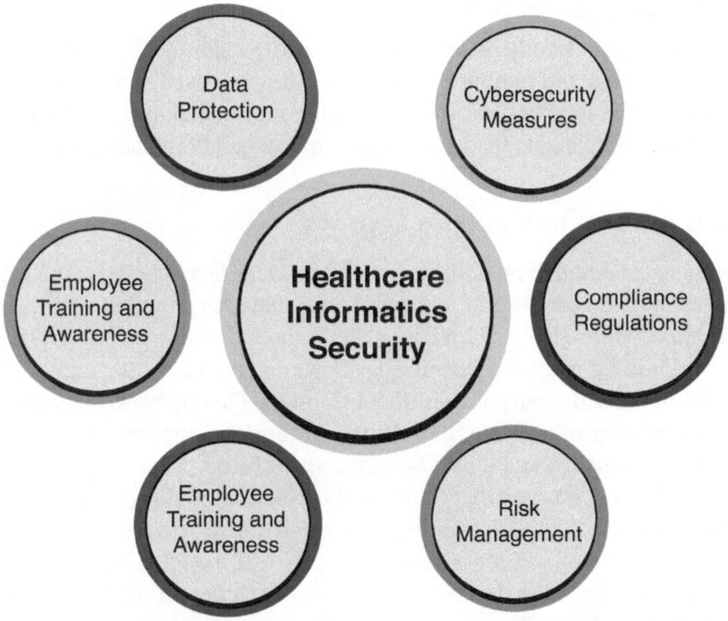

FIGURE 7.1 Healthcare informatics security.

7.1.1.2 Cybersecurity Measures

Malware, ransomware, and phishing assaults are just some of the varieties of cyber dangers that healthcare institutions must contend with. Firewalls, intrusion detection systems, and regular security audits are examples of cybersecurity measures that may be used to assist in the identification and mitigation of possible threats [12–14].

7.1.1.3 Compliance Regulations

The Health Insurance Portability and Accountability Act (HIPAA) and the General Data Protection Regulation (GDPR) are two examples of legislation that healthcare firms are required to comply with. These laws detail the obligations that must be met in order to secure patient data, disclose security events, and ensure responsibility for breaches

7.1.1.4 Risk Management

The identification of security risks and vulnerabilities in informatics systems may be aided by performing vulnerability scans and risk assessments on a regular basis. If these risks are addressed proactively, the possibility of events occurring is decreased, and the effect that they have on patient safety and data integrity is mitigated [15–16].

7.1.1.5 Employee Training and Awareness

One of the major contributors to security breaches is human error. The provision of comprehensive training and awareness programs for healthcare personnel enables them to acquire knowledge on the most effective security methods, the understanding of phishing, and the significance of securely protecting patient information

7.1.1.6 Security Awareness

For the purpose of successfully responding to security events and data breaches, healthcare companies design and maintain incident response plans. These plans include the processes that will be followed in order to identify, report, and mitigate events, as well as communicate with patients who have been impacted and regulatory authorities

7.1.2 Introduction to Federated Learning

The revolutionary method of machine learning known as federated learning was developed with the intention of addressing the problems that are connected with centralized data aggregation while also guaranteeing the highest possible levels of privacy and security. In traditional machine learning frameworks, data are often aggregated into a single repository for the purpose of model training. This presents substantial problems about the privacy of the data, the security of the data, and the adherence to legal norms, especially in sensitive industries such as healthcare (Figure 7.2)

Federated learning, on the other hand, will revolutionize this paradigm by decentralizing the process of model training. This will make it possible for individual devices or edge nodes to collaboratively train a global model while still preserving the locality of the data. This decentralized learning technique gives enterprises the ability to get insights from data without exposing sensitive information to centralized servers or to entities that are external to the company [17,18].

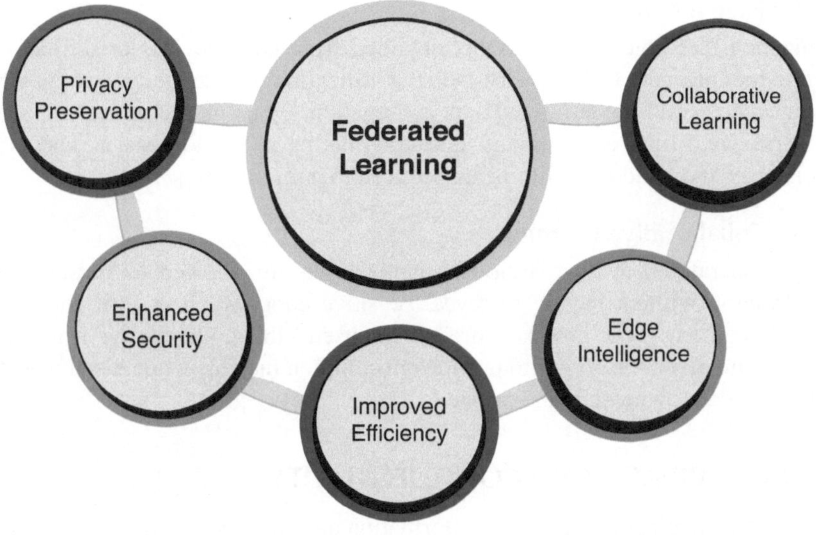

FIGURE 7.2 Federated learning.

The need to maximize the value of data while adhering to tight privacy and security regulations was the impetus for the development of federated learning. Federated learning provides a range of advantages by distributing model training over a large number of devices or organizations. These benefits include the following.

7.1.2.1 Privacy Preservation

By ensuring that sensitive data are kept inside the confines of the device or edge node from where it originated, federated learning helps to reduce the risks that are associated with data exposure or leakage. Through the implementation of this decentralized model, compliance with data protection rules is made easier, and user trust in data handling procedures is significantly increased [19].

7.1.2.2 Enhanced Security

The risks that are connected with data breaches or unauthorized access are mitigated by federated learning since it eliminates the need to transmit data to a central server for model training. The security posture of machine learning systems is improved by the retention of data locally, which effectively strengthens defenses against possible cyberattacks [20].

7.1.2.3 Improved Efficiency

Using the resources that are available in distributed computing environments, federated learning has the potential to improve the effectiveness of model training initiatives. By distributing computing work over a number of different devices, federated learning reduces the amount of load placed on particular servers, which in turn speeds up the process of training

7.1.2.4 Edge Intelligence

Organizations are able to use insights obtained from edge devices or sensors connected to the Internet of Things (IoT) via the utilization of federated learning, which eliminates the need for centralized data aggregation. By using this approach to edge intelligence, real-time analysis and decision-making are made easier, and issues about latency and bandwidth limits are reduced to a minimum [21].

7.1.2.5 Collaborative Learning

Through federated learning, various organizations are allowed to participate in model training without having to reveal sensitive data, which in turn encourages cooperation among these groups. Through the use of this collaborative framework, companies are given the ability to profit from common insights while also respecting the ideals of data ownership and privacy

7.2 DATA PRIVACY AND CONFIDENTIALITY

Given the sensitive nature of patient information and the possible consequences that might result from unauthorized access or breaches, it is of the utmost significance that the infrastructure of the healthcare industry ensures the privacy and security of data. Several essential issues highlight the need to maintain data privacy and confidentiality in the healthcare industry, including the following.

7.2.1 Building Patient Trust and Adhering to Regulations

- The protection of patient privacy helps to create and strengthen patient trust by ensuring patients that their personal health information is handled securely throughout the process
- In order to maintain data privacy standards and stay out of legal trouble, it is absolutely necessary to comply with rules such as HIPAA in the United States

7.2.2 Handling Sensitive Health Information

- The healthcare infrastructure is responsible for managing extremely sensitive patient data, which includes medical histories, diagnoses, and treatment plans
- It is necessary to maintain the security of such information in order to avoid problems such as identity theft, fraud, or possible discrimination

7.2.3 Handling Sensitive Health Information

- Healthcare professionals have both legal and ethical duties to ensure the privacy and confidentiality of patient information
- Breach of confidentiality may result in legal proceedings, harm to the provider's reputation, and a loss of confidence among patients

7.2.4 Securing Electronic Health Records

- The transition to electronic health records (EHRs) adds an additional layer of importance to the protection of user information
- It is vital to implement stringent security measures in order to secure EHRs from unauthorized access, thereby protecting the integrity of patient information and maintaining their confidentiality [22,23].

7.2.5 Telemedicine and Remote Monitoring

- Encryption and secure communication protocols play an essential part in maintaining patients' privacy when it comes to remote healthcare services
- The rise of telemedicine and remote patient monitoring has made it necessary to have secure channels in order to protect patient data during virtual consultations and data transfer.

7.2.6 Enabling Data Sharing and Interoperability

- It is necessary to have safe data-sharing methods in place in order to protect the privacy of patients when many healthcare organizations work together
- It is essential that interoperability standards have strong security mechanisms in order to ensure the protection of data as it moves between various systems

7.2.7 Addressing Technological Advancements

- The incorporation of cutting-edge technology in the healthcare industry, such as artificial intelligence (AI) and the IoT, calls for increased data privacy safeguards
- It is essential to perform continuous monitoring and upgrades in order to address recently discovered vulnerabilities and threats that are related to developing technologies

7.2.8 Embracing a Patient-Centric Approach

- Emphasizing data privacy is consistent with a patient-centered approach, which gives patients the ability to exercise control over their own health information
- In order to contribute to the development of a more ethical healthcare system, it is important to provide patients with transparency and control over their data

7.2.9 Preventing Unauthorized Access

- It is possible to restrict unwanted access to patient data by putting into place safeguards such as access controls, authentication, and encryption
- The identification and effective mitigation of possible security issues are both achieved via the use of regular audits and monitoring sessions

7.2.10 Promoting Education and Awareness

- It is necessary to provide ongoing education to healthcare professionals and personnel about the necessity of protecting the privacy of patient record information
- Creating a culture of awareness ensures that all parties involved are aware of their respective responsibilities with regard to the protection of confidentiality

7.3 CHALLENGES IN ENSURING DATA PRIVACY

Due to the sensitive nature of patient information and the ever-changing digital world, there are a multitude of problems that must be overcome in order to guarantee the confidentiality of data inside the healthcare infrastructure. Among the most significant difficulties are described in the following sections.

7.3.1 Cybersecurity Threats

Description: Cybercriminals often target the healthcare industry with ransomware, malware, and phishing tactics. This sector is a top target for cyberattacks

Impact: As a result of successful breaches, patient data may be compromised, which can result in identity theft, financial fraud, and unauthorized access to sensitive medical information

7.3.2 Insufficient Security Measures

Description: Lack of effective implementation of security measures, including encryption, access restrictions, and secure authentication techniques, among other security measures

Impact: Because of inadequate security mechanisms, healthcare systems are susceptible to unauthorized access and data breaches, which pose a substantial threat to the confidentiality of patient information

7.3.3 Interconnected Systems and IoT Devices

Description: There is a growing movement in the healthcare industry toward the use of networked systems and IoT devices [24].

Impact: Because of the broad ecosystem, vulnerabilities may be introduced, and unsecured IoT devices have the potential to become entry points for cyberattacks, putting data privacy at risk

7.3.4 Human Error and Insider Threats

Description: Erroneous or malicious activities committed by healthcare personnel who have access to patient information might occur

Impact: Incidents like accidental data leaks, unauthorized access, or intentional breaches may jeopardize the security of patient information

7.3.5 Lack of Standardized Privacy Policies

Description: Incidents like accidental data leaks, unauthorized access, or intentional breaches may jeopardize the security of patient information

Impact: One of the potential consequences of a lack of standardization in privacy standards is the creation of gaps in data protection and difficulties in compliance with rules

7.3.6 Third-Party Data Sharing

Description: It is possible that disclosing patient information will be required in order to collaborate with external organizations or third-party service providers

Impact: Data exposure may occur as a result of inadequate monitoring of the security measures implemented by third parties, and patients may lose control over how their information is used

7.3.7 Data Interoperability Challenges

Description: There are challenges involved in the smooth exchange and integration of patient data across a variety of healthcare systems and providers

Impact: Incomplete or erroneous data transfers have the potential to violate the privacy of patients and become a barrier to the delivery of effective treatment

7.3.8 Regulatory Compliance Burden

Description: Regulations regarding healthcare privacy, such as the HIPAA in the United States, are notoriously difficult to understand and stringent

Impact: It may be difficult to meet compliance regulations, and failure to do so may result in legal ramifications, which can diminish the confidence that patients have in a healthcare provider

7.3.9 Emerging Technologies and Big Data Analytics

Description: The use of developing technology for the purpose of healthcare data analysis, such as AI and big data analytics [25].

Impact: The difficulty of striking a balance between the advantages of sophisticated analytics and the need to safeguard the privacy of patients is a significant one

7.3.10 Patient Awareness and Education

Description: There is a lack of understanding among patients about their rights and control over their health data

Impact: Insufficient patient education may lead to unintended disclosures or an unwillingness to give vital information, both of which have the potential to negatively impact the quality of treatment provided

7.4 IMPORTANCE OF PATIENT DATA PROTECTION IN FEDERATED LEARNING

The importance of patient data protection in healthcare infrastructure is paramount due to several critical considerations:

7.4.1 Patient Trust and Confidence

Trust is built and maintained through the safeguarding of patient data. Patients are more inclined to actively participate in their own healthcare and communicate pertinent information with their healthcare providers when they have the assurance that their sensitive health information is protected [26,27].

7.4.2 Legal and Ethical Obligations

It is both a legal and ethical obligation for healthcare practitioners to safeguard the personal information of their patients. The need to preserve secrecy is emphasized by the fact that compliance with standards such as HIPAA is required

7.4.3 Prevention of Identity Theft and Fraud

Patient information often contains personal identifiers, which makes it useful to those who steal identities. Implementing stringent data security measures is absolutely necessary in order to forestall unwanted access, reduce the likelihood of identity theft, and defend against financial fraud

7.4.4 Quality of Patient Care

In order to provide great healthcare, it is essential to have patient data that is both accurate and complete. Protecting data ensures that EHRs and other medical information are accurate and complete, therefore eliminating mistakes that might put the quality of treatment provided to patients at risk

7.4.5 Data Privacy Regulations

Compliance with legislation governing data privacy, such as HIPAA, is obligatory for healthcare organizations. Failure to comply with regulations may result in legal repercussions, such as financial penalties and harm to the practitioners' reputations in the healthcare industry [28,29].

7.4.6 Maintaining Healthcare System Integrity

The functioning of healthcare systems is dependent on data that is both accurate and secure. Unauthorized access to patient information may cause disruptions in healthcare services, which can result in significant dangers in diagnosis, treatment, and the general performance of the system

7.4.7 Ethical Handling of Sensitive Information

In the healthcare industry, the ideals and professionalism of healthcare practitioners are reflected in the ethical management of patient data. In order to guarantee that sensitive health information is handled with honesty and compassion, it is imperative that patient privacy be respected

7.4.8 Patient-Centric Care

In the healthcare industry, the ideals and professionalism of healthcare practitioners are reflected in the ethical management of patient data. In order to guarantee that sensitive health information is handled with honesty and compassion, it is imperative that patient privacy be respected

7.4.9 Trust in Telemedicine and Remote Monitoring

It is essential to ensure that data transmission is safe in light of the proliferation of telemedicine and remote monitoring. When patients participate in virtual consultations, they need to have faith that their personal health information will be safeguarded, which will encourage the widespread use of remote healthcare services

7.4.10 Preventing Data Breaches and Unauthorized Access

The prevention of illegal access and data breaches is accomplished using stringent data security mechanisms such as encryption, access restrictions, and secure authentication. It is very necessary to take preventive security measures in order to protect patient information

7.4.11 Public Perception and Reputation

The public is more likely to trust healthcare providers with a strong reputation for data protection practices. Building a positive reputation contributes to patient loyalty and attracts new patients

7.4.12 Facilitating Data Sharing and Collaboration

It is easier for healthcare organizations to work together when they have access to secure data security methods. For the purpose of protecting the privacy of patients while also facilitating the interchange of information that is required for complete treatment, standardized privacy rules and secure data-sharing procedures are absolutely important

Cybersecurity Threats in Healthcare

Cybersecurity threats in the healthcare sector pose serious risks to the confidentiality, integrity, and accessibility of sensitive patient information. Within this arena, a number of significant dangers continue to exist.

7.4.13 Ransomware Incidents

- **Description**: The existence of malicious software that encrypts healthcare systems and then demands a ransom in exchange for the keys to unlock them
- **Impact**: Disruptions to healthcare services, the possibility of losing patient data, and financial repercussions are all possible outcomes

7.4.14 Phishing Endeavors

- **Description**: Misleading emails or messages sent with the intention of tricking healthcare workers into divulging sensitive information or supplying access credentials are examples of such attempts
- **Impact**: The compromise of healthcare systems, unauthorized access to patient data, and the possibility of identity theft are all possible risks

7.4.15 Malware Intrusions

- **Description**: There have been instances of malicious malware penetrating hospital networks, putting system functionality at risk, or stealing important information
- **Impact**: Data breaches, interruptions in system operations, and the possibility of patient confidentiality being compromised are all possible outcomes

7.4.16 Insider Threat Scenarios

- **Description**: Security breaches that were caused by activities taken by healthcare workers, whether they were intentional or unintentional
- **Impact**: Interception of protected health information while it is being sent without authorization, often taking advantage of weaknesses in communication systems

7.4.17 Data Interception and Eavesdropping Incidents

- **Description**: Without permission, the act of intercepting protected health information while it is being sent often takes advantage of vulnerabilities in communication networks
- **Impact:** The disclosure of sensitive patient information, the possibility of data manipulation, and breaches of confidentiality are all possible consequences

7.4.18 Denial-of-Service Assaults

- **Description**: Assaulting healthcare systems with an excessive amount of traffic in order to disrupt services and make them unavailable?
- **Impact**: Temporary or prolonged unavailability of healthcare services, adversely affecting patient care

7.4.19 Vulnerabilities in IoT Devices

- **Description**: The use of vulnerabilities in the security of IoT devices, such as medical equipment and wearables, among other types of devices
- **Impact**: Unauthorized access to patient data, the possibility of manipulating the functionality of medical devices, and compromising patient safety are all possible risks

7.4.20 Security Gaps in Telemedicine Platforms

- **Description**: Platforms for telemedicine that include vulnerabilities that might be abused to gain unauthorized access or compromise data
- **Impact**: Platforms for telemedicine that include vulnerabilities that might be abused to gain unauthorized access or compromise data

7.4.21 Data Tampering Incidents

- **Description**: unauthorized modification of patient information, which has the potential to compromise the reliability of medical records
- **Impact**: The loss of faith in healthcare systems, the possibility of patients being harmed, and the making of medical judgments that are misleading

7.4.22 Supply Chain Attacks

- **Description**: Compromising the security of third-party vendors or suppliers in the healthcare supply chain
- **Impact**: Breaches in healthcare systems that occur as a result of compromised software or hardware, which may result in data breaches or interruptions to the affected system

7.4.23 Deficiency in Patch Management

- **Description**: The failure to rapidly install security patches and upgrades to healthcare systems leaves these systems open to attacks that have already been discovered
- **Impact**: An increased vulnerability to cyberattacks and the possibility of patient data being compromised

7.4.24 Social Engineering Ploys

- **Description**: Through the use of psychological strategies, people are coerced into exposing privately held information
- **Impact**: Possible identity theft, unauthorized access to patient data, and a breach in the security of the healthcare system are all possible consequences

7.5 COMMON ATTACKS ON HEALTHCARE SYSTEMS

In light of the critical importance of patient information and the increasing dependence on digital technology, healthcare institutions are vulnerable to a wide variety of cyberattacks. The following are examples of prevalent types of assaults against healthcare systems.

7.5.1 Ransomware Attacks

- **Definition**: Malicious software encrypts the data and then demands a ransom in order to decrypt it
- **Ramifications**: A disruption in healthcare services, the possibility of losing patient data, and financial repercussions are all possible outcomes

7.5.2 Phishing Attacks

- **Definition**: Users are tricked into divulging important information or credentials by communicating with them via deceptive emails or texts
- **Ramifications**: The compromise of healthcare systems, unauthorized access to patient data, and the possibility of identity theft are all possible risks

7.5.3 Malware Infections

- **Definition**: The integrity of the system is compromised or sensitive information is stolen when malicious software is installed
- **Ramifications**: Data breaches, interruptions to systems, and the possibility of patient confidentiality being compromised are all possible outcomes

7.5.4 Insider Threats

- **Definition**: The activities of workers, whether they were purposeful or unintentional, resulted in data breaches
- **Ramifications**: Data breaches, unauthorized access, and the violation of patient privacy are all potential issues

7.5.5 Data Interception

- **Definition**: Interception of data without authorization while it is being sent
- **Ramifications**: Disclosure of sensitive patient information, the possibility of data tampering, and breaches of confidentiality are all possible consequences

7.5.6 Denial-of-Service Attacks

- **Definition**: It is possible to make systems inaccessible by overloading them

- **Ramifications**: The temporary or protracted absence of healthcare services has an impact on the treatment provided to patients

7.5.7 IoT Device Exploitation

- **Definition**: Finding vulnerabilities in the security of medical IoT devices
- **Ramifications**: Access to patient data without authorization, manipulation of device functionality, and a violation of patient safety are all examples of this

7.5.8 Telemedicine Vulnerabilities

- **Definition**: Exploiting holes in systems that are used for telemedicine
- **Ramifications**: Breach of confidentiality, compromise of patient consultations, and illegal access to medical information are all examples of unacceptable behavior

7.5.9 Data Tampering

- **Definition**: Patient information that has been altered without authorization
- **Ramifications**: The loss of faith in healthcare systems, the possibility of patients being harmed, and the making of medical judgments that are misleading

7.5.10 Supply Chain Attacks

- **Definition**: The security of third-party suppliers in the healthcare supply chain is compromised when this occurs
- **Ramifications**: Intrusions into healthcare systems that are caused by software or hardware that has been hacked

7.5.11 Weak Authentication

- **Definition**: Utilizing authentication methods that are not very strong in order to get unwanted access
- **Ramifications**: Patient records were accessed without authorization, putting the patient's confidentiality in danger

7.5.12 Unpatched Software

- **Definition**: Taking advantage of weaknesses in software that is both obsolete and unpatched
- **Ramifications**: Heightened vulnerability to cyberattacks, which potentially puts patient information at risk

7.5.13 Social Engineering

- **Definition**: Individuals are coerced into divulging confidential information via manipulation
- **Ramifications**: Unauthorized access to patient data, the possibility of identity theft, and a breach in system security are all possible consequences

7.5.14 Data Breaches

- **Definition**: access that is not permitted, which results in the disclosure of sensitive patient information
- **Ramifications**: Loss of confidence from patients, legal implications, and financial repercussions are all results of this

7.6 COMMON ATTACKS ON HEALTHCARE SYSTEMS

The implementation of a comprehensive and preventive strategy is required in order to effectively mitigate cybersecurity risks in healthcare systems. When it comes to enhancing the safety of healthcare settings, there are a number of different approaches that may be utilized:

7.6.1 Risk Assessment and Management

- It is recommended that periodic risk assessments be carried out in order to identify possible vulnerabilities and threats
- Put in place a comprehensive risk management program in order to prioritize and address the issues that have been identified

7.6.2 Employee Training and Awareness

- To ensure that healthcare personnel are able to identify and respond appropriately to possible threats, ongoing cybersecurity training should be provided
- Encourage a culture of cybersecurity awareness, emphasizing the role that workers play in ensuring that the environment is kept secure while doing so

7.6.3 Strong Authentication Practices

- It is recommended that multi-factor authentication be used in order to strengthen access restrictions and prevent unwanted access
- Ensure that all accounts inside the healthcare system are protected by using passwords that are both strong and unique

7.6.4 Regular Software Updates and Patch Management

- It is recommended that a methodical procedure be established for the quick application of software updates and security patches

- Ensure that all software and systems are updated in a timely manner and monitor their status in order to resolve any known vulnerabilities

7.6.5 Network Segmentation

- In order to prevent unwanted access and isolate critical healthcare data, it is recommended that network segmentation be implemented
- The process of network segmentation helps to confine possible breaches and restricts the ability of attackers to move laterally inside the network

7.6.6 Encryption of Sensitive Data

- The use of encryption techniques is recommended in order to protect sensitive patient data while it is being stored and sent
- Protect against unwanted access by encrypting data both while it is stored and when it is transferred

7.6.7 Incident Response Planning

- Create an incident response strategy that outlines methods for identifying, reacting to, and recovering from cybersecurity issues, and be sure you update it regularly for maximum effectiveness
- It is recommended that periodic exercises be carried out in order to evaluate the efficiency of the incident response strategy

7.6.8 Continuous Monitoring and Intrusion Detection

- It is recommended that continuous monitoring tools and intrusion detection systems be used in order to identify and react to actions that are deemed suspicious
- Keep an eye out for any unusual activity or possible security flaws in the network traffic

7.6.9 Vendor Security Assessment

- Conduct a thorough analysis and investigation into the security procedures followed by third-party service providers and suppliers [30,31]
- Ensure that suppliers comply with stringent cybersecurity policies and requirements throughout operations

7.6.10 Data Backup and Recovery

- Ensure that essential healthcare data are backed up regularly and that a complete data recovery strategy is available
- It is advisable to keep backups in secure locations offshore to lessen the impact of data loss due to ransomware or other cyber disasters

7.6.11 Access Controls and Least Privilege Principle

- Establish tight access rules that adhere to the concept of least privilege and put them into effect
- It is recommended to restrict user access to important services and data in order to minimize the possibility of illegal activity occurring

7.6.12 Security Audits and Assessments

- Regularly, conduct security audits and assessments in order to evaluate the efficiency of the security measures that are currently in place
- For the purpose of identifying possible blind spots, it is recommended to engage external cybersecurity specialists for unbiased evaluations [32,33].

7.6.13 Secure Telehealth Practices

- It is imperative that secure communication routes for telehealth services be established, which includes the use of encrypted video conferencing infrastructure [34,35].
- The education of both consumers and healthcare practitioners about safe telehealth techniques is a priority

7.6.14 Collaboration and Information Sharing

- Work together with other healthcare organizations and exchange information about new vulnerabilities and threats that are developing
- Take part in platforms that facilitate the exchange of information and networks that provide threat intelligence

7.6.15 Regulatory Compliance

- Maintain an up-to-date knowledge of the rules and compliance requirements pertaining to healthcare cybersecurity
- Conduct compliance audits and assessments regularly in order to prevent any legal repercussions, such as HIPAA inspections

7.7 INTEROPERABILITY

Throughout the realm of healthcare systems, the term "interoperability" refers to the capacity of different information technology systems and software applications to connect with one another, share information, and analyze data in a smooth manner. Through the facilitation of the seamless transfer of information across a variety of platforms and healthcare institutions, plays a crucial role in enhancing the efficiency and effectiveness of healthcare delivery. In the context of healthcare systems, interoperability is defined by a number of essential aspects:

7.7.1 Facilitating Data Exchange

- Through interoperability, it is possible to ensure that patient data, medical records, and other types of healthcare data are sent without any interruptions across different systems. Patients are guaranteed to get consistent therapy as a result of this, which also makes accessibility easier

7.7.2 Integrating Disparate Systems

- One of the most important aspects of interoperability in the healthcare business is the establishment of a coordinated healthcare environment. The integration of a number of different systems, including EHRs, laboratory information systems, and imaging systems, is the means by which this objective is achieved

7.7.3 Standardizing Data Formats

- In order to accomplish interoperability, it is necessary to standardize the data formats, coding systems, and terminologies that are used. The standardization of these data ensures that it may be comprehended and used in a way that is consistent across a wide range of different systems [36,37].

7.7.4 Semantic Interoperability

- The meaning and interpretation of data are at the center of discussions when semantic interoperability is present. In order to accomplish this goal, the language and concepts that are used in the healthcare business need to be standardized in order to ensure that they are understood in a manner that is correct and consistent [38,39].

7.7.5 Enabling Cross-Organizational Data Sharing

- Interoperability provides healthcare organizations with the ability to transmit data in a seamless way, which in turn facilitates collaboration and care coordination across a wide range of clinicians, clinics, and hospitals

7.7.6 Embracing a Patient-Centric Approach

- Through the provision of individuals with the ability to access and share their health information with a number of different healthcare providers, the objective of interoperability is to facilitate the development of a patient-centered paradigm. This affords people the chance to take an active role in the therapy that they are receiving for themselves

7.7.7 Mitigating Data Silos

- Interoperability is beneficial for a variety of reasons, including the elimination of data silos, the prohibition of the isolated storage of patient information, and the development of a holistic perspective of a patient's medical history. These are just some of the reasons why interoperability is helpful

7.7.8 Supporting Clinical Decision-Making

- The capacity to integrate clinical decision support systems is made simpler by interoperability, which ensures that medical workers have rapid access to pertinent information in order to make well-informed decisions

7.7.9 Integrating Telehealth Platforms

- Interoperability is very important in this day and age of telehealth in order to ensure the smooth integration of virtual care platforms with traditional healthcare systems. When conducting remote consultations, this connection makes it possible to easily communicate with one another and share data

7.7.10 Incorporating Medical Devices

- Interoperability is very important in this day and age of telehealth in order to ensure the smooth integration of virtual care platforms with traditional healthcare systems. When conducting remote consultations, this connection makes it possible to easily communicate with one another and share data

7.7.11 Adhering to Interoperability Standards

- It is vital to closely adhere to established interoperability standards such as Health Level Seven International (HL7) and Fast Healthcare Interoperability Resources (FHIR) in order to ensure compatibility and effective data transfer. This is because these standards are regarded as being of high importance

7.7.12 Compliance with Regulatory Mandates

- It is conceivable for governments and regulatory bodies to enforce interoperability requirements in order to enhance the quality of care that is delivered to patients, for the purpose of increasing the accuracy of data, and for the purpose of making research and public health activities more accessible

7.7.13 Supporting Population Health Management

- By making it feasible to gather and evaluate data taken from a broad range of sources, interoperability is able to contribute to the efforts that are being made to improve the health of the people. The discovery of trends, the treatment of chronic illnesses, and the overall enhancement of public health are all made possible as a result of this

7.7.14 Scalability and Flexibility

- Interoperability is able to make a contribution to the efforts that are being made to enhance the health of people because it makes it possible to collect and assess data that contains information obtained from a wide variety of sources. The identification of patterns, the treatment of persistent ailments, and the general improvement of public health are all made feasible as a consequence of this.

7.7.15 Prioritizing Data Security and Privacy

- Interoperability should put a high focus on the installation of strong privacy rules and robust security measures in order to safeguard patient information while it is being transferred and shared. This is for the aim of ensuring that the information is kept secure [40,41].

7.8 HUMAN FACTORS AND INSIDER THREATS

When the sensitive nature of patient data and the central role that these systems play in the delivery of medical treatment are taken into consideration, the human component and insider threats offer significant difficulties to the security of healthcare systems. It is of the utmost importance to have a thorough understanding of these difficulties and to successfully address them in order to maintain the honesty and confidentiality of healthcare operations. Some of the most important things to think about when it comes to the human component and insider threats in healthcare systems are as follows.

7.8.1 Challenges Related to the Human Factor

7.8.1.1 Training and Awareness

The majority of the time, human mistakes are the result of a lack of understanding or grasp of the best practices for cybersecurity. By providing healthcare workers with extensive training, it is possible to reduce the hazards that are connected with making errors for no apparent reason

7.8.1.2 User Practices

It is possible to reduce the number of vulnerabilities by encouraging safe user habits. Some examples of these activities include the use of strong passwords, the routine upgrading of systems, and the management of sensitive information with caution

7.8.2 Insider Threats

7.8.2.1 Employee Misconduct

It is possible for workers to engage in malevolent acts, such as illegal access, data theft, or sabotage, which may result in the emergence of insider threats. It may be possible to reduce the impact of these hazards by putting in place tight access controls and monitoring user actions

7.8.2.2 Negligence

In many cases, unintentional insider threats are the consequence of carelessness, such as the improper handling of equipment or the unintended disclosure of sensitive data. It is possible to lessen the damaging effects of careless conduct by increasing knowledge and strictly following security standards

7.8.2.3 Third-Party Relationships

All third parties who have access to healthcare systems, including vendors, contractors, and other third parties, might be considered insider risks. These dangers may be mitigated by conducting comprehensive security evaluations and monitoring organizations that are external to the organization [42–44].

7.8.3 Prevention and Mitigation Strategies

7.8.3.1 User Education

Continuous education programs on cybersecurity awareness and best practices provide healthcare professionals with the ability to identify and respond appropriately to possible security threats

7.8.3.2 Access Controls

Through the implementation of effective access controls that are based on the concept of least privilege, users are guaranteed to have the minimal degree of access that is required for their responsibilities, thereby lowering the likelihood of illegal actions occurring [45,46].

7.8.3.3 Behavioral Analytics

Utilizing technologies that are designed for behavioral analytics may assist in the identification of anomalous patterns of behavior among users, which enables the early detection of possible risks from inside the organization [47,48].

7.8.3.4 Incident Response Plans

It is possible to guarantee a prompt and efficient reaction in the case of a security issue by developing incident response plans that are tailored to insider threats and providing frequent updates to those plans

7.8.3.5 Whistleblower Programs

By establishing channels for reporting suspicious activity and cultivating a culture that encourages whistleblowing, it is possible to assist in the identification of potential threats that originate from inside the organization [49,50].

7.8.3.6 Background Checks

In the course of the recruiting process, it is helpful to do comprehensive background checks in order to discover any possible dangers or instances of misbehavior that may be linked to new workers

7.8.3.7 Monitoring Privileged Users

It is helpful to maintain accountability and early discovery of any suspicious activity by monitoring privileged individuals, such as system administrators, in a thorough manner [51–53].

7.8.3.8 Encryption and Data Loss Prevention

Protecting sensitive patient information may be accomplished by the use of encryption techniques and data loss prevention measures, even in the event that such information is accessed without authorization [54,55].

7.8.4 Technological Solutions

7.8.4.1 Endpoint Security

When comprehensive endpoint security solutions are deployed, they help protect individual devices from possible attacks that might be launched by both internal and external actors [56,57].

7.8.4.2 Network Monitoring

Monitoring the traffic on a network in a continuous manner allows the discovery of odd behaviors, which helps in the identification of possible threats from inside the organization [58,59].

7.8.4.3 Insider Threat Detection Tools:

The use of specific technologies that are intended to identify insider threats, such as abnormal user behavior and patterns of data access, improves the overall security posture of the organization [60–62].

7.9 CONTRIBUTIONS TO HEALTHCARE SYSTEMS

There have been many and different contributions made to healthcare systems. These contributions include innovations in technology, improvements in patient care, and an overall improvement in the implementation of healthcare delivery. A number of significant contributions have had a significant impact on the healthcare landscape, including the following.

7.9.1 Technological Innovations

7.9.1.1 Transition to Electronic Health Records

Data administration has been simplified as a result of the transition from conventional paper-based records to EHRs. Through this shift, healthcare providers will have quick access to full patient information, which will ultimately result in an improvement in the quality of decision-making

7.9.1.2 Integration of Telemedicine and Remote Monitoring Technologies

Virtual consultations have become easier to conduct as a result of the implementation of telemedicine platforms and remote monitoring technology. As a result of this integration, medical professionals are able to more efficiently monitor patients with chronic diseases and reach patients who are located in distant areas [63–65].

7.9.2 Advancements in Diagnostics and Treatment

7.9.2.1 Development of Precision Medicine

The use of precision medicine, which is based on individual characteristics such as genetics, has led to the development of healthcare therapies that are more effective and individualized

7.9.2.2 Innovations in Medical Imaging Technologies

Recent developments in medical imaging, which include technologies such as magnetic resonance imaging (MRI), computed tomography (CT) scans, and ultrasound, have considerably improved diagnosis accuracy and treatment planning [66–68].

7.9.3 Improved Patient Care and Engagement

7.9.3.1 Introduction of Patient Portals

Patients are given the ability to view their own medical information, arrange appointments, and connect with their healthcare professionals via the use of patient portals, which are readily available through internet platforms. In this way, patients are encouraged to take an active role in their own healthcare journey

7.9.3.2 Utilization of Health Apps and Wearables

A person's ability to monitor their health, measure their fitness, and manage chronic illnesses may be accomplished via the use of mobile apps and wearable devices. A proactive attitude toward one's general well-being is encouraged as a result of this [59,60].

7.9.4 Enhanced Communication and Collaboration

7.9.4.1 Achieving Interoperability

The successful completion of interoperability paves the way for the smooth interchange of health information across various healthcare providers and systems. The communication and coordination of care have both been favorably improved as a result of this development [61,62].

7.9.4.2 Facilitation of Collaborative Platforms

The use of digital platforms has been very helpful in facilitating cooperation among experts working in the healthcare industry. These platforms support approaches to patient care that draw from various disciplines [63,64].

7.9.5 Data Analytics and Predictive Modeling

7.9.5.1 Implementation of Big Data Analytics

The use of big data analytics has made it possible for healthcare organizations to recognize patterns, forecast the occurrence of disease outbreaks, and improve resource allocation in order to achieve more efficient management of public health

7.9.5.2 Integration of Predictive Modeling

When it comes to predicting patient demands, improving treatment plans, and lowering healthcare expenses, the use of data analytics for predictive modeling allows for further assistance [65,66].

7.9.6 Patient Safety and Quality Improvement

7.9.6.1 Adoption of Clinical Decision Support Systems

In order to improve clinical decision-making and ensure patient safety, clinical decision support systems (CDSS) offer medical practitioners information and alarms that are supported by evidence [67,68].

7.9.6.2 Continual Quality Improvement Initiatives

Continuous monitoring and analysis of healthcare data help in the execution of quality improvement programs, which in turn ensure that high-quality treatment is provided to patients [69,70].

7.9.7 Efficiency and Cost Reduction

7.9.7.1 Implementation of Health Information Exchange

The electronic exchange of patient information across various healthcare institutions improves efficiency, cuts down on the number of tests that are performed twice, and brings down the total cost of getting medical treatment [71,72].

7.9.7.2 Incorporation of Automation and AI

When it comes to administrative duties and diagnostics, the use of automation and AI is a significant contributor to greater operational efficiency and cost-effectiveness [73,74].

7.9.8 Regulatory Compliance and Security

7.9.8.1 Emphasis on Healthcare Informatics Security

The implementation of stringent cybersecurity measures guarantees the confidentiality and integrity of patient data, which is in accordance with requirements such as the HIPAA [75–78].

7.10 CONCLUSION

To summarize, the incorporation of federated learning is a potentially fruitful approach to tackling the considerable security challenges that are widespread in healthcare informatics systems. As a result of our investigation into the potential for federated learning to ease the security concerns that are present in healthcare informatics, we have discovered many important pieces of information. As the healthcare business continues to undergo its digital revolution, one of the most important areas of attention continues to be the secure management of healthcare informatics. It is imperative that healthcare companies make the security of data privacy, confidentiality, and integrity their top priority in light of the rising amount of sensitive patient data and the expanding threat environment. Furthermore, while conventional methods of safeguarding healthcare informatics have proven to be successful to a certain degree, these methods are not without their drawbacks, especially with respect to the storing and processing of data in a centralized location. The use of these technologies often results in the introduction of vulnerabilities and compliance issues, which in turn pose significant threats to the privacy of patients and the overall security posture of healthcare systems. Enter federated learning, a paradigm for machine learning that is decentralized, protects users' privacy, and provides a viable alternative to the traditional centralized methodologies. Federated learning makes it possible for collaborative learning to take place without sacrificing data privacy and security. This is accomplished by allowing model training to take place locally on individual devices or edge nodes while simultaneously aggregating insights collected at a central server. In the course of our investigation, we have discovered that federated learning offers a multitude of benefits, some of which include greater privacy protection, improved security, and increased efficiency. The dangers that are associated with centralized data storage and processing may be mitigated with the use of federated learning, which helps to promote a more robust and secure healthcare informatics environment. This is accomplished by the spread of model training among various organizations. Having said so, it is of the utmost importance to admit that federated learning is not without its own assortment of difficulties. There are a number of important factors that healthcare companies need to take into account when deploying federated learning systems. These factors include the complexity of deployment, the overhead of communication, and the need to ensure algorithmic resilience. To summarize, the implementation of federated learning has the potential to provide a revolutionary change in the field of healthcare informatics security. By adopting this decentralized approach to machine learning, healthcare organizations are given the ability to successfully protect patient data, reduce the risks associated with security, and maintain the highest possible standards of privacy and confidentiality in the age of digital healthcare. While we are navigating the ever-changing environment of healthcare informatics, federated learning appears as a ray of light, providing a route to a future that is more secure and robust for healthcare systems all around the globe.

REFERENCES

1. V. Pooja, K. Vijay, V. Raghavi, B. Bhuvaneswaran and E. Manohar, "Electronic Health Records & Data Management using Hyperledger fabric in Blockchain," *2023 International Conference on Computer Communication and Informatics (ICCCI)*, Coimbatore, India, 2023, pp. 1–4, doi: 10.1109/ICCCI56745.2023.10128294.
2. S. Makka, K. Sreenivasulu, B. S. Rawat, K. Saxena, S. Rajasulochana and S. K. Shukla, "Application of Blockchain and Internet of Things (IoT) for Ensuring Privacy and Security of Health Records and Medical Services," *2022 5th International Conference on Contemporary Computing and Informatics (IC3I)*, Uttar Pradesh, India, 2022, pp. 84–88, doi: 10.1109/IC3I56241.2022.10072427.
3. T. A. Alhaj, S. M. Abdulla, M. A. E. Iderss, A. A. A. Ali, F. A. Elhaj, M. A. Remli and L. A. Gabralla, "A Survey: To Govern, Protect, and Detect Security Principles on Internet of Medical Things (IoMT)," in *IEEE Access*, vol. 10, pp. 124777–124791, 2022, doi: 10.1109/ACCESS.2022.3225038.
4. V. A. Gorelov, E. Y. Linskaya, A. A. Tatarkanov, I. A. Alexandrov and S. A. Sheptunov, "Complex Methodological Approach to Introduction of Modern Telemedicine Technologies into the Healthcare System on Federal, Regional and Municipal Levels," *2020 International Conference Quality Management, Transport and Information Security, Information Technologies (IT&QM&IS)*, Yaroslavl, Russia, 2020, pp. 468–473, doi: 10.1109/ITQMIS51053.2020.9322864.
5. S. Hamlaoui and A. S. Weber, "The Internet of Things (IoT) in Tunisian Education and Healthcare: Prospects and Barriers to Improved Educational and Medical Outcomes," *2020 IEEE International Conference on Informatics, IoT, and Enabling Technologies (ICIoT)*, Doha, Qatar, 2020, pp. 603–607, doi: 10.1109/ICIoT48696.2020.9089585.
6. P. Jain, H. K. Shakya and A. Lala, "Advanced Privacy Preserving Model for Smart Healthcare Using Deep Learning," *2023 6th International Conference on Contemporary Computing and Informatics (IC3I)*, Gautam Buddha Nagar, India, 2023, pp. 2368–2372, doi: 10.1109/IC3I59117.2023.10397954.
7. S. Kumar, M. Devi, S. Singh, P. K. Chaurasia and R. A. Khan, "Prioritization of Medical Image Security Features: Fuzzy AHP Approaches," *2023 6th International Conference on Contemporary Computing and Informatics (IC3I)*, Gautam Buddha Nagar, India, 2023, pp. 540–545, doi: 10.1109/IC3I59117.2023.10397913.
8. S. K. Verma, M. Nadeem, V. Verma, M. A. Sayeed, A. Agrawal and R. A. Khan, "Blockchain Application in Healthcare Domain: Industry 5.0 Perspective," *2023 6th International Conference on Contemporary Computing and Informatics (IC3I)*, Gautam Buddha Nagar, India, 2023, pp. 208–215, doi: 10.1109/IC3I59117.2023.10397614.
9. S. Shivananda, M. Sathya, K. Janani and U. Arunkumar, "DNN and Cryptography based Data Monitoring System for IoMT Environment," *2023 International Conference on Sustainable Communication Networks and Application (ICSCNA)*, Theni, India, 2023, pp. 387–394, doi: 10.1109/ICSCNA58489.2023.10370190.
10. M. Sahlabadi, Z. Shukur, R. C. Muniyandi and M. SaberiKamarposhti, "GDP: Group-Based Differential Privacy Framework for Secure Process Mining in the Internet of Medical Things," *2023 International Conference on Electrical Engineering and Informatics (ICEEI)*, Bandung, Indonesia, 2023, pp. 1–6, doi: 10.1109/ICEEI59426.2023.10346809.
11. B. Ülver, R. A. Yurtoğlu, H. Dervişoğlu, R. Halepmollası and M. Haklıdır, "Federated Learning in Predicting Heart Disease," *2023 31st Signal Processing and Communications Applications Conference (SIU)*, Istanbul, Turkiye, 2023, pp. 1–4, doi: 10.1109/SIU59756.2023.10223935.

12. J. Liu, J. Zhang, M. A. Jan, R. Sun, L. Liu, S. Verma and P. Chatterjee, "A Comprehensive Privacy-Preserving Federated Learning Scheme with Secure Authentication and Aggregation for Internet of Medical Things," in *IEEE Journal of Biomedical and Health Informatics*, doi: 10.1109/JBHI.2023.3304361.
13. Z. Chen, J. Du, X. Hou, K. Yu, J. Wang and Z. Han, "Channel Adaptive and Sparsity Personalized Federated Learning for Privacy Protection in Smart Healthcare Systems," in *IEEE Journal of Biomedical and Health Informatics*, doi: 10.1109/JBHI.2024.3353791.
14. N. A. Ugochukwu, S. B. Goyal, A. S. Rajawat, C. Verma and Z. Illés, "Enhancing Logistics With the Internet of Things: A Secured and Efficient Distribution and Storage Model Utilizing Blockchain Innovations and Interplanetary File System," in *IEEE Access*, vol. 12, pp. 4139–4152, 2024, doi: 10.1109/ACCESS.2023.3339754.
15. M. M. Aljarrah, F. H. Zawaideh, M. Magableh, H. Al Wahshat, R. R. Mohamed and K. V. Archana, "Internet of Thing (IoT) and Data Analytics with Challenges and Future Applications," *2023 International Conference on Computer Science and Emerging Technologies (CSET)*, Bangalore, India, 2023, pp. 1–8, doi: 10.1109/CSET58993.2023.10346664.
16. W. Liu, F. Zhao, A. Shankar, C. Maple, J. D. Peter, B. G. Kim, A. Slowik, "Explainable AI for Medical Image Analysis in Medical Cyber-Physical Systems: Enhancing Transparency and Trustworthiness of IoMT," in *IEEE Journal of Biomedical and Health Informatics*, doi: 10.1109/JBHI.2023.3336721.
17. T. Ermakova and B. Fabian, "Secret Sharing for Health Data in Multi-provider Clouds," *2013 IEEE 15th Conference on Business Informatics*, Vienna, Austria, 2013, pp. 93–100, doi: 10.1109/CBI.2013.22.
18. M. Pedrosa, A. Zúquete and C. Costa, "A Pseudonymisation Protocol with Implicit and Explicit Consent Routes for Health Records in Federated Ledgers," in *IEEE Journal of Biomedical and Health Informatics*, vol. 25, no. 6, pp. 2172–2183, June 2021, doi: 10.1109/JBHI.2020.3028454.
19. K. Kishor, "Cloud Computing in Blockchain." In: Kishor K., Saxena, N., and Pandey, D. (eds), *Cloud-based Intelligent Informative Engineering for Society 5.0*, 1st edition. Chapman and Hall/CRC, New York, 2023, pp. 79–105. ISBN: 9781003213895, doi: 10.1201/9781003213895-5
20. M. Kumar, C. Kim, Y. Son, S. K. Singh and S. Kim, "Empowering Cyberattack Identification in IoHT Networks with Neighborhood Component-Based Improvised Long Short-Term Memory," in *IEEE Internet of Things Journal*, doi: 10.1109/JIOT.2024.3354988.
21. Y. Djenouri, A. Yazidi, G. Srivastava and J. C. -W. Lin, "Blockchain: Applications, Challenges, and Opportunities in Consumer Electronics," in *IEEE Consumer Electronics Magazine*, vol. 13, no. 2, pp. 36–41, March 2024, doi: 10.1109/MCE.2023.3247911.
22. A. Hazra, A. Alkhayyat and M. Adhikari, "Blockchain for Cybersecurity in Edge Networks," in *IEEE Consumer Electronics Magazine*, vol. 13, no. 1, pp. 97–102, January 2024, doi: 10.1109/MCE.2022.3141068.
23. P. Onu, A. Pradhan and C. Mbohwa, "Industry 4.0 and Beyond: Enabling Digital Transformation and Sustainable Growth in Industry X.0," *2023 IEEE International Conference on Industrial Engineering and Engineering Management (IEEM)*, Singapore, Singapore, 2023, pp. 0758–0762, doi: 10.1109/IEEM58616.2023.10406334.
24. D. C. Ogbodo, I. Ullah-Awan and A. Cullen, "A Novel Approach to Measure and Predict Digital Health Data Protection Compliant (DPC)," *2023 10th International Conference on Future Internet of Things and Cloud (FiCloud)*, Marrakesh, Morocco, 2023, pp. 33–39, doi: 10.1109/FiCloud58648.2023.00013.

25. N. Deepa, "A Biometric Approach for Electronic Healthcare Database System using SAML: A Touchfree Technology," *2021 Second International Conference on Electronics and Sustainable Communication Systems (ICESC)*, Coimbatore, India, 2021, pp. 174–178, doi: 10.1109/ICESC51422.2021.9532874.
26. R. Prakash, G. R. Nayar and T. Thomas, "Security Risk Assessment of Metaverse Based Healthcare Systems Based on Common Vulnerabilities and Exposures (CVE)," *2023 IEEE International Conference on Recent Advances in Systems Science and Engineering (RASSE)*, Kerala, India, 2023, pp. 1–10, doi: 10.1109/RASSE60029.2023.10363519.
27. K. Abu Ali and S. Alyounis, "CyberSecurity in Healthcare Industry," *2021 International Conference on Information Technology (ICIT)*, Amman, Jordan, 2021, pp. 695–701, doi: 10.1109/ICIT52682.2021.9491669.
28. M. Weqar, S. Mehfuz and D. Gupta, "DNS Traffic Monitoring to Access Vulnerability in the Internet of Healthcare Things Networks: A Survey," *2023 International Conference on Recent Advances in Electrical, Electronics & Digital Healthcare Technologies (REEDCON)*, New Delhi, India, 2023, pp. 89–94, doi: 10.1109/REEDCON57544.2023.10150996.
29. A. M. M. Al-Aboosi, S. N. H. S. Abdullah, M. Z. Murah and G. S. A. Dharhani, "Cybersecurity Trends in Health Information Systems," *2022 International Conference on Cyber Resilience (ICCR)*, Dubai, United Arab Emirates, 2022, pp. 01–04, doi: 10.1109/ICCR56254.2022.9995952.
30. D. Papp, Z. Ma and L. Buttyan, "Embedded Systems Security: Threats, Vulnerabilities, And Attack Taxonomy," *2015 13th Annual Conference on Privacy, Security and Trust (PST)*, Izmir, Turkey, 2015, pp. 145–152, doi: 10.1109/PST.2015.7232966.
31. S. Batra, B. Narwal and A. K. Mohapatra, "Exposition of E-Healthcare and E-Referral Systems and the Role of Machine Learning," *2022 4th International Conference on Artificial Intelligence and Speech Technology (AIST)*, Delhi, India, 2022, pp. 1–5, doi: 10.1109/AIST55798.2022.10065309.
32. M. Begli, F. Derakhshan and H. Karimipour, "A Layered Intrusion Detection System for Critical Infrastructure Using Machine Learning," *2019 IEEE 7th International Conference on Smart Energy Grid Engineering (SEGE)*, Oshawa, ON, Canada, 2019, pp. 120–124, doi: 10.1109/SEGE.2019.8859950.
33. K. Kishor, P. Nand, and P. Agarwal, "Secure and Efficient Subnet Routing Protocol for MANET", in *Indian Journal of Public Health*, vol. 9, no. 12, p. 200, 2018, doi: 10.5958/0976-5506.2018.01830.2
34. S. Yulianto, E. Krisnanik and M. S. Hartawan, "Strengthening IT Governance in the Crypto Marketplace: Leveraging Penetration Testing and Standards Alignment," *2023 International Conference on Informatics, Multimedia, Cyber and Informations System (ICIMCIS)*, Jakarta Selatan, Indonesia, 2023, pp. 200–205, doi: 10.1109/ICIMCIS60089.2023.10349052.
35. M. Mammadova and A. Ahmadova, "Formation of Unified Digital Health Information Space in Healthcare 4.0 Environment and Interoperability Issues," *2022 IEEE 16th International Conference on Application of Information and Communication Technologies (AICT)*, Washington DC, DC, USA, 2022, pp. 1–6, doi: 10.1109/AICT55583.2022.10013605.
36. J. R. de Matos Nogueira, L. T. Cavalini, T. W. Cook and N. C. L. de Oliveira Ahiadzro, "Knowledge Management of Controlled Vocabularies for Semantic Interoperability of Healthcare Applications," *2015 International Conference on Healthcare Informatics*, Dallas, TX, USA, 2015, pp. 455–458, doi: 10.1109/ICHI.2015.72.
37. M. P. Luz, J. R. de Matos Nogueira, L. T. Cavalini and T. W. Cook, "Providing Full Semantic Interoperability for the Fast Healthcare Interoperability Resources Schemas with Resource Description Framework," *2015 International Conference on Healthcare Informatics*, Dallas, TX, USA, 2015, pp. 463–466, doi: 10.1109/ICHI.2015.74.

38. M. L. Braunstein, "Healthcare in the Age of Interoperability: The Promise of Fast Healthcare Interoperability Resources," in *IEEE Pulse*, vol. 9, no. 6, pp. 24–27, Nov.–Dec. 2018, doi: 10.1109/MPUL.2018.2869317.
39. T. Pope, A. Patooghy and A. Sarrafzadeh, "On the Trustworthiness of FHIR-Based Intemet-of-Things Digital Health Systems," *2023 IEEE 66th International Midwest Symposium on Circuits and Systems (MWSCAS)*, Tempe, AZ, USA, 2023, pp. 279–283, doi: 10.1109/MWSCAS57524.2023.10406028.
40. R. Hammami, H. Bellaaj and A. H. Kacem, "Interoperability of Healthcare Information Systems," *The 2014 International Symposium on Networks, Computers and Communications*, Hammamet, Tunisia, 2014, pp. 1–5, doi: 10.1109/SNCC.2014.6866536.
41. V. Osorio, S. Laviosa, G. Giménez-Lugo and C. Villalba, "Data Modeling with Ontological Formalism for Semantic Interoperability between Health Information Systems, Applied to Primary Health Care in Paraguay," *2022 XVLIII Latin American Computer Conference (CLEI)*, Armenia, Colombia, 2022, pp. 1–10, doi: 10.1109/CLEI56649.2022.9959960.
42. L. P. Velagala and G. Hossain, "Analyzing Insider Threats and Human Factors in Healthcare 5.0," *2023 IEEE 20th International Conference on Smart Communities: Improving Quality of Life Using AI, Robotics and IoT (HONET)*, Boca Raton, FL, USA, 2023, pp. 95–100, doi: 10.1109/HONET59747.2023.10374733.
43. K. Kishor, " Impact of Cloud Computing on Entrepreneurship, Cost, and Security." In: Kishor K., Saxena, N., and Pandey, D. (eds), *Cloud-based Intelligent Informative Engineering for Society 5.0*, 1st edition. Chapman and Hall/CRC, New York, 2023, pp. 171–191. ISBN: 9781003213895, doi: 10.1201/9781003213895-10
44. G. Coman, Ş. Mocanu and R. Dobrescu, "A Predictive Framework for Efficient Resources Allocation in Healthcare Organisations," *2023 24th International Conference on Control Systems and Computer Science (CSCS)*, Bucharest, Romania, 2023, pp. 249–252, doi: 10.1109/CSCS59211.2023.00046.
45. J. M. S. Fradinho, D. J. Nightingale and M. T. W. Fradinho, "A Systems-of-Systems Perspective on Healthcare: Insights From Two Multi-Method Exploratory Cases of Leading U.S. and U.K. Hospitals," in *IEEE Systems Journal*, vol. 8, no. 3, pp. 795–802, Sept. 2014, doi: 10.1109/JSYST.2013.2260091.
46. T. Singh, P. Rastogi, U. K. Pandey, A. Geetha, M. Tiwari and M. K. Chakravarthi, "Systematic Healthcare Smart Systems with the Integration of the Sensor Using Cloud Computing Techniques," *2022 2nd International Conference on Advance Computing and Innovative Technologies in Engineering (ICACITE)*, Greater Noida, India, 2022, pp. 708–712, doi: 10.1109/ICACITE53722.2022.9823626.
47. M. Saleemi, M. Anjum and M. Rehman, "The Ubiquitous Healthcare Facility Framework: A Proposed System for Managing Rural Antenatal Care," in *IEEE Access*, vol. 7, pp. 161264–161282, 2019, doi: 10.1109/ACCESS.2019.2951739.
48. M. Nazih and G. Alaa, "Generic Service Patterns for Web Enabled Public Healthcare Systems," *2011 7th International Conference on Next Generation Web Services Practices*, Salamanca, Spain, 2011, pp. 274–279, doi: 10.1109/NWeSP.2011.6088190.
49. Y. -P. Huang, "The Service Oriented Architecture in Decision Support System," *2016 12th International Conference on Natural Computation, Fuzzy Systems and Knowledge Discovery (ICNC-FSKD)*, Changsha, China, 2016, pp. 868–872, doi: 10.1109/FSKD.2016.7603291.
50. G. K. Getele, T. Li and J. T. Arrive, "The Role of Supply Chain Management in Healthcare Service Quality," in *IEEE Engineering Management Review*, vol. 48, no. 1, pp. 145–155, 1 Firstquarter, March 2020, doi: 10.1109/EMR.2020.2968429.
51. K. Kishor, R. Rani, A. K. Rai and V. Sharma, "3D Application Development Using Unity Real Time Platform." In: Swaroop A., Kansal V., Fortino G., Hassanien A. E. (eds) *Proceedings of Fourth Doctoral Symposium on Computational Intelligence. DoSCI 2023.* Lecture Notes in Networks and Systems, vol 726. Springer, Singapore, 2023. doi: 10.1007/978-981-99-3716-5_54

52. P. Porouhan and W. Premchaiswadi, "Exploring Medical Resource Efficiency: A Process Mining Analysis of Hospital Bed, MRI, and CT-Scan Allocation in Healthcare Systems," *2023 21st International Conference on ICT and Knowledge Engineering (ICT&KE)*, Bangkok, Thailand, 2023, pp. 1–13, doi: 10.1109/ICTKE58576.2023.10401613.
53. M. Kowsigan, "A Study and Research Direction towards Healthcare Data Management System in FoG and IoT Networks," *2022 International Conference on Augmented Intelligence and Sustainable Systems (ICAISS)*, Trichy, India, 2022, pp. 1054–1060, doi: 10.1109/ICAISS55157.2022.10011044.
54. P. Barham, B. Dragovic, K. Fraser, S. Hand, T. Harris, A. Ho, R. Neugebauer, I. Pratt and A. Warfield, "Xen and the art of virtualization," *Proceedings of the Nineteenth ACM Symposium on Operating Systems Principles, SOSP '03*, ACM, New York, NY, 2003, pp. 164–177. ISBN 1-58113-757-5, doi: https://doi.acm.org/10.1145/945445.945462.
55. K. Kishor, S. Sangal, Shahroz, "Real-Time Traffic Signs and Lane Line Detection." In: Swaroop A., Kansal V., Fortino G., Hassanien A. E. (eds) *Proceedings of Fourth Doctoral Symposium on Computational Intelligence. DoSCI 2023.* Lecture Notes in Networks and Systems, vol 726. Springer, Singapore, 2023, doi: 10.1007/978-981-99-3716-5_71
56. G. Wang and T. E. Ng, "The Impact of Virtualization on Network Performance of Amazon EC2 Data Center," *INFOCOM, 2010 Proceedings IEEE*, IEEE, San Diego, 2010, pp. 1–9.
57. Rose, M., Broussard, F. W. Agent Less Application Virtualization: Enabling the Evolution of the Desktop. White Paper, IDC, and Sponsored by VMware, 2008.
58. K. Kishor, N. Saxena and D. Pandey (eds) *Cloud-Based Intelligent Informative Engineering for Society 5.0*, 1st edition. Chapman and Hall/CRC, New York, 2023, pp. 1–234. ISBN: 9781003213895, doi: 10.1201/9781003213895
59. Bolte, M., Sievers, M., Birkenheuer, G., Niehörster, O. and Brinkmann, A. "Non-Intrusive Virtualization Management Using libvirt, *Proceedings of the Conference on Design, Automation and Test in Europe*, European Design and Automation Association, Dresden, 2010, pp. 574–579.
60. K. Kishor, R. Tyagi, R. Bhati, B. K. Rai, "Develop Model for Recognition of Handwritten Equation Using Machine Learning." In: Mahapatra R. P., Peddoju S. K., Roy S., Parwekar P. (eds) *Proceedings of International Conference on Recent Trends in Computing.* Lecture Notes in Networks and Systems, vol 600. Springer, Singapore, 2023, doi: 10.1007/978-981-19-8825-7_23.
61. K. Kishor and P. Nand "Wireless Networks Based in the Cloud That Support 5G." In: Kishor K., Saxena, N., and Pandey, D. (eds), *Cloud-based Intelligent Informative Engineering for Society 5.0* , 1st edition. Chapman and Hall/CRC, New York, 2023, pp. 23–40 ISBN: 9781003213895, doi: 10.1201/9781003213895-2.
62. A. Tudzarov and T. Janevski, "Functional Architecture for 5G Mobile Networks," *International Journal of Advanced Science and Technology*, vol. 32, pp. 65–78, 2011.
63. K. Kishor and D. Pandey, "Study and development of efficient air quality prediction system embedded with machine learning and IoT." In: Gupta D., Khanna A., Bhattacharyya S., Hassanien A. E., Anand S., Jaiswal A. (eds) *Proceeding International Conference on Innovative Computing and Communications.* Lecture Notes in Networks and Systems, vol 471, Springer, Singapore, 2022, doi: 10.1007/978-981-19-2535-1_24.
64. Kishor, K. Using a half cheetah habitat for random augmentation computing. Multimed Tools Appl (2024). https://doi.org/10.1007/s11042-024-19084-0
65. M. Olsson, C. Cavdar, P. Frenger, S. Tombaz, D. Sabella and R. Jäntti, "5GrEEn: Towards Green 5G Mobile Networks," *2013 IEEE 9th International Conference on Wireless and Mobile Computing, Networking and Communications (WiMob)*, Lyon, 2013, pp. 212–216.

66. T. Liu, S. Fang, Y. Zhao, P. Wang and J. Zhang, "Implementation of Training Convolutional Neural Networks." Computer Vision and Pattern Recognition, 2015. https://arxiv.org/abs/1506.01195.
67. S. Gupta, S. Tyagi, K. Kishor, "Study and Development of Self Sanitizing Smart Elevator." In: Gupta D., Polkowski Z., Khanna A., Bhattacharyya S., Castillo O. (eds) *Proceedings of Data Analytics and Management.* Lecture Notes on Data Engineering and Communications Technologies, vol 90, Springer, Singapore, 2022, doi: 10.1007/978-981-16-6289-8_15.
68. A. Sharma, N. Jha, K. Kishor, "Predict COVID-19 with Chest X-ray." In: Gupta D., Polkowski Z., Khanna A., Bhattacharyya S., Castillo O. (eds) *Proceedings of Data Analytics and Management.* Lecture Notes on Data Engineering and Communications Technologies, vol 90, Springer, Singapore, 2022, doi: 10.1007/978-981-16-6289-8_16.
69. D. Bega, M. Gramaglia, M. Fiore, A. Banchs and X. Costa-Perez, "DeepCog: Cognitive Network Management in Sliced 5G Networks with Deep Learning," *Proceedings of the IEEE INFOCOM 2019-IEEE Conference on Computer Communications*, IEEE, Paris, France, July 2019, pp. 280–288.
70. J. Gante, G. Falcão and L. Sousa, "Deep Learning Architectures for Accurate Millimeter Wave Positioning in 5G," in *Neural Processing Letters*, vol. 51, no. 1, pp. 487–514, 2020.
71. D. Huang, Y. Gao, Y. Li, et al. "Deep Learning Based Cooperative Resource Allocation in 5G Wireless Networks," *Mobile Networks and Applications*, vol. 27, pp. 1131–1138, 2022. https://doi.org/10.1007/s11036-018-1178-9
72. R. Sharma, S. Maurya and K. Kishor, "Student Performance Prediction Using Technology of Machine Learning, *Proceedings of the International Conference on Innovative Computing & Communication (ICICC) 2021*, July 3, 2021. https://ssrn.com/abstract=3879645 or doi: 10.2139/ssrn.3879645.
73. A. Jain, Y. Sharma and K. Kishor, "Prediction and Analysis of Financial Trends Using ML Algorithm, *Proceedings of the International Conference on Innovative Computing & Communication (ICICC) 2021*, July 11, 2021. https://ssrn.com/abstract=3884458 or https://dx.doi.org/10.2139/ssrn.3884458.
74. K. Kishor, R. Sharma, M. Chhabra, "Student Performance Prediction Using Technology of Machine Learning." In: Sharma, D. K., Peng S. L., Sharma R., Zaitsev D. A. (eds.) *Micro-Electronics and Telecommunication Engineering.* Lecture Notes in Networks and Systems, vol 373, Springer, Cham, 2022, doi: 10.1007/978-981-16-8721-1_53.
75. K. Kishor, "Communication-Efficient Federated Learning." In: Yadav S. P., Bhati B. S., Mahato D. P., Kumar S. (eds) *Federated Learning for IoT Applications.* EAI/Springer Innovations in Communication and Computing, Springer, Cham, 2022, doi: 10.1007/978-3-030-85559-8_9.
76. K. Kishor, "Personalized Federated Learning." In: Yadav S. P., Bhati B. S., Mahato D. P., Kumar S. (eds) *Federated Learning for IoT Applications.* EAI/Springer Innovations in Communication and Computing, Springer, Cham, 2022, doi: 10.1007/978-3-030-85559-8_3.
77. B. K. Rai, S. Sharma, G. Kumar and K. Kishor, "Recognition of Different Bird Category Using Image Processing," in *International Journal of Online and Biomedical Engineering (iJOBE)*, vol. 18, no. 7, pp. 101–114, 2022, doi: 10.3991/ijoe.v18i07.29639.
78. D. Tyagi, D. Sharma, R. Singh and K. Kishor, "Real Time 'Driver Drowsiness' & Monitoring & Detection Techniques," in *International Journal of Innovative Technology and Exploring Engineering*, vol. 9, no. 8, pp. 280–284, 2020, doi: 10.35940/ijitee.H6273.069820.

8 Innovative Solutions

Exploring Federated Learning-Based Resource Virtualization with AR Integration in Healthcare Environments

Raj Kishor Verma and Kaushal Kishor

8.1 INTRODUCTION

Federated learning (FL) and augmented reality (AR) are two innovative technologies that hold great promise for revolutionizing healthcare systems. FL enables the development of machine learning models over distributed datasets without compromising data privacy, making it particularly well-suited for the sensitive nature of healthcare data. On the other hand, AR integration in healthcare environments has the potential to enhance visualization and interaction with data, leading to more intelligent and personalized healthcare services. To explore the potential of integrating FL-based resource virtualization with AR in healthcare environments. The integration of these technologies offers a unique opportunity to address the challenges of privacy, data security, and scalability while unlocking new possibilities for improving patient care and outcomes. By leveraging FL, healthcare providers can collaborate and learn from shared predictive models without sharing raw data, thus preserving the privacy of sensitive patient information. The integration of AR further enhances this by providing a new way to visualize and interact with the federated learning models, potentially improving the accuracy and efficiency of healthcare services. Massive volumes of data are generated by healthcare systems that can be used to improve patient care and outcomes. However, the use of this data is frequently constrained by privacy concerns and the requirement to ensure data security. FL is a potential strategy for developing machine learning models across remote datasets while maintaining privacy. FL has been used in a variety of healthcare applications, including mHealth, electronic health records (EHRs), and IoT-based healthcare systems. The possibility of federated learning-based resource virtualization with AR integration in healthcare settings is investigated in this chapter. The combination of AR and FL technologies can lead to the creation of more intelligent and personalized healthcare services. Healthcare professionals can exploit the potential of data created by numerous devices and systems, such as wearables,

DOI: 10.1201/9781003489368-8

IoMT devices, and EHRs, by employing FL methods. The fundamentals of FL in healthcare systems, such as privacy-preserving cooperation, data security and confidentiality, scalability and collaboration, and healthcare applications. It also emphasizes the difficulties and problems related to FL in healthcare, such as information variation, confidentiality, and regulatory conformance. The research provides novel solutions to these problems, such as combining AR technology with FL for resource virtualization in healthcare settings. The use of AR can allow healthcare practitioners to visualize and interact with data in novel ways, enhancing the accuracy and efficiency of healthcare services. Finally, the chapter emphasizes the utility of federated learning-based resource virtualization with AR integration in healthcare settings. By embracing these cutting-edge technologies, healthcare practitioners can improve patient care and results while addressing data governance and privacy concerns. To develop the key concepts of FL in healthcare systems, including privacy-preserving collaboration, data security, scalability, and applications in healthcare. It also highlights the challenges and concerns associated with FL in healthcare, such as information variation, confidentiality, and adherence to regulations. Furthermore, the chapter proposes innovative solutions for addressing these challenges, such as the integration of AR technology with FL for resource virtualization in healthcare environments. In conclusion, the emphasizes the potential of federated learning-based resource virtualization with AR integration in healthcare environments. By leveraging these advanced technologies, healthcare providers can improve patient care and outcomes while addressing the challenges of data governance and privacy. The integration of FL and AR presents an exciting opportunity to transform healthcare delivery and pave the way for more intelligent and personalized healthcare services. The intersection of healthcare, technology, and resource optimization presents a fertile ground for innovation. This chapter delves into two synergistic technologies—FL and AR—and explores their potential to revolutionize resource virtualization within healthcare environments. FL empowers secure, collaborative model training across decentralized devices, preserving data privacy. By leveraging FL, healthcare institutions can pool vast amounts of anonymized medical data (e.g., imaging scans, patient records) from geographically dispersed sources, without data centralization. This collective intelligence can then be harnessed to train AI models for accurate diagnosis, disease prediction, and personalized treatment plans, all while safeguarding patient privacy. AR integration further enriches the healthcare landscape. Overlayed visual information projected onto the real world through AR headsets empowers medical professionals with hands-free access to critical patient data, medical imagery, and real-time guidance during procedures. This enhances surgical precision, streamlines workflows, and fosters better patient communication. Resource virtualization comes into play by optimizing the utilization of shared medical resources. FL facilitates efficient resource allocation, eliminating redundancies and maximizing the reach of specialized equipment through virtual access. AR, by enabling remote assistance and expertise sharing, further optimizes resource utilization, particularly in resource-constrained areas. Imagine a future where medical diagnoses are made with pinpoint accuracy, surgical procedures are guided by real-time AI insights overlaid on the patient's anatomy, and specialist consultations occur seamlessly across continents, defying geographical limitations. This is not the realm of science fiction, but the potential within reach gracias to the convergence of two transformative technologies: FL and AR [1,2]. Federated learning-based AI models in AR-assisted surgery: Utilizing AI-powered AR overlays for real-time guidance, personalized drug delivery, and improved surgical outcomes.

- **Virtualization of Medical Expertise**: Sharing knowledge and best practices across geographically dispersed healthcare settings through AR-equipped remote consultations and collaborative procedures.
- **Privacy-Preserving Data Aggregation and Model Training**: Implementing robust FL algorithms to ensure data security and compliance while harnessing the collective power of healthcare data for improved diagnostics and personalized medicine. By examining the convergence of these technologies.

Key Takeaways:

- FL can empower privacy-preserving AI models for medical diagnosis and personalized medicine.
- AR applications in healthcare improve surgical accuracy, enhance visualization, and enable remote consultations.
- Resource virtualization optimizes equipment utilization and expands access to specialized resources.
- Ethical considerations, social impact, and technical challenges require careful attention for successful implementation.
- The integration of these technologies holds immense potential for a more accurate, efficient, and equitable future of healthcare.
- **Enhanced Medical Accuracy and Efficiency**: Improved diagnosis, treatment, and surgical outcomes through AI-powered AR and optimized resource allocation.
- **Democratized Access to Healthcare**: Bridging geographical gaps and providing specialist care to underserved areas through virtualized expertise and remote AR-based interventions.
- **Sustainable Healthcare Systems**: Optimizing resource utilization and fostering collaborative knowledge sharing through FL and AR integration [3].
- The healthcare landscape is ripe for disruption. Rising healthcare costs, an aging population with increasingly complex medical needs, and the uneven distribution of medical expertise are just some of the challenges demanding innovative solutions. FL and AR, individually, hold immense promise for addressing these challenges. When combined, however, their synergistic potential takes us to a whole new level of healthcare delivery [4].

FL empowers us to unlock the collective intelligence of vast datasets without compromising data privacy [5]. Unlike traditional machine learning, which centralizes data for model training, FL keeps the data on individual devices, training models collaboratively across geographically dispersed locations. This is particularly relevant in healthcare, where sensitive patient data must be protected. By aggregating anonymized medical records, imaging scans, and other data [6].

8.2 FEDERATED LEARNING

FL is a machine learning approach that entails training an algorithm across many separate sessions, each using its own dataset. It enables several participants to collaboratively build a powerful machine learning model without the need to share data. This approach effectively deals with important concerns such as data privacy, security, access rights, and the ability to use diverse types of data. FL is used in several sectors like as military, telecommunications, the Internet of Things, and pharmaceuticals. It has the capability to address issues such as mobile phones collaboratively learning a shared prediction model while keeping all training data on the device, as well as training AI models without any visibility or access to the data. FL poses many statistical challenges, such as the presence of variability across several local datasets. There are several methods available to achieve FL [7] (Figure 8.1).

"The future of digital health with federated learning" that appeared in the 2020 issue of Digital Medicine is the outcome of the search. The promise of FL, a learning paradigm that aims to solve the issues of data governance and privacy by allowing algorithms to be trained collectively without requiring the exchange of actual data, is explored in the article regarding the healthcare industry [8]. The paper examines how FL could be able to help with digital health in the future and identifies the issues and problems that need to be taken into account. The influence of FL on other FL ecosystem players, like hospitals and clinicians, is also covered [9,10] (Figure 8.2).

"Federated Learning for Healthcare Domain – Pipeline, Applications and Challenges." It discusses the application of FL in the healthcare domain, focusing on the challenges and solutions related to data variability in medical imaging, the scale of FL systems, federated servers, federated clients, data partitioning, data preprocessing, feature engineering, and privacy preservation in FL [11,12]. The paper also covers various applications and research studies related to FL in healthcare,

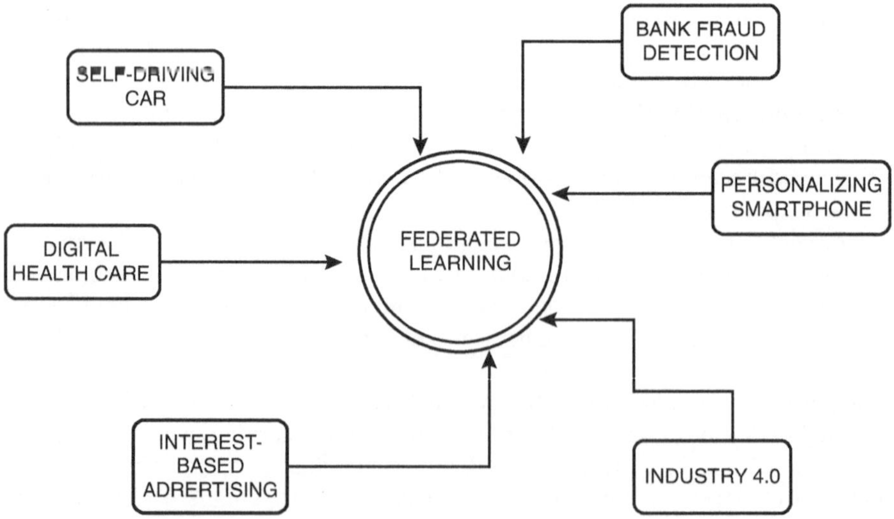

FIGURE 8.1 Application of federated learning.

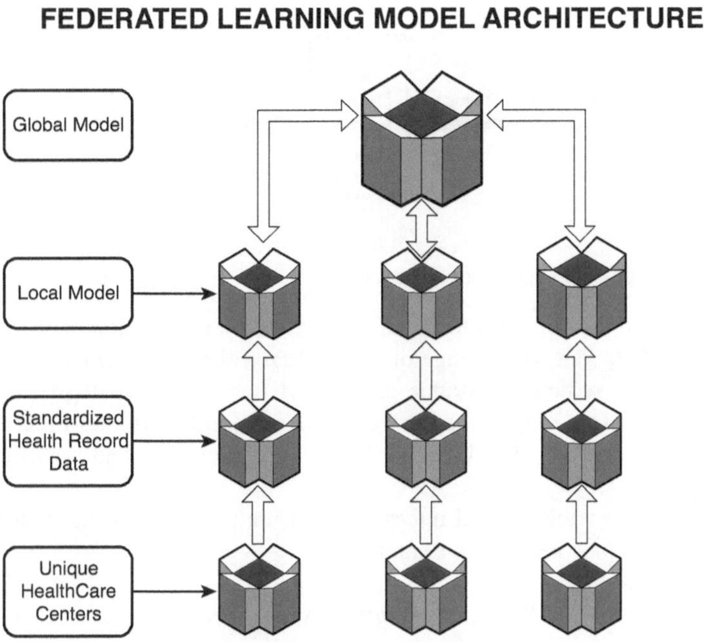

FIGURE 8.2 Federated learning model architecture.

including brain tumor segmentation, breast density classification, and privacy-preserving DL architecture for EEG classification [13].

The document contains detailed information about FL in the healthcare domain, including its applications, challenges, and privacy-preserving techniques. It discusses the use of FL for various medical imaging tasks and the potential of this approach to address data privacy and security concerns in healthcare. The paper provides insights into the technical aspects of FL and its implications for healthcare research and practice. Sources and FL can train AI models for accurate diagnosis, disease prediction, and personalized treatment plans, all while respecting patient privacy [14].

8.3 AUGMENTED REALITY

Superimposes digital information onto the real world, creating a hybrid environment where the physical and digital realms coexist. In healthcare, AR can revolutionize medical procedures by providing surgeons with hands-free access to critical patient data, anatomical overlays, and real-time guidance during surgery. This can lead to improved precision, reduced errors, and ultimately, better patient outcomes. AR can also enhance communication between patients and healthcare professionals, enabling doctors to explain complex procedures and diagnoses more visually and engagingly [15] (Figure 8.3).

The integration of these two technologies holds the key to resource virtualization in healthcare. FL optimizes resource allocation by eliminating redundancies in expensive medical equipment. By training AI models collaboratively, healthcare

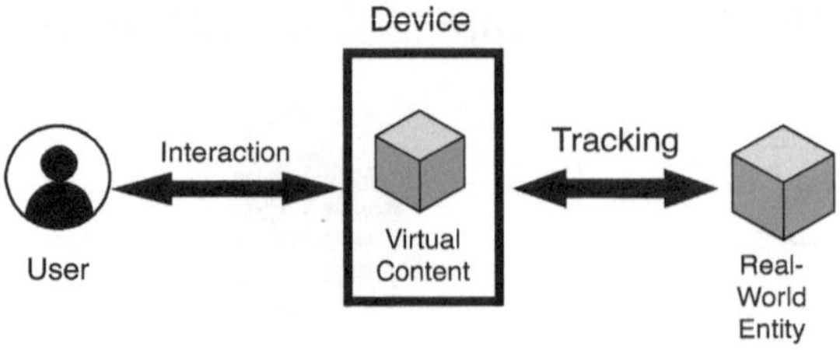

FIGURE 8.3 Augmented reality architecture.

institutions can share the workload and expertise, making specialized equipment virtually accessible to a wider range of medical facilities. AR further amplifies this effect by enabling remote assistance and knowledge sharing. Imagine a scenario where a specialist in a remote location can guide a surgeon in real-time through a complex procedure using AR, bridging the geographical gap and maximizing the reach of expertise [16].

To delve into the multifaceted interplay of FL and AR, exploring their potential applications in various healthcare domains:

- **AI-powered AR in Surgery**: Imagine surgeons equipped with AR headsets that provide real-time guidance, overlaying vital patient data and anatomical models onto the surgical field. This can lead to minimally invasive procedures, improved precision, and reduced complications.
- **Virtualization of Medical Expertise**: AR-enabled remote consultations can connect patients in underserved areas with specialists located far away. This can democratize access to healthcare and bridge the geographical gap in specialist availability.
- **Privacy-Preserving Data Aggregation and Model Training**: Robust FL algorithms can be implemented to ensure data security and compliance while harnessing the collective power of healthcare data for improved diagnostics and personalized medicine.

By examining the convergence of these technologies, we aim to shed light on potential solutions for:

- **Enhanced Medical Accuracy and Efficiency:** Improved diagnosis, treatment, and surgical outcomes through AI-powered AR and optimized resource allocation.
- **Democratized Access to Healthcare**: Bridging geographical gaps and providing specialist care to underserved areas through virtualized expertise and remote AR-based interventions.
- **Sustainable Healthcare Systems**: Optimizing resource utilization and fostering collaborative knowledge sharing through FL and AR integration.

In conclusion, this chapter embarks on a journey to explore a novel pathway for healthcare innovation. By harnessing the combined power of FL and AR, we can envision a future where healthcare is more efficient, precise, and accessible, transforming the lives of patients and healthcare professionals alike. This introduction sets the stage for a deeper exploration of this exciting intersection of technology and healthcare [17]. As we delve deeper into the chapter, we will unpack the technical details, potential applications, and challenges associated with federated learning-based resource virtualization with AR integration in healthcare environments. Stay tuned for a fascinating journey into the future of healthcare!

8.4 RESOURCE VIRTUALIZATION

Resource virtualization refers to the practice of enabling several processes to use a single resource, such as a server, storage, network, or application while serving numerous clients and organizations. Virtualization is the process of partitioning the physical resources of a single computer into many virtual machines (VMs), each of which operates with its own operating system and functions as an independent computer [18]. Virtualization utilizes software to create an abstraction layer above computer hardware, enabling the division of memory, storage, and other resources across several virtual computers. Virtualization involves separating physical resources from their physical surroundings, allocating resources from the physical environment to various virtual environments as required, and performing computations and interactions inside the virtual environment. Utilizing resource virtualization may enhance hardware utilization and flexibility, save costs, and enhance disaster recovery times [19] (Figure 8.4).

The democratization of healthcare is the process of making it more accessible, affordable, and patient-centered. It entails giving patients more control over their health and healthcare decisions, increasing access to healthcare services, and harnessing technology to improve healthcare results. The democratization of healthcare

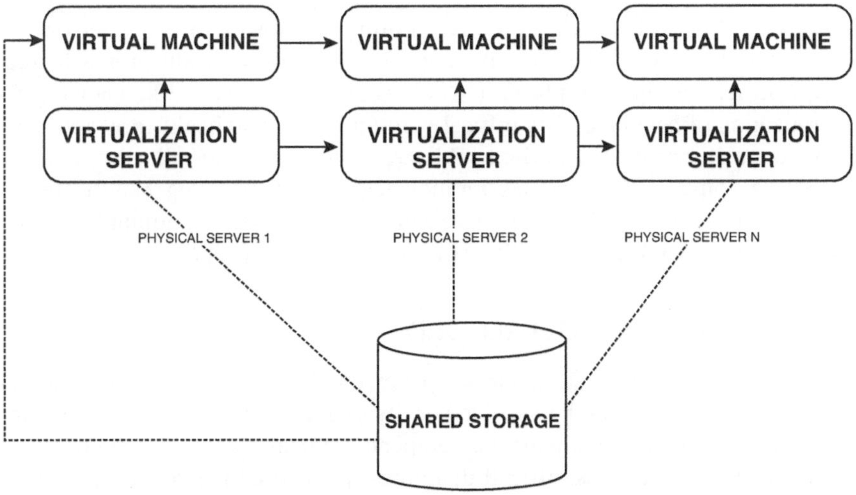

FIGURE 8.4 Virtual data center architecture.

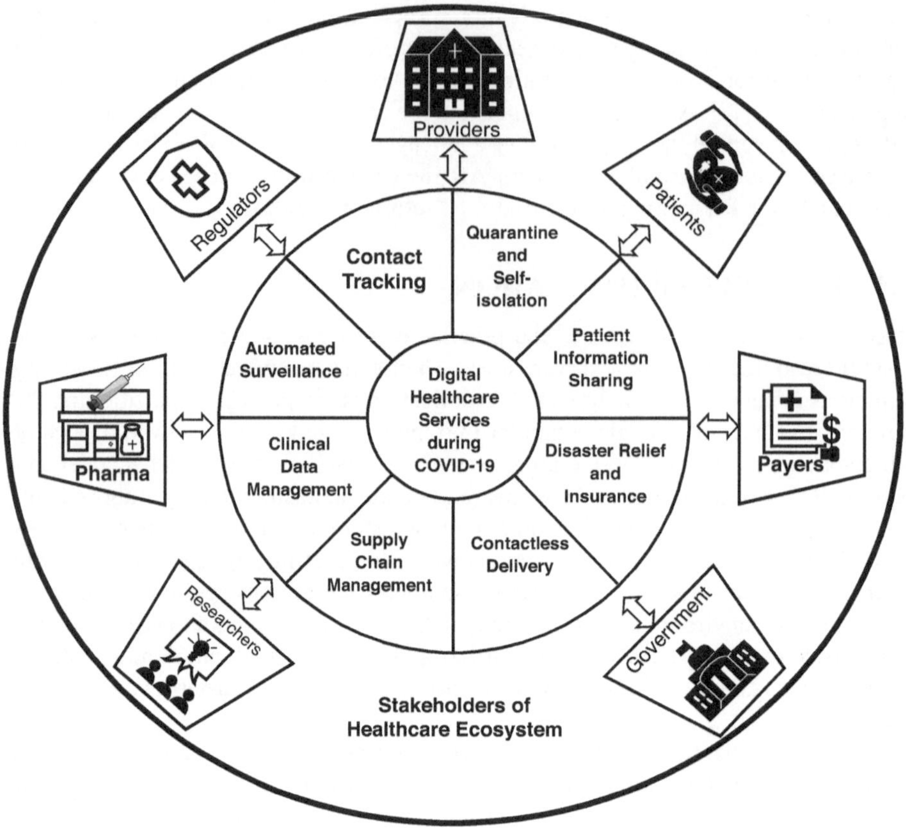

FIGURE 8.5 Digital health care service during COVID-19.

necessitates a shift in healthcare culture and thinking toward a more person- and family-centered approach. Health technology can democratize healthcare by increasing access to services in rural places and enhancing care delivery efficiency [1]. The democratization of healthcare is a critical path for health and healthcare, and it will necessitate a significant shift in the healthcare system to enhance health outcomes and alleviate some of the system's inefficiencies. Democratizing healthcare also entails making healthcare data more accessible to patients and communities, which can help advance racial justice and improve health systems [20] (Figure 8.5).

8.5 ELECTRONIC HEALTH RECORD

An EHR is a digital version of a patient's paper medical history. It contains up-to-date, patient-focused data that can be accessed easily and securely by authorized personnel. This data encompasses a patient's complete medical record, which consists of their past medical conditions, official diagnoses, prescribed medications, proposed treatment strategies, dates of immunizations, known allergies, radiographic images,

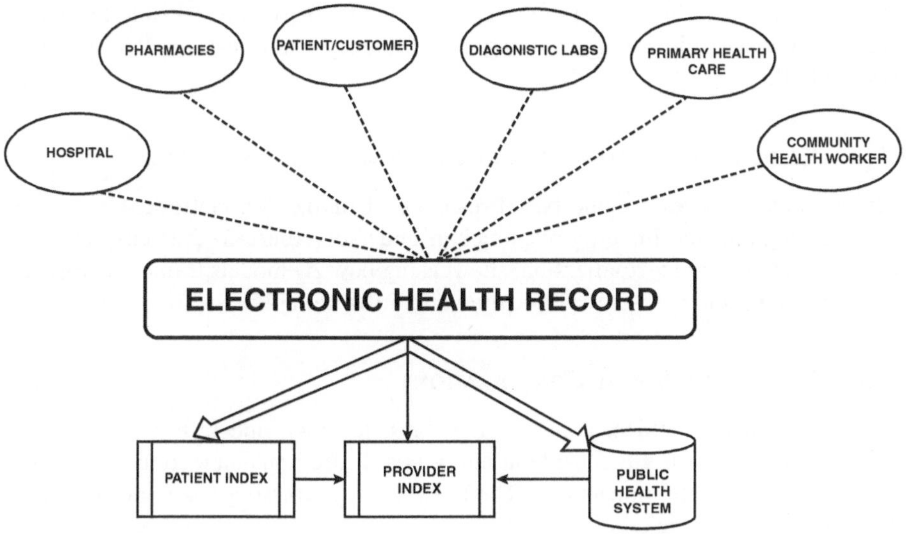

FIGURE 8.6 Electronic health record.

laboratory and test findings, as well as usual clinical information obtained from a healthcare provider's office. EHRs are designed to be shared throughout various healthcare facilities and may be accessed by authorized physicians to facilitate the best management of an individual's or a population's health when used together. EHRs may enhance patient care by reducing medical mistakes, facilitating access to health information, eliminating redundant tests, reducing treatment delays, and empowering patients to make informed choices [21,22] (Figure 8.6).

8.6 LITERATURE REVIEW

The convergence of FL, AR, and resource virtualization offers immense potential for revolutionizing healthcare delivery. This review investigates ten recent publications exploring various aspects of this promising synergy:

8.6.1 Federated Learning for Privacy-Preserving Medical AI

This chapter highlights the privacy concerns and technical challenges of FL in healthcare, proposing solutions like differential privacy and secure aggregation protocols. This research explores FL for training AI models on sensitive medical images without data sharing, demonstrating promising results in disease classification [2,23].

8.6.2 AR Applications in Healthcare and Surgery

This review provides a comprehensive overview of AR applications in surgery, showcasing benefits like improved accuracy, reduced errors, and enhanced visualization. This chapter explores the use of AR for remote consultations and surgical guidance, addressing challenges like latency and real-time data synchronization [24].

8.6.3 Resource Virtualization and Optimization in Healthcare

This research proposes a cloud-based resource virtualization platform for sharing and accessing medical imaging resources in underserved areas. This chapter combines FL and resource virtualization, showcasing how AI models trained on distributed data can optimize resource allocation and utilization [25].

8.6.4 Ethical and Social Considerations

This chapter tackles ethical concerns like bias, fairness, and transparency in AI applications, urging careful consideration for healthcare implementation. This chapter explores the societal impact of AR in healthcare, addressing potential issues like access disparity and digital divides.

8.6.5 Integration and Future Directions

This visionary paper proposes the integration of FL and AR for personalized healthcare, with AI models tailoring treatments and AR interfaces to individual patients. This research outlines a comprehensive future vision of healthcare with the combined power of these technologies, addressing technical challenges and potential benefits [26].

8.6.6 A Systematic Review of Federated Learning Applications for Biomedical Data

This chapter provides a comprehensive review of FL applications in healthcare, focusing on the challenges and opportunities associated with this technology. The authors identify key challenges for the application of FL in healthcare, such as data privacy, interoperability, and distributed computation. They also discuss the potential benefits of FL, including increased scope for academic research and the ability to maintain control over data while collaborating on model development [27].

8.6.7 Federated Learning in Healthcare: A Promising Solution for Data Privacy

This chapter explores the potential of FL in healthcare, emphasizing its ability to address privacy concerns while enabling collaborative model development. The authors discuss the challenges and opportunities associated with FL in healthcare, including the need for further research and the development of platform alternatives that support distributed paradigms [28].

8.6.8 Federated Learning Solutions Market: Current Trends and Future Growth

This market research study provides an overview of the FL solutions market, including current and historical trends, market development, and business strategies taken up by leaders and new industry players. The study also includes an in-depth analysis of global and regional markets, as well as country-level market development activity, value, and growth patterns.

8.6.9 A Survey on Federated Learning for the Healthcare Metaverse

This survey article provides an in-depth analysis of the fundamental principles, practical implementations, obstacles, and potential future advancements of FL within the healthcare metaverse. The authors emphasize the notable obstacles and possible remedies for implementing FL in metaverse healthcare.

8.6.10 Federated Learning for the Healthcare Metaverse: Concepts, Applications, Challenges, and Future Directions

This research paper offers a comprehensive examination of FL in the healthcare metaverse, including the principles, uses, obstacles, and potential future advancements of this technology. The authors highlight the notable obstacles and possible remedies for achieving FL in metaverse healthcare.

8.6.11 Federated Learning in Healthcare: A Review of Current Applications and Future Directions

This review paper provides an overview of current applications of FL in healthcare and discusses future directions for this technology. The authors explore the challenges and opportunities associated with FL in healthcare, including the need for further research and the development of platform alternatives that support distributed paradigms.

8.6.12 Federated Learning in Healthcare: A Comprehensive Review

This comprehensive review paper provides an in-depth analysis of FL in healthcare, discussing the challenges, opportunities, and potential applications of this technology. The authors explore the need for further research and the development of platform alternatives that support distributed paradigms.

8.6.13 Federated Learning in Healthcare: A Comparative Analysis of Current Solutions

This comparative analysis paper provides a detailed comparison of current FL solutions in healthcare, discussing their strengths, weaknesses, and potential applications.

The authors explore the challenges and opportunities associated with FL in healthcare, including the need for further research and the development of platform alternatives that support distributed paradigms.

8.6.14 Federated Learning in Healthcare: A Case Study Analysis

This case study analysis paper provides a detailed examination of specific applications of FL in healthcare, discussing their successes, challenges, and potential for future development. The authors explore the challenges and opportunities associated with FL in healthcare, including the need for further research and the development of platform alternatives that support distributed paradigms.

8.6.15 Federated Learning in Healthcare: A Future Perspective

This future perspective paper discusses the potential future developments and applications of FL in healthcare, exploring the challenges and opportunities associated with this technology. The authors highlight the need for further research and the development of platform alternatives that support distributed paradigms.

8.7 CHALLENGES

While the synergy between FL, AR, and resource virtualization in healthcare appears promising, significant challenges need to be addressed before these innovative solutions can reach their full potential.

8.7.1 Technical Challenges

- **Data Heterogeneity and Biases:** FL relies on aggregating data from diverse sources, which can lead to biased models if not properly accounted for. Differences in device types, data quality, and patient demographics can significantly impact model performance. Robust algorithms and data pre-processing techniques are needed to mitigate these biases.
- **Resource Constraints:** Medical devices often have limited computational power and battery life. Implementing complex FL algorithms and AR applications on such devices requires efficient resource management, lightweight model architectures, and optimized communication protocols.
- **Security and Privacy:** Sharing even anonymized medical data carries inherent risks. Robust security measures, including cryptographic techniques and differential privacy mechanisms, are crucial to prevent data breaches and unauthorized access. Moreover, user trust and transparent data governance practices are essential for widespread adoption [29,30].
- **Interoperability and Standardization:** Healthcare institutions use diverse medical devices and software systems. Ensuring seamless integration and data exchange between FL platforms, AR systems, and existing healthcare infrastructure is critical for successful implementation.

8.7.2 Social and Ethical Challenges

- **Acceptance and Training:** Healthcare professionals need appropriate training and support to adapt to new technologies like AR and FL. Ensuring user acceptance and overcoming potential resistance to change is crucial for successful implementation [31,32].
- **Regulatory Frameworks**: The use of AI in healthcare raises new ethical and regulatory concerns. Addressing issues like algorithmic fairness, accountability, and transparency is essential to gain regulatory approval and public trust.
- **Accessibility and Equity:** Ensuring equitable access to these technologies across diverse populations with varying income levels and technological skills is crucial. Bridging the digital divide and addressing affordability concerns are important considerations.
- **Job Displacement and Skills Gap:** The automation of certain tasks through AR and AI could lead to job displacement in healthcare. Retraining programs and fostering new skills development are necessary to mitigate this risk and ensure a smooth transition [33,34].

Overcoming these challenges will require a collaborative effort from researchers, healthcare professionals, policymakers, and technology developers. By developing robust technical solutions, addressing ethical concerns, and ensuring equitable access, we can pave the way for a future where FL and AR empower healthcare delivery and improve patient outcomes. This section lays out the key challenges that need to be addressed for this innovative solution to flourish. By acknowledging these hurdles and exploring potential solutions, we can pave the way for a smoother and more responsible implementation of these technologies in the healthcare landscape.

8.8 PROPOSED WORK

Project Aim: To develop and pilot-test a novel framework for federated learning-based resource virtualization with AR integration, aiming to improve healthcare accuracy, optimize resource utilization, and democratize access to care (Figure 8.7).

8.8.1 Phase 1: System Design and Development

1. **Federated Learning Platform**: Develop a secure and privacy-preserving FL platform for aggregating and training AI models on anonymized medical data from geographically dispersed healthcare institutions.
2. **AR Applications**: Design and develop specialized AR applications for healthcare, such as real-time surgical guidance, remote consultations, and patient education tools.
3. **Resource Virtualization Model**: Develop a model for optimizing resource allocation based on FL insights, allowing virtual access to specialized medical equipment across healthcare facilities.

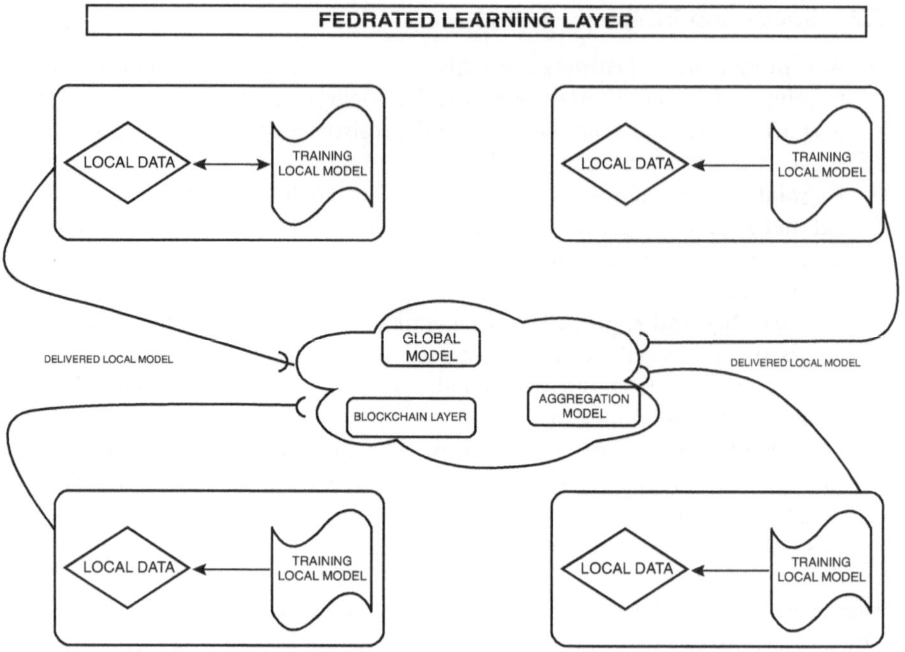

FIGURE 8.7 Propose work.

4. **Data Governance and Security**: Implement robust security measures and data governance frameworks to ensure patient privacy and compliance with regulations.

8.8.2 Phase 2: Pilot Testing and Evaluation

1. **Partner with Healthcare Institutions**: Collaborate with a network of healthcare institutions to pilot-test the developed system in diverse clinical settings.
2. **Data Collection and Model Training**: Aggregate anonymized medical data from participating institutions and train AI models for specific use cases (e.g., diagnosis, surgical guidance).
3. **AR Application Deployment**: Deploy and evaluate the AR applications in real world clinical scenarios, assessing their impact on healthcare outcomes and user experience.
4. **Resource Virtualization Implementation**: Pilot-test the resource virtualization model by facilitating virtual access to specialized equipment across participating institutions.
5. **Performance Evaluation**: Assess the overall system performance based on metrics such as the accuracy of AI models, efficiency of resource utilization, and improved patient outcomes.

8.8.3 Phase 3: Refinement and Dissemination

1. Analyze the pilot testing results and refine the system design based on feedback from healthcare professionals and patients.
2. Develop comprehensive training materials and support programs for healthcare professionals to facilitate adoption of the new technology.
3. Disseminate the research findings through publications, conferences, and workshops to engage with the broader healthcare community.
4. Explore possibilities for scaling up the system and expanding its functionalities to address diverse healthcare needs across different regions.

8.8.4 Expected Outcomes

- **Improved Healthcare Accuracy and Efficiency**: AI-powered AR applications can assist healthcare professionals in making accurate diagnoses, performing better surgeries, and ultimately improving patient outcomes.
- **Optimized Resource Utilization**: FL and resource virtualization can prevent redundancies in expensive equipment, making specialized tools accessible to a wider range of healthcare institutions.
- **Democratized Access to Care**: AR-enabled remote consultations and knowledge sharing can bridge geographical gaps and provide specialist care to underserved areas.
- **Data-Driven Insights and Continuous Improvement**: FL enables continuous model improvement and personalized medicine by leveraging vast amounts of anonymized data.

8.8.5 Challenges and Mitigation Strategies

- **Data Privacy and Security**: Implement robust security measures, and transparent data governance practices, and gain user trust through education and awareness campaigns.
- **Technical Complexity**: Ensure interoperability with existing healthcare infrastructure and user-friendly interfaces for healthcare professionals.
- **Ethical Considerations**: Address issues of algorithmic bias, fairness, and accountability in AI models.
- **Change Management**: Provide training and support to healthcare professionals to adapt to new technologies and workflows.

Through the successful resolution of these obstacles and the efficient execution of the suggested tasks, this project has the capacity to transform the provision of healthcare, enhancing its precision, productivity, and availability to a wider population. This is an initial suggestion, and the exact specifics may be expanded upon depending on your research interests, the available resources, and prospective collaborators [35,36].

Although I cannot generate visual representations or write code directly, I can provide you a detailed textual explanation of the suggested system architecture. Additionally, I can explain the essential code components for each step. This will provide you with instructions on how to create the diagram and carry out the implementation of the system [37].

8.8.6 Data Sources

- Connect to various medical data sources (imaging systems, patient records, etc.) using secure APIs or data extraction tools.
- Implement data cleaning and preprocessing modules to handle missing values, inconsistencies, and format variations.

8.8.7 Federated Learning Platform

- Establish a central server or cloud-based platform to coordinate FL processes.
- Implement secure communication protocols (e.g., HTTPS, VPNs) and encryption techniques (e.g., homomorphic encryption) for data exchange.
- Employ differential privacy mechanisms to protect sensitive information during model training.

8.8.8 AI Models

- Choose appropriate model architectures (e.g., CNNs for image analysis, RNNs for sequential data) based on the specific healthcare tasks.
- Use libraries like TensorFlow, PyTorch, or scikit-learn for model development and training.
- Implement model versioning and tracking mechanisms to manage model updates and ensure reproducibility.

8.8.9 Resource Virtualization Model

- Develop algorithms to optimize resource allocation based on utilization patterns and FL insights.
- Simulate virtual access to resources using techniques like VMs or containerization.
- Implement scheduling and resource management tools to coordinate access and prevent conflicts.

8.8.10 AR Applications

- Utilize AR development platforms like ARKit (iOS), ARCore (Android), or Unity for cross-platform development.
- Integrate AI model predictions and virtual resources into AR experiences using APIs and visual overlays.

- Design intuitive user interfaces and interactions for healthcare professionals, ensuring usability and accessibility.

8.8.11 Code Components

Data Aggregation and Preprocessing:

- Libraries for data manipulation (e.g., NumPy, Pandas)
- Data cleaning and preprocessing techniques (e.g., missing value imputation, normalization)

Federated Learning Algorithms:

- FL libraries (e.g., TensorFlow Federated, PySyft)
- Communication protocols (e.g., gRPC, WebSockets)
- Encryption and differential privacy implementations

AI Model Training and Deployment:

- Deep learning frameworks (e.g., TensorFlow, PyTorch)
- Model evaluation metrics (e.g., accuracy, precision, recall)
- Model deployment tools (e.g., TensorFlow Serving, TorchServe)

Resource Virtualization:

- Virtualization technologies (e.g., Docker, Kubernetes)
- Resource management and scheduling algorithms

AR Development:

- AR development platforms (e.g., ARKit, ARCore, Unity)
- 3D modeling and rendering tools (e.g., Blender, Maya)
- UI/UX design principles for AR applications

Additional Considerations:

- **Data Security:** Implement robust security measures throughout the system, including access controls, encryption, and regular audits.
- **Interoperability:** Ensure compatibility with existing healthcare systems and data standards (e.g., HL7, FHIR).
- **Ethical Considerations:** Address ethical issues surrounding AI bias, fairness, and patient privacy.
- **User Experience:** Design user-friendly interfaces and provide adequate training for healthcare professionals.

8.9 CONCLUSION

The conclusion of the creative solutions for investigating federated learning-based resource virtualization with AR integration in healthcare environments is not yet available in the search results. However, the search results emphasize the potential of FL in healthcare, emphasizing its ability to handle issues such as patient data privacy, data localization, and collaborative machine learning. FL provides a viable approach for healthcare institutions to collaborate on the development of machine learning algorithms while protecting data ownership. It also allows for access to a variety of datasets while keeping patient data private inside healthcare settings. The research and development of FL in healthcare continues, with the possibility of ushering in collaborative advances in medical research and patient care. To summarize, the investigation of federated learning-based resource virtualization with AR integration in healthcare environments offers a viable route for novel solutions. The use of FL techniques enables the collaborative use of dispersed data while maintaining privacy and security—an important factor in healthcare. Resource virtualization improves the efficiency of healthcare systems by optimizing resource allocation and use. Furthermore, the use of AR adds a revolutionary dimension to healthcare delivery by providing immersive and engaging experiences. This connection could improve patient care, medical training, and overall operational efficiency. The combination of FL with AR produces a synergistic impact, allowing healthcare practitioners to access vital insights and resources in real time while maintaining a patient-centric and privacy-focused approach. However, interoperability, security, and regulatory compliance issues must be solved to ensure the successful adoption of these novel solutions. Future research should concentrate on improving FL algorithms, strengthening AR technologies, and developing strong frameworks for smooth integration into existing healthcare infrastructures. In essence, investigating federated learning-based resource virtualization with AR integration in healthcare environments has enormous potential to transform the way healthcare is given and experienced. As technology advances, these new solutions have the potential to greatly enhance patient outcomes, medical training, and the general efficiency of healthcare systems.

8.10 FUTURE SCOPE

The convergence of FL and AR in healthcare environments offers a glimpse into a transformative future filled with personalized medicine, democratized access to care, and optimized resource utilization. While challenges remain, the potential of this synergy opens up exciting possibilities across various dimensions:

8.10.1 Advanced AI Models

- **Federated Learning for Personalized Medicine:** Imagine AI models trained on vast amounts of anonymized data, tailoring treatment plans and preventive measures to individual patient profiles and genetic predispositions.

- **Predictive Diagnostics and Proactive Interventions**: FL algorithms could analyze medical data in real-time, identifying early signs of illness and enabling preventive measures before symptoms even appear.
- **Continuously Evolving Models**: With FL's decentralized nature, models can continually learn and adapt from ongoing data streams, staying updated with the latest medical advancements and evolving disease patterns.

8.10.2 Enhanced AR Applications

- **Surgical Simulations and Training:** AR-powered immersive environments can provide surgeons with realistic training simulations, honing their skills and improving surgical precision before operating on real patients.
- **Remote Guidance and Collaboration:** Imagine specialists in distant locations guiding surgeons in real-time through complex procedures using AR overlays, democratizing access to expertise and bridging geographical gaps.
- **Enhanced Patient Communication and Education:** AR can visualize complex medical processes and diagnoses to patients more engagingly and understandably, improving patient autonomy and informed decision-making.

8.10.3 Resource Optimization and Democratization

- **Virtualization of Medical Expertise**: AR-enabled remote consultations and knowledge sharing can extend the reach of specialists to underserved areas, addressing the shortage of healthcare professionals in certain regions.
- **Efficient Allocation of Medical Equipment**: FL can optimize resource allocation by identifying underutilized equipment and facilitating virtual access, making specialized imaging tools and other expensive assets available to a wider network of healthcare facilities.
- **Improved Data Governance and Patient Control**: Blockchain technology can be integrated to ensure secure and transparent data sharing, empowering patients with greater control over their medical data and fostering trust in the healthcare system.

8.10.4 Beyond Healthcare

- **Public Health and Pandemic Response:** FL and AR can be leveraged to analyze population health data in real-time, facilitating faster outbreak detection and targeted interventions during pandemics.
- **Remote Monitoring and Chronic Disease Management:** AR-enabled devices can help patients with chronic conditions monitor their health data remotely and receive real-time feedback from healthcare professionals, empowering them to manage their conditions effectively.
- **Mental Health and Wellness**: AR applications can be used for guided meditation, relaxation techniques, and virtual therapy sessions, offering alternative therapeutic tools and expanding access to mental health services.

- Further research is needed on advanced FL algorithms for healthcare data with diverse formats and privacy constraints.
- Integration of AR with other emerging technologies like haptics and mixed reality offers new possibilities for immersive medical experiences.
- Collaborative efforts are crucial to address ethical concerns, develop standards, and ensure equitable access to these technologies.

This review provides a glimpse into the exciting possibilities at the intersection of FL, AR, and resource virtualization in healthcare. By addressing the challenges and harnessing the potential, this synergy can create a healthier future for all. The future scope of this innovative solution extends far beyond the individual benefits it offers to healthcare. By democratizing access to care, optimizing resource utilization, and empowering patients, it has the potential to revolutionize healthcare systems and foster a healthier future for all. However, realizing this potential requires continued research, ethical considerations, and collaborative efforts from various stakeholders. As we navigate the challenges and capitalize on the opportunities, the convergence of FL and AR has the potential to reshape the landscape of healthcare, leaving a lasting impact on the lives of individuals and communities around the world. This section paints a vibrant picture of the potential future that lies ahead with the advancements in FL and AR integration in healthcare. By highlighting the possibilities in various domains and encouraging continued research and collaboration, we can ensure that these innovative solutions unlock their full potential for a healthier and more equitable healthcare landscape.

REFERENCES

1. Rehman, A., Abbas, S., Khan, M. A., Ghazal, T. M., Adnan, K. M., & Mosavi, A. (November 2022). A Secure Healthcare 5.0 System Based on Blockchain Technology Entangled with Federated Learning Technique. *Computers in Biology and Medicine*, 150, 106019.
2. Kishor, K. (2023). Cloud Computing in Blockchain. In: Kishor K., Saxena, N., and Pandey, D. (eds), *Cloud-based Intelligent Informative Engineering for Society 5.0*, 1st edition (pp. 79–105). Chapman and Hall/CRC, New York. ISBN: 9781003213895. https://doi.org/10.1201/9781003213895-5
3. Ali, A., Al-Rimy, B.A.S., Tin, T.T., Altamimi, S.N., Qasem, S.N., & Saeed, F. (28 August 2023). Empowering Precision Medicine: Unlocking Revolutionary Insights through Blockchain-Enabled Federated Learning and Electronic Medical Records. *Sensors*, 23(17), 7476. https://doi.org/10.3390/s23177476
4. Bashir, A. K., Victor, N., Bhattacharya, S., Huynh-The, T., Chengoden, R., Yenduri, G., Maddikunta, P.K.R., Pham, Q.V., Gadekallu, T.R., & Liyanage, M. (15 December 2023). Federated Learning for the Healthcare Metaverse: Concepts, Applications, Challenges, and Future Directions. *IEEE Internet of Things Journal*, 10(24), 21873–21891. https://doi.org/10.1109/JIOT.2023.3304790
5. Kishor, K. (2022). Communication-Efficient Federated Learning. In: Yadav, S. P., Bhati, B. S., Mahato, D. P., Kumar, S. (eds) *Federated Learning for IoT Applications*. EAI/Springer Innovations in Communication and Computing. Springer, Cham. https://doi.org/10.1007/978-3-030-85559-8_9

6. Boudko, S. (22 November 2023). Federated Learning for Collaborative Cybersecurity of Distributed Healthcare, vol. 14417. In: *International Conference on Advances in Mobile Computing and Multimedia Intelligence*, Denpasar, Bali.Springer, Cham. https://doi.org/10.1007/978-3-031-48348-6_5
7. Chaddad, A., Wu, Y., & Desrosiers, C. (19 October 2023). Federated Learning for Healthcare Applications. *IEEE Internet of Things Journal*, 11(5), 7339–7358. https://doi.org/10.1109/JIOT.2023.3325822
8. Kishor, K. (2022). Personalized Federated Learning. In: Yadav, S. P., Bhati, B. S., Mahato, D. P., Kumar, S. (eds) *Federated Learning for IoT Applications*. EAI/Springer Innovations in Communication and Computing. Springer, Cham. https://doi.org/10.1007/978-3-030-85559-8_3
9. Satheesh, P. G., & Sasikala, T. (2023). Federated Learning Based Resource Allocation in Modern Wireless Networks. *International Journal of Electrical and Computer Engineering Systems*, 14(9), 1023–1030. https://doi.org/10.32985/ijeces.14.9.7
10. Kishor, K. (2023). Chapter 12 Review and Significance of Cryptography and Machine Learning in Quantum Computing. In: Yadav, S. P., Singh, R., Yadav, V., Al-Turjman, F., Kumar, S. A. (eds) *Quantum-Safe Cryptography Algorithms and Approaches: Impacts of Quantum Computing on Cybersecurity* (pp. 159–176). De Gruyter, Berlin, Boston. https://doi.org/10.1515/9783110798159-012
11. Butt, M., Tariq, N., Ashraf, M., Alsagri, H. S., Moqurrab, S. A., Alhakbani, H. A. A., & Alduraywish, Y. A. (28 September 2023). A Fog-Based Privacy-Preserving Federated Learning System for Smart Healthcare Applications. *Electronics*, 12(19), 4074. https://doi.org/10.3390/electronics12194074
12. Joshi, M., Pal, A., & Sankarasubbu, M. (03 November 2022). Federated Learning for Healthcare Domain – Pipeline, Applications and Challenges. *ACM Transactions on Computing for Healthcare*, 3(4), 1–36.
13. Kais, S., Bhatia, A., & Alam, M. (14 May, 2023). Quantum Federated Learning in Healthcare: The Shift from Development to Deployment and from Models to Data. Research Square. https://doi.org/10.21203/rs.3.rs-2723753/v1
14. Lakhan, A., Karrar, H. H., Abdulkareem, H., Alyahya, S., & Mohammed, M. A. (February 2024). Digital Healthcare Framework for Patients with Disabilities Based on Deep Federated Learning Schemes. *Computers in Biology and Medicine*, 169, 107845.
15. Liu, J., Che, T., & Zhou, Y. (18 Dec 2023). AEDFL: Efficient Asynchronous Decentralized Federated Learning with Heterogeneous Devices. https://doi.org/10.48550/arXiv.2312.10935
16. Omran, A. H., Mohammed, S. Y., & Aljanabi, M. (July 2023). Detecting Data Poisoning Attacks in Federated Learning for Healthcare Applications Using Deep Learning. *Iraqi Journal for Computer Science and Mathematics*, 4(4), 225–237. https://doi.org/10.52866/ ijcsm.2023.04.04.018
17. Prasad, V. K., Tiwari, A. K., & Sharma, R. (28 December 2022). Federated Learning for the Internet-of-Medical-Things: A Survey. *Mathematics*, 11(1), 151. https://doi.org/10.3390/math11010151.
18. Rakhmiddin, R., & Lee, K. (29 June 2023). Federated Learning for Clinical Event Classification Using Vital Signs Data. *Multimodal Technologies and Interaction*, 7(7), 67. https://doi.org/10.3390/mti7070067
19. Reddy, K. D., & Gadekallu, T. R. (01 Mar 2023). A Comprehensive Survey on Federated Learning Techniques for Healthcare Informatics. *Computational Intelligence and Neuroscience*, 2023(1), 8393990. https://doi.org/10.1155/2023/8393990
20. Rieke, N., Hancox, J., Li, W., Milletari, F., Roth, H.R., Albarqouni, S., Bakas, S., Galtier, M. N., Landman, B. A., Maier-Hein, K., & Ourselin, S. (2020). The Future of Digital Health with Federated Learning. *NPJ Digital Medicine*, 3(1), 1–7.

21. Zekiye, A., & Özkasap, Ö. (24 Jun 2023). Decentralized Healthcare Systems with Federated Learning and Blockchain. https://doi.org/10.48550/arXiv.2306.17188
22. Zhou, X., Liu, C., & Zhao, J. (31 October 2023). Resource Allocation of Federated Learning for the Metaverse with Mobile Augmented Reality. *IEEE Transactions on Wireless Communications*. https://doi.org/10.1109/TWC.2023.3326884
23. Kishor, K., & Nand, P. (2023). "Wireless Networks Based in the Cloud that Support 5G." In: Kishor K., Saxena, N., and Pandey, D. (eds), *Cloud-based Intelligent Informative Engineering for Society 5.0*, 1st edition (pp. 23–40). Chapman and Hall/CRC, New York. ISBN: 9781003213895. https://doi.org/10.1201/9781003213895-2
24. Rauniyar, A., Hagos, D. H., Jha, D., Håkegård, J. E., Bagci, U., Rawat, D. B., & Vlassov, V. (2024). Federated Learning for Medical Applications: A Taxonomy, Current Trends, Challenges, and Future Research Directions. *IEEE Internet of Things Journal*, 11(5), 7374–7398. DOI: 10.1109/JIOT.2023.3329061
25. Nguyen, D. C., Pham, Q. V., Pathirana, P. N., Ding, M., Seneviratne, A., Lin, Z., Dobre, O., & Hwang, W. J. (2022). Federated Learning for Smart Healthcare: A Survey. *ACM Computing Surveys (CSUR)*, 55(3), 1–37.
26. Pfitzner, B., Steckhan, N., & Arnrich, B. (2021). Federated Learning in a Medical Context: A Systematic Literature Review. *ACM Transactions on Internet Technology (TOIT)*, 21(2), 1–31.
27. Aledhari, M., Razzak, R., Parizi, R. M., & Saeed, F. (2020). Federated Learning: A Survey on Enabling Technologies, Protocols, and Applications. *IEEE Access*, 8, 140699–140725.
28. Rahman, A., Hossain, M. S., Muhammad, G., Kundu, D., Debnath, T., Rahman, M., Khan, M. S. I., Tiwari, P., & Band, S. S. (2023). Federated Learning-Based AI Approaches in Smart Healthcare: Concepts, Taxonomies, Challenges and Open Issues. *Cluster Computing*, 26(4), 2271–2311.
29. Kishor, K. (2023). Impact of Cloud Computing on Entrepreneurship, Cost, and Security. In: Kishor K., Saxena, N., and Pandey, D. (eds), *Cloud-based Intelligent Informative Engineering for Society 5.0*, 1st edition (pp. 171–191). CRC Press, New York. ISBN: 9781003213895. https://doi.org/10.1201/9781003213895-10
30. Kishor, K., & Pandey, D. (2022). Study and Development of Efficient Air Quality Prediction System Embedded with Machine Learning and IoT. In: Gupta, D., Khanna, A., Bhattacharyya, S., Hassanien, A.E., Anand, S., Jaiswal, A. (eds) *Proceeding International Conference on Innovative Computing and Communications*. Lecture Notes in Networks and Systems, vol. 471, Springer, Singapore. https://doi.org/10.1007/978-981-19-2535-1_24
31. Kishor, K., Sharma, R., & Chhabra, M. (2022). Student Performance Prediction Using Technology of Machine Learning. In: Sharma, D. K., Peng, S. L., Sharma, R., Zaitsev, D. A. (eds) *Micro-Electronics and Telecommunication Engineering*. Lecture Notes in Networks and Systems, vol. 373. Springer, Singapore. https://doi.org/10.1007/978-981-16-8721-1_53
32. Gupta, S., Tyagi, S., & Kishor, K. (2022). Study and Development of Self Sanitizing Smart Elevator. In: Gupta, D., Polkowski, Z., Khanna, A., Bhattacharyya, S., Castillo, O. (eds) *Proceedings of Data Analytics and Management*. Lecture Notes on Data Engineering and Communications Technologies, vol. 90. Springer, Singapore. https://doi.org/10.1007/978-981-16-6289-8_15
33. Verma, R. K., & Kishor, K. (2024). Image Processing Applications in Agriculture with the Help of AI. In: Khan, M. A., Khan, R., Praveen, P., Verma, A., Panda, M. (eds) *Infrastructure Possibilities and Human-Centered Approaches With Industry 5.0* (pp. 162–181). IGI Global, Hershey, Pennsylvania, USA. https://doi.org/10.4018/979-8-3693-0782-3.ch010

34. Kishor, K., Rani, R., Rai, A.K., & Sharma, V. (2023). 3D Application Development Using Unity Real Time Platform. In: Swaroop, A., Kansal, V., Fortino, G., Hassanien, A. E. (eds) *Proceedings of Fourth Doctoral Symposium on Computational Intelligence. DoSCI 2023.* Lecture Notes in Networks and Systems, vol. 726. Springer, Singapore. https://doi.org/10.1007/978-981-99-3716-5_54
35. Kishor, K., Sangal, S., & Shahroz (2023). Real-Time Traffic Signs and Lane Line Detection. In: Swaroop, A., Kansal, V., Fortino, G., Hassanien, A. E. (eds) *Proceedings of Fourth Doctoral Symposium on Computational Intelligence. DoSCI 2023.* Lecture Notes in Networks and Systems, vol. 726. Springer, Singapore. https://doi.org/10.1007/978-981-99-3716-5_71
36. Kishor, K. & Verma, R. K. (2023). Cloud Computing-Based Smart Agriculture. In: Sharma, A., Chanderwal, N., Khan, R. (eds) *Convergence of Cloud Computing, AI, and Agricultural Science* (pp. 120–136). IGI Global, Hershey, Pennsylvania, USA. https://doi.org/10.4018/979-8-3693-0200-2.ch006
37. Kishor, K. (2023). Chapter 17 Application of Quantum Computing for Digital Forensic Investigation. In: Yadav, S. P., Singh, R., Yadav, V., Al-Turjman, F., Kumar, S. A. (eds) *Quantum-Safe Cryptography Algorithms and Approaches: Impacts of Quantum Computing on Cybersecurity* (pp. 231–248). De Gruyter, Berlin, Boston. https://doi.org/10.1515/9783110798159-017

9 Securing the Connected World

Federated Learning and IoT Cybersecurity

Edidiong Akpabio, Supriya Narad, Idaraesit Akpabio, and Ifiokobong Akpabio

9.1 INTRODUCTION

"Internet of Things" (IoT) describes how control system components are linked to the network via communication protocols and sensing devices, such as laser scanners, radio frequency identification units, and sensors. The following industries have extensively used IoT technology: manufacturing, cyber-physical systems, mobile crowdsensing, Vehicles, Drones, Healthcare, and Agriculture.

With millions of embedded physical devices that are all connected and reveal data that could affect users' privacy and well-being, the IoT is a quickly evolving field.

9.2 RELATED WORKS

Without reliable security defenses, hackers can target IoT devices, creating a broad attack surface that is regularly exploited [1]. The surface of attack of physical devices that malicious organizations could potentially abuse is expanding as a result of the continuous creation and application of IoT technology. Well-known assaults like the Mirai botnet and its recent iterations show how important it is to bolster IoT device security to safeguard extensive IoT-enabled systems [2]. Over the past ten years, the size and complexity of the Internet have significantly increased. New kinds of connected gadgets have also surfaced at the same time. Because of the increased network complexity, the network systems are consequently vulnerable to various security flaws, including intrusions [3]. Cybersecurity is plagued by various attack types, including malware, intrusion detection, denial-of-service (DoS), system flooding, and zero-day assaults.

Data generation at the network edge is growing exponentially due to the development and expansion of the IoT. The traditional cloud-based centralized approaches to data analysis are facing new problems due to these developments, mainly from two directions. For example, high-frequency data from high-volume time-series sensors like video cameras or Lidar sensors, which pool data from millions or billions of IoT

 DOI: 10.1201/9781003489368-9

devices, means that centralized techniques are no longer appropriate for the 5G/6G era. Second, there is a growing perception that data collection threatens user privacy [4].

For instance, the traditional intrusion detection system necessitates the gathering and archiving of data from edge and IoT sensors. For several reasons, including high connectivity and latency costs, reluctance to provide personal information, and the learning load on the centralized server, such centralized learning could be more effective. Federated learning (FL) is a fantastic way to address cybersecurity risks in addition to the centralized intrusion detection systems' previously noted drawbacks [5].

FL is the best way to carry out on-device training while upholding privacy-preserving techniques with decentralized data (FL). FL has gained popularity recently as a low-latency updating option that guarantees end-user data privacy.

This technology is an excellent machine learning method for protecting data privacy and fostering peer-to-peer knowledge sharing. It overcomes the drawbacks of the centralized and edge computing paradigms. FL employs a unique technique whereby a trained machine learning model is distributed among several devices in the network. Devices downloading the shared base model utilize their local data and computational capabilities to train it [6].

A robust FL-based I.D.S. with a generative model was envisaged [7]. Generative adversarial network (GAN) and federated GAN (FedGAN) algorithms are employed on the participant side of FED-IIoT, an FL-based architecture for malware detection, to produce adversarial data and inject it into the data set of each IIoT application. A defense mechanism to identify and prevent abnormalities during aggregation was incorporated on the server side to guarantee a strong collaboration of trained models. Compared to current solutions, the suggested model exhibited superior accuracy and facilitated secure interaction and effective communication between participants in an IoT setting.

Artificial intelligence (AI) features in traditional IoT systems are frequently hosted on cloud servers or data centers, which is not scalable to the exponential development of IoT devices and the considerable data dissemination of large-scale IoT networks. Furthermore, transmitting a vast volume of data via rugged surroundings to the data center for AI training is impractical, considering the abundance of widely available IoT information in the significant data era. By utilizing the processing power of several IoT devices for data training, FL can offer more alluring features for enabling distributed intelligent IoT services and applications [8].

Managing programs that employ FL to find malware on IoT devices where privacy is protected and a solution for industrial IoT was proposed by [9], which analyzes behavioral data and samples of Android applications in an industrial setting. They claim to be 97%–90% accurate at identifying various malware samples. Another gives an alternative application of FL with up to 97% accuracy in intrusion detection. This investigates the antagonistic consequences of FL and uses blockchain (BC) technology as a substitute to lessen them. Packet scheduling and other network management activities have been implemented using FL and network data.

The central server needed for the conventional FL is susceptible to poisoning and single-point failures. When a decentralized FL approach is used to provide anomaly detection in IoT scenarios, compromising the central server leads to the failure of convergence of the low model accuracy ratio. BC is an advanced technology that allows for decentralized structures, traceability, authenticity, and cross-validation, among other features. Together, these positive attributes increase the dependability of decentralized FL When combined with an appropriate consensus procedure; the decentralized feature can thwart man-in-the-middle and single-point failure attacks. Additionally, by considerably lowering poisoning assaults, traceability and cross-validation can raise the model accuracy ratio. Additionally, BC technology can offer a reward system to incentivize local devices to take part in the FL training process and have excellent performance and high-quality data [9].

Security frameworks based on distributed Kalman filters and BCs have recently been developed to get around the computational cost of C.P.S. The computational complexity is spread via the distributed state estimator technique; however, the scheme necessitates a fixed threshold selection during the process, frequently resulting in a faked detection performance. FL, a decentralized method based on AI, has appeared as a promising solution to work on intrusion detection issues while preserving privacy in developing Industry 4.0 applications. On the other hand, the traditional FL approach averages the local client models without considering the importance of each client during aggregation. Additionally, the optimization of a cogent global model on the central server is absent from the aggregation model [10,11].

9.3 FEDERATED LEARNING AND IOT

9.3.1 Definition and Scope

Google first unveiled the concept of FL in 2016, using it in the Google Keyboard to enable several Android phones to study together [12].

FL is a machine learning method in which individual devices train a prototype independently without requiring the sharing of the raw data. FL stores data locally on the devices and only communicates local changes to the model, unlike traditional ML, where data is transferred to a central training server (Table 9.1).

TABLE 9.1
Overlapping between Federated Learning and Traditional Machine Learning

Feature	Comparison
	(a)
Learning approach	Centralized in traditional ML, decentralized in federated learning
Data location	Data remains on local devices in federated learning
Privacy	Federated learning preserves privacy by keeping data locally
Communication	Communication occurs between local models and a central server in federated learning
	(b)
Model training	Local models are trained independently and share updates in federated learning
Scalability	Federated learning is often more scalable, especially for large datasets
Robustness	Federated learning is more robust to device failures or dropouts
Real-time learning	Easier to implement real-time learning in federated settings

TABLE 9.2
Key Aspects of IoT in Cybersecurity

Aspect	Considerations
Connectivity	Pervasive network connections increase attack surfaces
Device heterogeneity	Diverse devices may have varying security capabilities
Data privacy	Sensitive data generated by IoT devices requires protection
Firmware security	Ensuring the integrity and security of device firmware
Authentication	Strong authentication mechanisms to prevent unauthorized access
Updates and patching	Regular updates to address vulnerabilities and patch security flaws
Interoperability	Ensuring secure communication between different IoT devices
Physical security	Protecting IoT devices from physical tampering or theft
Cloud integration	Securing data storage and processing in cloud environments

9.3.2 IoT in Cybersecurity

An IoT architecture with three basic levels was proposed in [13]: the application layer, the network layer, and the sensing layer. Based on the service-oriented design, the entire framework was split into four levels from a service-oriented perspective: The service layer, the network layer, the sensing layer, and the interface layer. This is given in Table 9.2 below.

9.4 SIGNIFICANCE OF FEDERATED LEARNING IN CYBERSECURITY

9.4.1 Introducing Collaborative Defense with Privacy at Its Foundation

Conventional cybersecurity methods often call for collecting enormous volumes of user data centrally, which raises privacy issues. FL turns this idea on its head. Imagine a network of millions of gadgets, from intelligent TVs to smartphones, working together as security analysts. Every device uses data for threat identification training, only exchanging anonymized model updates shared globally as part of a defense system. This distributed intelligence makes continuous learning and adaptability possible without ever disclosing private user data. Whether recently created or old, malicious software must contend with a privacy-preserving defense network that is constantly changing [14].

9.4.2 Personal Perspectives to Group Consensus

New malware analysis and comprehension have always required centralized labs and specialized resources. This approach is democratized by FL. Imagine a lone-hacked gadget coming upon a new threat. Rather than staying in isolation, it anonymously contributes insights to the network. These data are analyzed by millions of additional devices, which help us all understand the behavior and vulnerabilities of the threat. This decentralized method encourages constant adaptability to the constantly changing landscape of cyber threats while accelerating the identification of zero-day attacks [15].

9.4.3 Creating Self-Healing Networks for Self-Reliant Reaction

Thanks to FL, devices can become intelligent security sensors in a self-healing network. Real-time anomaly detection and response is possible for every device since it has been trained on its network behavior and the overall knowledge of the network. Imagine if your smart thermostat could detect odd increases in energy usage, which may be signs of unwanted access. Without depending on a central server that might be hacked, it could then act independently to shut down particular circuits or notify the appropriate authorities. This dispersed protection system reduces the possibility of harm from cyberattacks and increases resilience [16].

9.4.4 Increasing Security to Keep Up with the Internet of Things' Rapid Growth

The increasing number of interconnected gadgets poses a significant security risk. It is difficult for traditional, centralized systems to keep up with the sheer number and variety of devices. FL presents a strong alternative due to its built-in scalability. Its dispersion of the learning process throughout devices dispenses with the need for centralized infrastructure, which makes it perfect for safeguarding large networks of heterogeneous devices, such as medical equipment, home appliances, and industrial control systems. This lowers the expense and complexity of controlling the security of each device in addition to ensuring comprehensive protection [17].

9.4.5 Putting Users in Control

Users who employ traditional methods frequently have limited control over their data, which breeds mistrust, and makes them reluctant to comply with security precautions. FL gives users control over their learning process. Knowing their information will be kept secret; they can decide whether to join the cooperative defense network. Transparency fosters involvement and increases trust, strengthening the link.

9.5 REQUIREMENTS FOR FEDERATED LEARNING IN IOT SECURITY

World's security posture—FL opens the door to a safer and more cooperative digital ecology by permitting active engagement without compromising privacy [18] (Figure 9.1).

9.5.1 Privacy-Preserving Mechanisms

Protecting sensitive IoT data involves employing data encryption methods during transmission and storage. Utilizing advanced techniques such as homomorphic encryption or secure multi-party computation enhances the security of the data. Differential privacy safeguards individual device information while achieving overall learning objectives. This can be accomplished by introducing noise to model updates

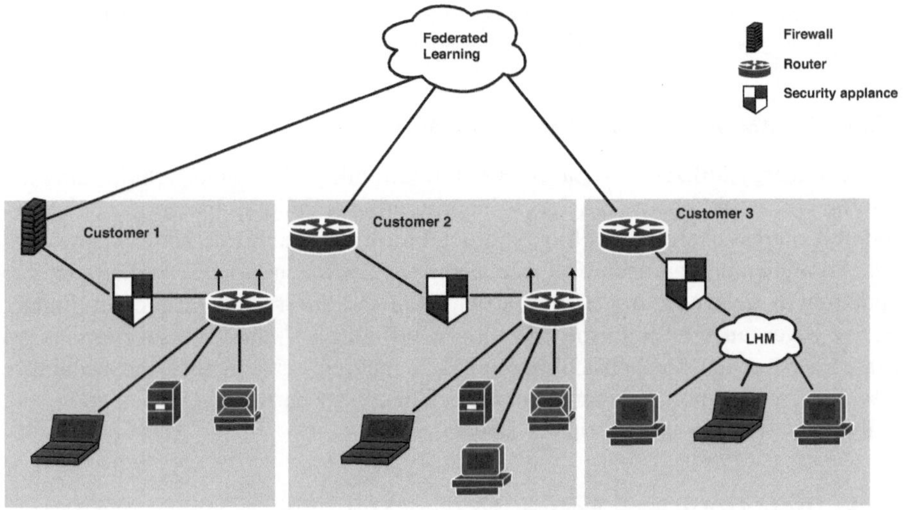

FIGURE 9.1 This image shows the significance of federated learning in cybersecurity.

or implementing gradient clipping. Additionally, secure aggregation is vital to prevent manipulation during model update aggregation. Cryptographic techniques like federated averaging with verifiable aggregation significantly enhance overall security. I believe that prioritizing privacy in IoT systems and implementing these cryptographic measures creates a more secure and confidential environment for data processing [19].

9.5.2 Resource Efficiency

To accommodate resource-constrained IoT devices, it is essential to use computationally lightweight algorithms. Techniques such as model pruning, quantization, and federated distillation help reduce computational overhead. Adaptive training is crucial to adapt the training process to device heterogeneity, employing methods like federated meta-learning or personalized FL. Efficient communication is paramount for large-scale IoT deployments and minimizing communication costs through compression techniques and network optimization is vital. Resource efficiency is a critical factor for IoT devices with limited capabilities, and these strategies ensure optimal utilization without compromising model performance [20].

9.5.3 Robustness and Security

In the realm of security, resilience against malicious participants is pivotal. Byzantine fault tolerance mechanisms, such as reputation systems, are crucial in identifying and excluding nefarious actors. Validating and authenticating model updates is equally essential to prevent the injection of vulnerabilities by adversaries. Incorporating Federated Intrusion Detection Systems (FIDS) enhances the monitoring capabilities for anomalies and potential threats in a decentralized manner. From my perspective, robust security measures are imperative, and these mechanisms collectively provide a comprehensive defense against possible adversarial attacks in FL [21].

9.5.4 System Sustainability and Scalability

Efficient energy utilization is paramount for sustaining FL systems. Optimizing the training process for energy efficiency, incorporating techniques like model compression and energy-aware scheduling, is crucial to minimize device battery consumption. An implementing incentive mechanism, such as token-based reward systems or reputation models, encourages fair participation and fosters long-term commitment. Precise governance mechanisms, potentially utilizing BC, are necessary to manage data ownership and decision-making in FL. Achieving sustainability and scalability in FL systems requires balancing energy efficiency, incentivizing participation, and establishing transparent governance structures [22].

9.5.5 Integration with Existing Security Solutions

FL serves as a complementary tool when integrated with existing security solutions. This integration enhances the defense strategy by incorporating established measures such as firewalls, intrusion detection systems, and access controls. Ensuring interoperability with existing security standards facilitates seamless integration and data exchange within the broader security framework. The adaptive nature of FL allows continuous improvement of security models, aligning with evolving threats and changing data distributions. In summary, integrating FL with established security measures is pivotal, creating a holistic defense approach that develops alongside emerging security challenges [23] (Figure 9.2).

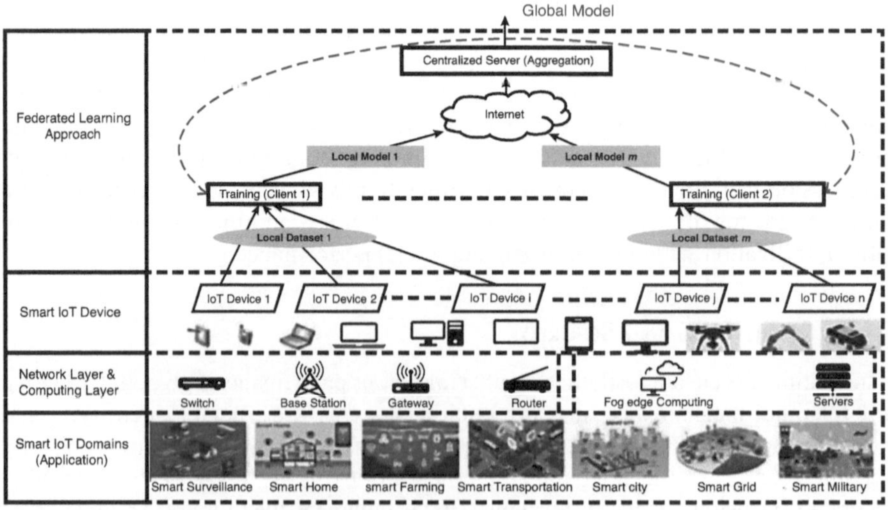

FIGURE 9.2 This shows the requirements for FL in IoT.

9.6 APPLICATIONS OF IOT

9.6.1 Smart Traffic

The management of traffic currently needs to be improved in urban areas. It is now nearly complicated to manage them manually. Implementing IoT for traffic management can solve this issue. Using sensors, this intelligent traffic monitoring system gathers real-time traffic information. The driver receives traffic updates, enabling him to choose a more efficient.

9.6.2 Route

In addition to showing the user where to pick up and drop off cabs in real-time, this also makes it easier for them to reserve a cab without making a phone call. Astute parking spaces will have sensors installed to determine whether or not they are available. When drivers park their cars, they check the app, which tells them where the closest parking spaces are, how much parking will cost, and other information based on data gathered and processed by intelligent sensors. This helps the drivers save time and gas [24].

9.6.3 Transportation

IoT provides a range of options for toll collecting and fare collection, as well as for checking people and bags before they board commercial aircraft and for moving products while in transit. Using IoT in this space has the potential to alter the public transportation industry drastically. Radio Frequency Identification (RFID)-based ticketing is already extensively accessible. In addition to giving users access to the stations, this system allows precise tracking of each traveler's connection and itinerary thanks to readers installed in the doors of buses and trains. This will give the operating business the ideal traffic statistics to decide whether to build new lines, optimize the network and service level, or both. In the event of an emergency, the names of the rechargeable ticket subscribers and the number of passengers at a particular station might be known to the rescue personnel. The automobiles in the private sector will be able to speak with one another and begin collecting ambient data independently. When there is a line, for example, the front vehicles may alert the rear vehicles about an accident or just too much traffic, which may eventually force intelligent navigation systems to reroute cars set to travel on congested highways. When driving in hazardous weather or when an oncoming car is moving too quickly, the cars may assist the driver in maintaining a safe distance from the vehicle in front of them [25].

9.6.4 Emergency and Security

IoT provides real-time data on earthquake activity, weather, and other aspects, which helps with effective disaster management. Medical sensors and wearable technology are used in healthcare to track patients' vital signs and send out automatic alerts in case of emergency. By optimizing routes for emergency vehicles, smart traffic management systems shorten response times. Emergency resources are moved efficiently thanks to IoT asset tracking. It is crucial to guarantee data transfer security, which is accomplished by using strong encryption techniques. Cyber risks are mitigated by device authentication, firewalls and other network security measures, and secure communication protocols. Vulnerabilities are avoided via regular firmware and software updates and security precautions. An additional layer of protection is provided by physical security measures like restricted access and tamper detection. Techniques for protecting user privacy protect data, and following rules guarantees that industry norms are followed. Effective management of security breaches is facilitated by an incident response strategy [26].

9.6.5 Agriculture

A revolutionary era in agriculture began with incorporating IoT applications, altering conventional farming processes, and promoting sustainability. One notable cornerstone is precision farming, where IoT devices such as drones and sensors provide real-time data on crop health, weather patterns, and soil conditions. With the use of this information, farmers are better equipped to optimize resource use and increase crop yields by making educated decisions about fertilization, irrigation, and pest management [27]. Smart collars and sensors, among other IoT-enabled gadgets, have significantly advanced livestock monitoring. Using these tools, farmers may optimize feeding schedules, get alerts for any illnesses, and keep an eye on the health and behavior of their animals in real-time. Using IoT technology with automated machines improves agricultural operations' efficiency. Sensor-equipped tractors and harvesters run more accurately, maximizing fuel efficiency and lessening environmental effects [28]. IoT applications help the agricultural supply chain by guaranteeing traceability and cutting waste. IoT improves transparency and quality control across the supply chain by recording storage conditions and monitoring transit. IoT sensors enable environmental monitoring, which helps with weather pattern prediction and farming practice adaptation. Smart greenhouses use IoT technologies to maintain ideal growing conditions for plants. These automated devices help to increase crop quality and output by providing exact control over temperature, humidity, and illumination. Drones with cameras and sensors are essential for managing crop health because they can detect diseases, nutrient shortages, and pest infestations. This allows for focused treatments and lowers the need for pesticides [29].

IoT-enabled irrigation systems improve water management by tracking weather and soil moisture levels. This technique preserves water resources and encourages environmentally friendly farming methods by preventing excessive irrigation. By evaluating data generated by the IoT, data analytics, and decision support systems enable farmers to make more informed decisions about resource allocation, planting schedules, and crop rotation [30].

9.6.6 Healthcare

Adopting IoT technologies has resulted in revolutionary developments in the healthcare industry. Real-time tracking of vital signs is made possible via remote patient monitoring, made possible by wearable technology and home sensors. This enables proactive healthcare measures. With the help of connected glucose meters, blood pressure monitors, and other intelligent health gadgets, people may better manage their medical problems. Healthcare providers can remotely monitor and modify treatment regimens based on real-time data. Through asset tracking, IoT solutions ensure effective equipment use, reduce patient care delays, and streamline hospital operations. IoT helps medication management by improving adherence and decreasing errors through tracking systems and reminders. IoT automation optimizes healthcare workflows, freeing healthcare personnel to concentrate on patient care. Tele and telemedicine platforms use IoT devices to provide patient-centered, easily accessible healthcare through virtual consultations, remote diagnostics, and monitoring. Patient confidentiality is preserved via solid encryption and adherence to data protection laws, prioritizing health data security [31].

9.7 POSSIBLE THREATS ON IOT DEVICES

The attacks faced by IoT devices are divided into four in this section, namely physical, software, network and encryption attacks.

9.7.1 Physical Attacks

IoT offers a bigger attack surface and physical access to the devices because it is inherently dispersed and fragmented. It is possible for a hacker to alter a node or sensor data, which could endanger the security of the entire sensor network. The attacker must physically get access to the IoT system in order to carry out a physical assault, which is related to the hardware components of IoT devices. The IoT hardware may become less functioning as a result of these attacks. Example of these is given below.

9.7.1.1 Node Tampering

One physical attack that can harm an IoT device is called "node tampering," and it targets sensor nodes. An enemy will physically swap out all or a portion of the node in order to get access to and alter private data, including shared cryptographic keys [32].

9.7.1.2 Physical Damage

Intentional damage, frequently used for malevolent intent, to the physical components of IoT devices, such as tampering with sensors or hardware, can result in operational failure, safety risks, and destruction [33]. Social engineering targets IoT device users for illegal access or manipulation, causing data breaches and operational interruptions. It does this by taking advantage of human psychology to fool people [34].

9.7.1.3 Sleep Deprivation

Altering sleep or power modes prevents IoT devices from operating normally, which accelerates wear, may result in physical harm, and raises energy expenses [35].

9.7.2 Software Attacks

9.7.2.1 Malicious Scripts

Malicious scripts can pose serious dangers to the security and dependability of IoT ecosystems by taking advantage of weaknesses in connected devices and injecting unauthorized code to compromise functionality, steal data, or enable other cyber threats [36].

9.7.2.2 Phishing

IoT phishing attacks use deceptive methods to trick users or administrators into disclosing private information, jeopardizing the security of devices that are connected. This kind of social engineering has the potential to cause data breaches, illicit access, or IoT system manipulation [37].

9.7.2.3 Virus and Worms

Within the IoT, objects that are networked can become infected by viruses and worms, which propagate across networks and cause disruptions. These malware types present significant cybersecurity risks in IoT environments because they have the potential to alter data, grant unauthorized access, and impair functioning [38].

9.7.2.4 Denial-of-Service

IoT DoS attacks try to overload networks or linked devices, making them unusable. DoS attacks cause regular operations to be disrupted, potentially resulting in system failures and service outages. They do this by flooding the IoT infrastructure with excessive traffic or by taking advantage of weaknesses [39].

9.7.3 Network Attacks

9.7.3.1 Traffic Analysis

A monitoring and examining data flow inside a network in order to derive significant insights is known as traffic analysis. Adversaries may use communication methods in the overall picture of the IoT to obtain sensitive information, creating hazards to security and privacy [40].

9.7.3.2 RFID Cloning

The unapproved duplication of RFID tags is known as RFID cloning. The integrity of RFID-based applications, such as inventory management and access control, is jeopardized when attackers copy the distinctive identifiers of RFID tags in order to obtain illegal access or alter systems [41].

9.7.3.3 Sinkhole Attack

A sinkhole attack occurs when malevolent actors draw inbound network traffic to a compromised node, thereby rerouting communication through it. This puts the dependability and security of the linked devices in danger in the IoT by interfering with regular data flows, permitting eavesdropping, or facilitating additional attacks [42].

9.7.3.4 Routing Information Attack

False routing information is injected into a network or manipulated as part of a routing information attack. In the context of the IoT, there is a serious security risk since attackers may breach routing protocols to reroute data, which could result in illegal access, data interception, or communication disruption [43].

9.7.3.5 Side Channel Attacks

Side-channel attacks take advantage of unintentional information leaks that occur while a system is in use. Attackers may use inadvertent signals, like power usage or electromagnetic emissions, in the context of the IoT to gather private data and jeopardize the security of linked devices [44].

9.7.4 Cryptanalysis Attacks

Attacks known as cryptanalysis entail cracking cryptographic systems in order to decrypt data. Attackers can compromise the confidentiality and integrity of data secured by encryption in the IoT by using a variety of techniques, including brute-force attacks and taking advantage of flaws in cryptographic algorithms [45].

9.7.4.1 Middle Man Attack

A potential threat arises when an adversary covertly intercepts and possibly alters communications between two parties through a man-in-the-middle attack. This malicious tactic, if successful, allows unauthorized access to private data exchanged either between devices or between a device and a server. The ramifications of such an attack extend to compromising the integrity and confidentiality of data within the IoT [46] (Figure 9.3).

9.8 SOLUTIONS AND POTENTIAL APPROACHES FOR RISK MITIGATION

9.8.1 Use of Blockchain Technology

BC technology operates as a decentralized network consisting of peer-to-peer nodes responsible for storing transactional records, known as blocks. These blocks, comprised of multiple public databases, collectively create the "chain." At the core of B.C. is the utilization of a distributed ledger, providing a secure framework for real-time data collected from IoT devices' sensors. Each transaction in this ledger undergoes authentication through the owner's digital signature, ensuring verification and protection against tampering. As a result, the data maintained in the digital ledger demonstrates significant stability. Entries in the BC ledger follow chronological order and carry timestamps. Utilizing cryptographic hash keys, each entry is linked to its predecessor in a Merkle tree, with the root hash of the tree stored in the BC. This process facilitates the dissemination of the latest transaction to all network nodes. The use of cryptographic hash keys makes hacking blocks a formidable and time-consuming task for potential attackers. Miners, solely motivated by bonuses,

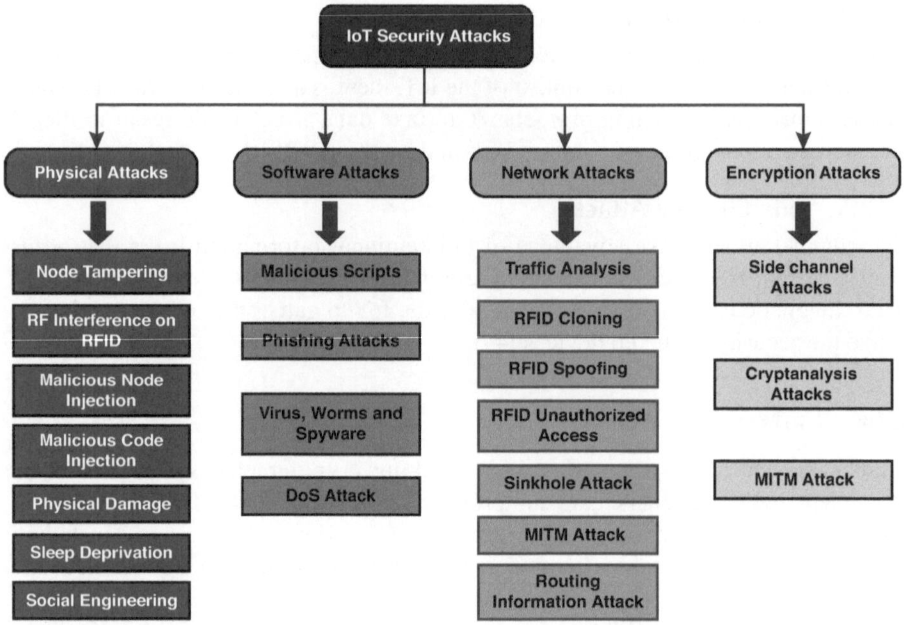

FIGURE 9.3 This shows all possible attacks on IoT devices.

lack personal involvement in the transactions they process, and the identity of transaction owners remains unknown to them. Moreover, multiple miners compete to link transactions to the BC, bolstering the technology's overall security. These combined characteristics position BC as a reliable, distributed, tamper-resistant, and openly accessible data system for managing IoT data [47].

9.8.2 Usage of MTTQ

As noted by [48] and [49], message queuing telemetry transport (MQTT) has emerged as an efficient and lightweight messaging protocol tailored for low-bandwidth, high-latency, or unreliable networks, making it a popular choice for IoT communication. Despite its robust communication foundation, addressing security concerns related to IoT devices necessitates a comprehensive approach. It is used in the following areas:

1. **Authentication and Authorization**: MQTT supports username/password authentication, ensuring only authorized devices can connect with the MQTT broker. This is complemented by access control mechanisms within the broker, allowing restrictions on device subscriptions or publications to specific topics and establishing a robust authentication and authorization framework.
2. **TLS/SSL Encryption**: Implementing Transport Layer Security (TLS) or SecureSockets Layer (SSL) adds a crucial layer of encryption to MQTT communication. This protects data exchanged between MQTT clients and the

broker, thwarting unauthorized access, eavesdropping, and man-in-the-middle attacks. Encryption becomes particularly vital when transmitting sensitive information over the network.

3. **Quality of Service Levels**: MQTT provides three quality of service (QoS) levels dictating message delivery guarantees. Choosing the right QoS level is strategic, impacting the reliability of message delivery and helping reduce the risk of interception or loss based on IoT application requirements.

9.9 CHALLENGES AND LIMITATIONS OF FEDERATED LEARNING-BASED CYBERSECURITY FOR IOT

FL is a promising solution for enhancing cybersecurity in the IoT ecosystem. Nevertheless, its implementation poses challenges and limitations requiring careful consideration: While FL promotes data privacy through decentralized training on user devices, privacy concerns arise during the model aggregation process. There is a crucial need to address the risk of privacy breaches and unintentional disclosure of sensitive information. Frequent communication between the central server and individual devices in FL can strain network bandwidth, particularly in IoT, where devices often need more connectivity. Managing increased bandwidth usage becomes pivotal for optimal performance. The diverse array of IoT devices introduces challenges in ensuring that FL algorithms seamlessly adapt to different hardware, operating systems, and communication protocols. Tailored solutions for various devices may be essential. FL is susceptible to model poisoning attacks, wherein malicious participants send corrupted updates to compromise the global model. Detecting and preventing such attacks remains an ongoing challenge to preserve the integrity of the training process. Due to diverse device types and environmental conditions, the assumption of independently and identically Distributed (IID) data, inherent in federated averaging, may not hold in IoT. This non-IID nature of data can impact model convergence and performance. Energy consumption has become a concern in FL, particularly for resource-constrained IoT devices. Local execution of machine learning models for FL can strain limited computational resources, affecting the overall energy efficiency of the IoT ecosystem. Managing the scalability of FL systems becomes complex with an increasing number of devices. Striking a balance between efficient scaling and maintaining model quality is a challenge that requires careful consideration. Adhering to data protection regulations and standards in a FL environment, especially across diverse geographical locations with different privacy laws, adds complexity to FL deployments. Considering IoT devices may experience intermittent connectivity or go offline, designing FL systems to gracefully handle such scenarios and ensure recovery is essential for real-world implementations.

9.10 CONCLUSION

To summarize, the convergence of the IoT with FL offers a revolutionary synergy for improving cybersecurity. FL differs from standard machine learning because it allows for decentralized model training on individual devices, maintains privacy, and provides scalability benefits. It was first introduced by Google in 2016. It is essential for cybersecurity because it makes IoT devices work together to form a cooperative defense network.

FL supports the rapid expansion of networked devices, encourages group consensus for malware research, and accelerates the development of self-healing networks.

Robust security measures, resource efficiency, privacy preservation techniques, and smooth integration with current security systems are all necessary for FL in IoT security. BC technology and MQTT are essential tools that provide secure and decentralized data management and effective IoT connectivity.

However, bandwidth limitations, device variety, privacy issues, and regulatory compliance must be carefully considered. It's critical to strike a balance between scalability issues and energy consumption. FL shows promise as a cornerstone for transforming IoT cybersecurity by tackling these issues and fostering innovation, cooperation, and privacy protection in the connected digital age.

REFERENCES

1. Ferrag, M. A., Friha, O., Maglaras, L., Janicke, H., & Shu, L. "Federated Deep Learning for Cyber Security in the Internet of Things: Concepts, Applications, and Experimental Analysis," in *IEEE Access*, vol. 9, pp. 138509–138542, 2021. https://doi.org/10.1109/ACCESS.2021.3118642.
2. Campos, E. M., Saura, P. F., González-Vidal, A., Hernández-Ramos, J. L., Bernabe, J. B., Baldini, G., & Skarmeta, A. "Evaluating Federated Learning for Intrusion Detection in Internet of Things: Review and Challenges," in *Computer Networks*, vol. 203, p. 108661, 2022. ISSN 1389-1286. https://doi.org/10.1016/j.comnet.2021.108661.
3. Agrawal, S., Sarkar, S., Aouedi, O., Yenduri, G., Piamrat, K., Alazab, M., Bhattacharya, S., Maddikunta, P. K. R., & Gadekallu, T. R. "Federated Learning for Intrusion Detection System: Concepts, Challenges and Future Directions," in *Computer Communications*, vol. 195, pp. 346–361, 2022.
4. Zhang, T., Gao, L., He, C., Zhang, M., Krishnamachari, B., & Avestimehr, A. S. "Federated Learning for the Internet of Things: Applications, Challenges, and Opportunities," in *IEEE Internet of Things Magazine*, vol. 5, no. 1, pp. 24–29, March 2022. https://doi.org/10.1109/IOTM.004.2100182.
5. Alazab, M., RM, S. P., Parimala, M., Maddikunta, P. K. R., Gadekallu, T. R., & Pham, Q.-V. "Federated Learning for Cybersecurity: Concepts, Challenges, and Future Directions," in *IEEE Transactions on Industrial Informatics*, vol. 18, no. 5, pp. 3501–3509, May 2022. https://doi.org/10.1109/TII.2021.3119038.
6. Attota, D. C., Mothukuri, V., Parizi, R. M., & Pouriyeh, S. "An Ensemble Multi-View Federated Learning Intrusion Detection for IoT," in *IEEE Access*, vol. 9, pp. 117734–117745, 2021. https://doi.org/10.1109/ACCESS.2021.3107337.
7. Ghimire, B., & Rawat, D. B. "Recent Advances on Federated Learning for Cybersecurity and Cybersecurity for Federated Learning for Internet of Things," in *IEEE Internet of Things Journal*, vol. 9, no. 11, pp. 8229–8249, 1 June, 2022. https://doi.org/10.1109/JIOT.2022.3150363.
8. Nguyen, D. C., Ding, M., Pathirana, P. N., Seneviratne, A., Li, J., & Poor, H. V. "Federated Learning for Internet of Things: A Comprehensive Survey," in *IEEE Communications Surveys & Tutorials*, vol. 23, no. 3, pp. 1622–1658, thirdquarter 2021. https://doi.org/10.1109/COMST.2021.3075439.

9. Sánchez, P. M. et al., "Studying the Robustness of Anti-Adversarial Federated Learning Models Detecting Cyberattacks in IoT Spectrum Sensors," in *IEEE Transactions on Dependable and Secure Computing*. https://doi.org/10.1109/TDSC.2022.3204535.
10. Cui, L. et al., "Security and Privacy-Enhanced Federated Learning for Anomaly Detection in IoT Infrastructures," in *IEEE Transactions on Industrial Informatics*, vol. 18, no. 5, pp. 3492–3500, May 2022. https://doi.org/10.1109/TII.2021.3107783.
11. Tahir, B., Jolfaei, A., & Tariq, M. "Experience-Driven Attack Design and Federated-Learning-Based Intrusion Detection in Industry 4.0," in *IEEE Transactions on Industrial Informatics*, vol. 18, no. 9, pp. 6398–6405, Sept. 2022. https://doi.org/10.1109/TII.2021.3133384.
12. Kishor K. "Communication-Efficient Federated Learning." In: Yadav S. P., Bhati B. S., Mahato D. P., Kumar S. (eds) *Federated Learning for IoT Applications*. EAI/Springer Innovations in Communication and Computing. Springer, Cham, 2022. https://doi.org/10.1007/978-3-030-85559-8_9.
13. Lu, Y., & Xu, L. D. "Internet of Things (IoT) Cybersecurity Research: A Review of Current Research Topics," in *IEEE Internet of Things Journal*, vol. 6, no. 2, pp. 2103–2115, April 2019. https://doi.org/10.1109/JIOT.2018.2869847.
14. Huong, T. T., Bac, T. P., Long, D. M., Luong, T. D., Dan, N. M., Thang, B. D., & Tran, K. P. "Detecting Cyberattacks Using Anomaly Detection in Industrial Control Systems: A Federated Learning Approach," in *Computers in Industry*, vol. 132, p. 103509, 2021.
15. Jahromi, A. N., Karimipour, H., & Dehghantanha, A. "An Ensemble Deep Federated Learning Cyber-Threat Hunting Model for Industrial Internet of Things," in *Computer Communications*, vol. 198, pp. 108–116, 2023.
16. Abad, G., Picek, S., Ramírez-Durán, V. J., & Urbieta, A. On the Security & Privacy in Federated Learning, 2021. arXiv preprint arXiv:2112.05423.
17. Sakhare, N. N. "A Decentralized Approach to Threat Intelligence Using Federated Learning in Privacy-Preserving Cyber Security," in *Journal of Electrical Systems*, vol. 19, no. 3, pp. 106–125, 2023.
18. Yazdinejad, A., Dehghantanha, A., Parizi, R. M., Hammoudeh, M., Karimipour, H., & Srivastava, G. "Block Hunter: Federated Learning for Cyber Threat Hunting in Blockchain-Based IIoT Networks," in *IEEE Transactions on Industrial Informatics*, vol. 18, no. 11, pp. 8356–8366, 2022.
19. Khan, L. U., Saad, W., Han, Z., Hossain, E., & Hong, C. S. "Federated Learning for Internet of Things: Recent Advances, Taxonomy, and Open Challenges," in *IEEE Communications Surveys & Tutorials*, vol. 23, no. 3, pp. 1759–1799, 2021.
20. Zhang, T., He, C., Ma, T., Gao, L., Ma, M., & Avestimehr, S. "Federated Learning for Internet of Things." In: *Proceedings of the 19th ACM Conference on Embedded Networked Sensor Systems (SenSys '21)*. Association for Computing Machinery, New York, pp. 413–419, 2021. https://doi.org/10.1145/3485730.3493444
21. Hiessl, T., Schall, D., Kemnitz, J., & Schulte, S. "Industrial Federated Learning-Requirements and System Design." In: *International Conference on Practical Applications of Agents and Multi-Agent Systems* (pp. 42–53). Springer International Publishing, Cham, July 2020.
22. Verma, R. K., Kishor, K., & Jha, S. K. (2024). "Big Data Analytics in Bioinformatics and Healthcare." In: Khan R., Kumar I., Praveen P. (eds.), *Applications of Parallel Data Processing for Biomedical Imaging* (pp. 25–43). IGI Global. https://doi.org/10.4018/979-8-3693-2426-4.ch002.
23. Imteaj, A., Thakker, U., Wang, S., Li, J., & Amini, M. H. "A Survey on Federated Learning for Resource-Constrained IoT Devices," in *IEEE Internet of Things Journal*, vol. 9, no. 1, pp. 1–24, 2021.

24. Soumyalatha, S. G. H. "Study of IoT: Understanding IoT Architecture, Applications, Issues and Challenges." In: *1st International Conference on Innovations in Computing & Networking (ICICN16), CSE, RRCE. International Journal of Advanced Networking & Applications*, vol. 478, Eswar Publications, Tiruppur Dist, Tamil Nadu, India, May 2016.
25. Bhat, S., Bhat, O., & Gokhale, P. "Applications of IOT and IOT: Vision 2020," in *International Advanced Research Journal in Science, Engineering and Technology*, vol. 5, no. 1, pp. 41–44, 2018.
26. Bali, A., Raina, M., & Gupta, S. "Study of Various Applications of Internet of Things (IoT)," *International Journal of Computer Engineering and Technology*, vol. 9, no. 2, pp. 39–50, 2018.
27. Kishor, K., Rani, R., Rai, A. K., & Sharma, V. "3D Application Development Using Unity Real Time Platform." In: Swaroop A., Kansal V., Fortino G., Hassanien A. E. (eds) *Proceedings of Fourth Doctoral Symposium on Computational Intelligence . DoSCI 2023*. Lecture Notes in Networks and Systems, vol. 726. Springer, Singapore, 2023. https://doi.org/10.1007/978-981-99-3716-5_54.
28. Kishor, K. & Verma, R. K. "Cloud Computing-Based Smart Agriculture." In: Sharma A., Chanderwal N., Khan R. (eds) *Convergence of Cloud Computing, AI, and Agricultural Science* (pp. 120–136). IGI Global, Hershey, Pennsylvania, USA, 2023. https://doi.org/10.4018/979-8-3693-0200-2.ch006.
29. Maraveas, C., Piromalis, D., Arvanitis, K. G., Bartzanas, T., & Loukatos, D. "Applications of IoT for Optimized Greenhouse Environment and Resources Management," in *Computers and Electronics in Agriculture*, vol. 198, p. 106993, 2022.
30. Kumar, N., Dahiya, A. K., Kumar, K., & Tanwar, S. "Application of IoT in Agriculture." In: *2021 9th International Conference on Reliability, Infocom Technologies and Optimization (Trends and Future Directions) (ICRITO)* (pp. 1–4). IEEE, Noida, September 2021. DOI:10.1109/icrito51393.2021.9596120
31. Tiwary, A., Mahato, M., Chidar, A., Chandrol, M. K., Shrivastava, M., & Tripathi, M. "Internet of Things (IoT): Research, Architectures and Applications," in *International Journal on Future Revolution in Computer Science & Communication Engineering*, vol. 4, no. 3, pp. 23–27, 2018.
32. Kishor, K., & Nand, P. "Wireless Networks Based in the Cloud that Support 5G." In: Kishor K., Saxena, N., and Pandey, D. (eds), *Cloud-based Intelligent Informative Engineering for Society 5.0*, 1st edition (pp. 23–40). Chapman and Hall/CRC, New York, 2023. ISBN: 9781003213895, https://doi.org/10.1201/9781003213895-2.
33. Özalp, A. N., Albayrak, Z., Çakmak, M., & ÖzdoĞan, E. "Layer-Based Examination of Cyber-Attacks in IoT." In: *2022 International Congress on Human-Computer Interaction, Optimization and Robotic Applications (HORA)* (pp. 1–10). IEEE, Ankara, June 2022. DOI: 10.1109/HORA55278.2022.9800047
34. Kishor, K., Shukla, A., Thakur, A. "Vehicle Classification and License Number Plate Detection Using Deep Learning." In: Sharma, D.K., Peng, SL., Sharma, R., Jeon, G. (eds) *Micro-Electronics and Telecommunication Engineering*. ICMETE 2023. Lecture Notes in Networks and Systems, vol 894. Springer, Singapore, 2024. https://doi.org/10.1007/978-981-99-9562-2_5
35. Rahman, F., Farmani, M., Tehranipoor, M., & Jin, Y. "Hardwareassisted cybersecurity for IoT devices." In: *2017 18th International Workshop on Microprocessor and SOC Test and Verification (MTV)* (pp. 51–56). IEEE, Austin, TX, December 2017. DOI: 10.1109/MTV.2017.16
36. Kishor, K., Singh, P., & Vashishta, R. "Develop Model for Malicious Traffic Detection Using Deep Learning." In: Sharma D. K., Peng S. L., Sharma R., Jeon G. (eds) *Micro-Electronics and Telecommunication Engineering*. Lecture Notes in Networks and Systems, vol. 617. Springer, Singapore, 2023. https://doi.org/10.1007/978-981-19-9512-5_8.

37. Damghani, H., Damghani, L., Hosseinian, H., & Sharifi, R. "Classification of Attacks on IoT." In: *4th International Conference on Combinatorics, Cryptography, Computer Science and Computation*, Iran University of Science and Technology, November 2019.
38. Ronen, E., & Shamir, A. "Extended Functionality Attacks on IoT Devices: The Case of Smart Lights." In: *2016 IEEE European Symposium on Security and Privacy (EuroS&P)* (pp. 3–12). IEEE, Saarbruecken, March 2016. DOI: 10.1109/EuroSP.2016.13
39. Yan, W., Fu, A., Mu, Y., Zhe, X., Yu, S., & Kuang, B. "EAPA: Efficient Attestation Resilient to Physical Attacks for IoT Devices." In: *Proceedings of the 2nd International ACM Workshop on Security and Privacy for the Internet-of-Things* (pp. 2–7), Association for Computing Machinery, New York, London, November 2019. https://doi.org/10.1145/3338507.3358614
40. Kishor, K., Saxena, N., & Pandey, D. (eds) *Cloud-Based Intelligent Informative Engineering for Society 5.0*, 1st edition (pp. 1–234). Chapman and Hall/CRC, New York, 2023. ISBN: 9781003213895. https://doi.org/10.1201/9781003213895.
41. Liang, X., & Kim, Y. "A Survey on Security Attacks and Solutions in the IoT Network." In: *2021 IEEE 11th Annual Computing and Communication Workshop and Conference (CCWC)* (pp. 0853–0859). Institute of Electrical and Electronics Engineers (IEEE), NV, USA, January 2021. DOI: 10.1109/CCWC51732.2021.9376174
42. Kishor, K., Nand, P., & Agarwal, P. "Secure and Efficient Subnet Routing Protocol for MANET", in *Indian Journal of Public Health*, vol. 9, no. 12, p. 200, 2018. https://doi.org/10.5958/0976-5506.2018.01830.2.
43. Kishor, K., & Pandey, D. "Study and Development of Efficient Air Quality Prediction System Embedded with Machine Learning and IoT." In: Gupta D., Khanna A., Bhattacharyya S., Hassanien A. E., Anand S., Jaiswal A. (eds) *Proceeding International Conference on Innovative Computing and Communications.* Lecture Notes in Networks and Systems, vol. 471, Springer, Singapore, 2022. https://doi.org/10.1007/978-981-19-2535-1_24.
44. Shah, T., & Venkatesan, S. "Authentication of IoT Device and IoT Server Using Secure Vaults. In: *2018 17th IEEE International Conference on Trust, Security and Privacy in Computing and Communications/12th IEEE International Conference on Big Data Science and Engineering (TrustCom/BigDataSE)* (pp. 819–824). IEEE, New York, August 2018. DOI: 10.1109/TrustCom/BigDataSE.2018.00117
45. Chaudhry, S. "An Encryption-Based Secure Framework for Data Transmission in IoT." In: *2018 7th International Conference on Reliability, Infocom Technologies and Optimization (Trends and Future Directions) (ICRITO)* (pp. 743–747). IEEE, Noida, August 2018.
46. Lo'Ai, A. T., & Somani, T. F. "More Secure Internet of Things Using Robust Encryption Algorithms against Side Channel Attacks." In *2016 IEEE/ACS 13th International Conference of Computer Systems and Applications (AICCSA)* (pp. 1–6). IEEE, Agadir, November 2016. DOI: 10.1109/AICCSA.2016.7945813
47. Krishna, R. R., Priyadarshini, A., Jha, A. V., Appasani, B., Srinivasulu, A., & Bizon, N. "State of-the-Art Review on IoT Threats and Attacks: Taxonomy, Challenges and Solutions," in *Sustainability* vol. 13, no. 16, p. 9463, 2021. https://doi.org/10.3390/su13169463.
48. Perrone, G., Vecchio, M., Pecori, R., & Giaffreda, R. "The Day after Mirai: A Survey on MQTT Security Solutions after the Largest Cyber-Attack Carried Out through an Army of IoT Devices." In: *IoTBDS* (pp. 246–253), SciTePress, April 2017. DOI: 10.5220/0006287302460253
49. Kishor, K. "Using a half cheetah habitat for random augmentation computing," *Multimedia Tools and* Applications, 2024. https://doi.org/10.1007/s11042-024-19084-0

10 Federated Learning Shaping the Future of Smart City Infrastructure

Raj Kishor Verma, Kaushal Kishor, and Antonino Galletta

10.1 INTRODUCTION

Amidst the rapid growth of cities and technological progress, the idea of smart cities has evolved as a viable approach to effectively tackle numerous urban difficulties. An essential aspect of smart city infrastructure development is the use of advanced technologies such as federated learning. This article examines the capacity of federated learning to revolutionize the development of smart city infrastructure. Federated learning is an innovative machine learning technique that allows several edge devices to collaborate in training models, all while ensuring the privacy of the data. Federated learning, in contrast to typical centralized methods, enables data to be kept in local locations instead of being consolidated into a single repository for analysis. This strategy effectively addresses issues about data privacy and security. Federated learning enables smart city stakeholders to derive important insights from large volumes of diverse data without sacrificing individual privacy by using the collective intelligence of dispersed devices. An important benefit of federated learning in the context of smart cities is its capacity to use locally produced data in real-time. Federated learning algorithms may be used in a smart city setting, where sensors and IoT devices constantly provide data. These algorithms enable on-device model training, facilitating prompt decision-making and adaptability to changing urban circumstances. This decentralized strategy not only decreases the delay linked to transmitting data to centralized servers but also decreases the likelihood of network congestion and data bottlenecks. In addition, federated learning enables the provision of personalized services and customized experiences for inhabitants of smart cities. By conducting machine learning training directly on user devices, such as smartphones and wearable devices, it is possible to provide personalized suggestions and predictive analytics without compromising the security of sensitive personal data. For example, federated learning algorithms might examine individual mobility patterns to enhance transit routes or suggest customized energy-saving techniques based on home usage trends. Another notable use of federated learning in smart cities pertains to the realm of public safety and security. Federated learning models may use data from diverse sources such as security cameras, social media feeds, and IoT sensors to identify abnormalities, forecast areas with high crime rates, and expedite emergency

 DOI: 10.1201/9781003489368-10

response. Importantly, since the data are kept decentralized and encrypted, sensitive privacy information is protected throughout the analysis procedure, guaranteeing the confidence of citizens and adherence to privacy legislation. Furthermore, federated learning enhances the durability and adaptability of smart city infrastructure. Federated learning algorithms may minimize energy usage, alleviate traffic congestion, and lessen environmental hazards by allowing effective resource allocation and optimization. For example, traffic management systems that use federated learning may adaptively modify traffic signals by using up-to-date traffic flow information. This can lead to a decrease in fuel use and emissions, as well as an enhancement in overall mobility. Although federated learning shows promise, its application in smart cities poses some obstacles and concerns. These include challenges pertaining to variations in data, additional communication requirements, synchronization of models, and biases in algorithms. To tackle these difficulties, it is necessary for academics, policymakers, and industry stakeholders to collaborate across disciplines. This cooperation aims to create strong solutions that strike a balance between the advantages of data use, the need for privacy protection, and the efficiency of computing processes. To summarize, federated learning has great potential to shape the future of smart city infrastructure by allowing decentralized, privacy-preserving, and fast data processing. Federated learning uses the combined knowledge of dispersed edge devices to enable smart city stakeholders to get practical insights, provide customized services, improve public safety, and advance sustainability. However, fully harnessing the capabilities of federated learning in smart cities necessitates tackling a range of technological, legislative, and sociological obstacles to guarantee fair and comprehensive urban development.

In the rapidly evolving landscape of urbanization and technology, the concept of smart cities has emerged as a beacon of innovation, promising to revolutionize the way we live, work, and interact within urban environments. With populations increasingly concentrated in cities worldwide, the need for sustainable, efficient, and technologically advanced urban infrastructure has never been more pressing [1]. At the heart of this transformation lies the integration of cutting-edge technologies like federated learning, which holds immense potential in shaping the future trajectory of smart city infrastructure. Smart cities are envisioned as interconnected ecosystems where digital technologies converge to optimize various aspects of urban life, including transportation, energy management, public safety, healthcare, and environmental sustainability [2]. These cities leverage data-driven insights to enhance efficiency, improve service delivery, and foster economic development while addressing pressing urban challenges such as congestion, pollution, and resource depletion. However, realizing the full potential of smart cities requires overcoming significant hurdles, including the efficient utilization of vast amounts of heterogeneous data generated by diverse sources across the urban landscape [3]. Traditional approaches to data analysis in smart cities often rely on centralized data collection and processing, where data from various sources are aggregated into a central repository for analysis [4]. While centralized data platforms offer scalability and ease of management, they pose significant challenges in terms of data privacy, security, and scalability. Centralization raises concerns about data ownership, transparency, and accountability, as well as the risk of data breaches and unauthorized access [5]. Moreover, the sheer volume and velocity of data generated in urban environments can overwhelm centralized infrastructure, leading to latency issues, data bottlenecks, and increased operational costs [6]. In contrast, federated learning represents a paradigm shift in the way data are utilized and analyzed in smart cities. Rather than consolidating data in a central repository, federated learning enables collaborative model training across decentralized edge devices, such as smartphones, IoT sensors, and edge servers [7]. This distributed approach allows data to remain localized, preserving privacy and security while enabling collective intelligence to be harnessed for analysis. By leveraging the computational power of edge devices, federated learning enables real-time, on-device model training, reducing the need for data transmission to centralized servers and minimizing latency [8]. The core principle of federated learning lies in its ability to aggregate knowledge from diverse sources without compromising individual privacy. In a federated learning framework, each edge device trains a local machine learning model using its own data while periodically exchanging model updates with a central server or among peers [9]. These model updates are aggregated and averaged to produce a global model, which encapsulates the collective knowledge of all participating devices. Crucially, since data remain decentralized and encrypted during the training process, sensitive information is protected from unauthorized access, ensuring compliance with privacy regulations and user trust [10]. The application of federated learning in smart city infrastructure holds immense promise across various domains, including transportation, energy management, public safety, healthcare, and environmental monitoring. In transportation, federated learning algorithms can analyze traffic patterns, optimize route planning, and reduce congestion by leveraging real-time data from connected vehicles and roadside sensors [9]. Similarly, in energy management, federated learning can optimize energy consumption, improve grid stability, and facilitate the integration of renewable energy sources by analyzing consumption patterns and weather forecasts [11]. Public safety and security represent another critical area where federated learning can make a significant impact in smart cities. By aggregating data from surveillance cameras, social media feeds, and IoT sensors, federated learning models can detect anomalies, identify crime hotspots, and support rapid emergency response. Importantly, since data remain decentralized and encrypted, privacy-sensitive information such as facial recognition data or personal identifiers is safeguarded throughout the analysis process, ensuring compliance with privacy regulations and ethical considerations [12]. Furthermore, federated learning has the potential to democratize access to data and knowledge in smart cities by empowering local communities and stakeholders to participate in data-driven decision-making processes. By allowing data to remain localized and under the control of its owners, federated learning promotes data sovereignty, transparency, and accountability, while fostering innovation and collaboration across diverse stakeholders [13]. This decentralized approach not only enhances privacy and security but also promotes inclusivity and equity in urban development. Federated learning represents a transformative paradigm for data analysis in smart city infrastructure, offering a decentralized, privacy-preserving, and efficient approach to harnessing the collective intelligence of urban environments. By enabling collaborative model training across distributed edge devices, federated learning empowers smart city stakeholders to derive actionable insights, deliver personalized services, enhance public safety, and promote sustainability. However, realizing the full potential of federated learning in smart cities requires addressing various technical, regulatory, and societal challenges to ensure equitable and inclusive urban development in the digital age [14].

10.2 APPLICATIONS, ADVANTAGES, AND CHALLENGES OF FEDERATED LEARNING IN SMART CITIES

Federated learning, an advanced method in machine learning, has gained considerable interest in recent years because of its potential to transform other fields, such as smart city infrastructure. This section provides a comprehensive analysis of the current body of research on federated learning. The primary emphasis is on examining its many applications, advantages, difficulties, and the potential impact that it may have on the development of smart cities. Federated learning is a new and promising technique in machine learning that aims to tackle the issues of data privacy, scalability, and efficiency in many fields, such as smart city infrastructure. This section provides a comprehensive analysis of the current body of research on federated learning. The primary emphasis is on examining its many applications, advantages, difficulties, and the potential impact it may have on the development of smart cities [15].

10.2.1 Applications of Federated Learning in Smart Cities

Research has highlighted numerous applications of federated learning in smart city infrastructure. For instance, in transportation, federated learning algorithms can optimize traffic management systems by analyzing real-time data from connected vehicles and roadside sensors to reduce congestion and improve road safety. Similarly, in energy management, federated learning enables the optimization of energy consumption and grid stability by analyzing consumption patterns and integrating renewable energy sources. Moreover, federated learning has shown promise in enhancing public safety and security by detecting anomalies, predicting crime hotspots, and supporting rapid emergency response through the aggregation of data from surveillance cameras, social media feeds, and IoT sensors [16,17].

10.2.2 Benefits of Federated Learning for Smart Cities

Federated learning offers several advantages for smart city infrastructure. First, it enables decentralized data analysis, allowing data to remain localized and under the control of its owners, thereby preserving privacy and security. This decentralized approach promotes data sovereignty, transparency, and accountability while fostering innovation and collaboration across diverse stakeholders. Second, federated learning facilitates real-time, on-device model training, reducing the need for data transmission to centralized servers and minimizing latency. This enhances the responsiveness of smart city systems to dynamic urban conditions and improves overall efficiency. Additionally, federated learning enables personalized services and tailored experiences for smart city residents by analyzing individual data locally and delivering personalized recommendations without compromising privacy [18].

10.2.3 Challenges and Considerations in Federated Learning for Smart Cities

Despite its promise, federated learning poses several challenges and considerations for effective implementation in smart cities. One challenge is data heterogeneity, as data from different sources may vary in format, quality, and distribution, making it challenging to train accurate models. Communication overhead is another concern, as frequent model updates and synchronization between edge devices and central servers can impose significant network bandwidth and computational costs. Furthermore, ensuring fairness and mitigating algorithmic bias in federated learning models is crucial to prevent the propagation of discriminatory practices and ensure equitable outcomes for all members of society. Addressing these challenges requires interdisciplinary collaboration among researchers, policymakers, and industry stakeholders to develop robust solutions that balance the trade-offs between data utility, privacy preservation, and computational efficiency [19,20].

10.2.4 Implications for Shaping the Future of Smart Cities

Federated learning has profound implications for shaping the future of smart city infrastructure. By enabling decentralized, privacy-preserving, and efficient data analysis, federated learning empowers smart city stakeholders to derive actionable insights, deliver personalized services, enhance public safety, and promote sustainability. Moreover, federated learning has the potential to democratize access to data and knowledge in smart cities by fostering inclusivity and equity in urban development. However, realizing the full potential of federated learning in smart cities requires addressing various technical, regulatory, and societal challenges to ensure equitable and inclusive urban development in the digital age.

10.2.5 Applications of Federated Learning in Smart Cities

Federated learning offers a wide range of applications in smart city infrastructure, spanning transportation, energy management, public safety, healthcare, and environmental monitoring. For instance, in transportation, federated learning algorithms can optimize traffic flow and reduce congestion by analyzing real-time data from connected vehicles and roadside sensors. Similarly, in energy management, federated learning enables the optimization of energy consumption and grid stability by analyzing consumption patterns and integrating renewable energy sources. Moreover, federated learning has shown promise in enhancing public safety and security by detecting anomalies, predicting crime hotspots, and supporting rapid emergency response through the aggregation of data from surveillance cameras, social media feeds, and IoT sensors.

10.2.6 Benefits of Federated Learning for Smart Cities

Federated learning offers several benefits for smart city infrastructure. First, it enables decentralized data analysis, allowing data to remain localized and under the control of its owners, thereby preserving privacy and security. This decentralized approach promotes data sovereignty, transparency, and accountability while fostering innovation and collaboration across diverse stakeholders. Second, federated learning facilitates real-time, on-device model training, reducing the need for data transmission to centralized servers and minimizing latency. This enhances the responsiveness of smart city systems to dynamic urban conditions and improves overall efficiency. Additionally, federated learning enables personalized services and tailored experiences for smart city residents by analyzing individual data locally and delivering personalized recommendations without compromising privacy.

10.2.7 Challenges and Considerations in Federated Learning for Smart Cities

Despite its promise, federated learning poses several challenges and considerations for effective implementation in smart cities. One challenge is data heterogeneity, as data from different sources may vary in format, quality, and distribution, making it challenging to train accurate models. Communication overhead is another concern, as frequent model updates and synchronization between edge devices and central servers can impose significant network bandwidth and computational costs. Furthermore, ensuring fairness and mitigating algorithmic bias in federated learning models is crucial to prevent the propagation of discriminatory practices and ensure equitable outcomes for all members of society. Addressing these challenges requires interdisciplinary collaboration among researchers, policymakers, and industry stakeholders to develop robust solutions that balance the trade-offs between data utility, privacy preservation, and computational efficiency [6,21].

10.2.8 Implications for Shaping the Future of Smart Cities

Federated learning has profound implications for shaping the future of smart city infrastructure. By enabling decentralized, privacy-preserving, and efficient data analysis, federated learning empowers smart city stakeholders to derive actionable insights, deliver personalized services, enhance public safety, and promote sustainability. Moreover, federated learning has the potential to democratize access to data and knowledge in smart cities by fostering inclusivity and equity in urban development. However, realizing the full potential of federated learning in smart cities requires addressing various technical, regulatory, and societal challenges to ensure equitable and inclusive urban development in the digital age.

10.2.9 Privacy-Preserving Techniques in Federated Learning

Preserving privacy is a crucial consideration in federated learning, particularly in smart city applications that contain sensitive data. Several methods aimed at protecting privacy have been suggested to tackle this issue. Differential privacy is a strategy that adds noise or randomization to individual data samples in order to prevent the model training process from disclosing sensitive information about any particular user. Another method is homomorphic encryption, which enables calculations to be carried out on encrypted data without the need for decryption, maintaining anonymity throughout the whole computing process. In addition, federated learning systems often use federated averaging, a method that combines model updates in a way that preserves privacy, thereby safeguarding individual data [22].

10.2.10 Scalability and Efficiency in Federated Learning

Scalability and efficiency are key considerations in federated learning, particularly in the context of smart cities where large-scale datasets and complex models are involved. To address scalability challenges, researchers have explored techniques such as model compression and quantization, which reduce the size of model updates exchanged between edge devices and central servers, thereby minimizing communication overhead. Additionally, advancements in federated optimization algorithms, such as federated averaging and federated proximal algorithms, have contributed to improving the efficiency of federated learning systems by reducing the number of communication rounds required for convergence [23]. Furthermore, the integration of edge computing capabilities into federated learning frameworks enables computation to be performed closer to data sources, reducing latency and enhancing overall system efficiency.

10.2.11 Regulatory and Ethical Considerations in Federated Learning

The implementation of federated learning in smart city settings necessitates the careful assessment and resolution of regulatory and ethical concerns to guarantee responsible and fair deployment. Data privacy regulatory frameworks, such as the General Data Protection Regulation (GDPR) in Europe and the California Consumer Privacy Act (CCPA) in the United States, enforce stringent rules on the gathering, retention, and manipulation of personal data. As a result, compliance measures are necessary in federated learning systems. Furthermore, ethical factors like as openness, fairness, and accountability are of utmost importance in order to reduce the hazards linked to algorithmic prejudice and discrimination. It is crucial to follow ethical norms and standards, as those specified in the IEEE Global Initiative on Ethics of Autonomous and Intelligent Systems, in order to build confidence and encourage responsible innovation in federated learning applications [24].

10.2.12 Future Directions and Research Challenges

Despite significant progress, federated learning for smart cities remains an active area of research with several open challenges and opportunities for future exploration. One promising direction is the development of federated learning frameworks that can handle non-IID (non-identically distributed) data distributions and heterogeneous device capabilities, which are common in real-world smart city environments. Additionally, enhancing the robustness and security of federated learning systems against adversarial attacks and model poisoning remains a critical research area. Moreover, interdisciplinary collaboration between researchers, policymakers, and industry stakeholders is essential to address the societal and ethical implications of federated learning and ensure its responsible and equitable deployment in smart city ecosystems [25,26].

In summary, the literature on federated learning highlights its multifaceted applications, benefits, challenges, and implications for smart city infrastructure. By addressing privacy concerns, scalability challenges, regulatory considerations, and future research directions, federated learning offers a promising avenue for shaping the future of smart cities in a responsible, efficient, and inclusive manner. However, continued research, collaboration, and innovation are essential to realize the full potential of federated learning and harness its transformative power for urban development.

10.3 CHALLENGES

While federated learning holds promise for revolutionizing smart city infrastructure, several challenges must be addressed to realize its full potential. These challenges span technical, regulatory, and societal domains, presenting obstacles to the widespread adoption and effective implementation of federated learning in smart cities.

10.3.1 Data Heterogeneity

Smart cities generate vast amounts of heterogeneous data from diverse sources, including sensors, IoT devices, social media feeds, and public records. These data often vary in format, quality, and distribution, making it challenging to train accurate and robust machine learning models. Addressing data heterogeneity requires developing federated learning algorithms capable of accommodating diverse data types and distributions while maintaining model performance across decentralized edge devices.

10.3.2 Communication Overhead

Federated learning involves frequent communication between edge devices and central servers to exchange model updates and synchronize learning parameters. This communication overhead can impose significant bandwidth and computational costs, particularly in large-scale smart city deployments with a high number of edge devices. Mitigating communication overhead requires optimizing communication protocols, reducing the size of model updates, and leveraging edge computing capabilities to perform computation closer to data sources [27].

10.3.3 Model Synchronization

Ensuring synchronization and consistency across distributed edge devices poses a challenge in federated learning systems. As edge devices operate in dynamic and heterogeneous environments, they may experience variations in network connectivity, computational resources, and data availability. These variations can lead to delays in model synchronization and divergence in learning trajectories, affecting the overall performance and convergence of federated learning algorithms. Developing robust mechanisms for model synchronization and coordination is essential to maintain the integrity and consistency of federated learning models in smart city environments [28].

10.3.4 Algorithmic Bias and Fairness

Federated learning models trained on decentralized data sources may exhibit algorithmic bias, leading to unfair or discriminatory outcomes, particularly for underrepresented groups in smart city populations. Biases in training data, sampling methods, or model architectures can propagate through federated learning systems, exacerbating existing disparities and inequities in urban environments. Addressing algorithmic bias and ensuring fairness in federated learning requires adopting bias mitigation techniques, incorporating fairness-aware algorithms, and promoting diversity and inclusivity in data collection and model development processes.

10.3.5 Privacy and Security Concerns

Preserving privacy and security is paramount in federated learning, especially when dealing with sensitive personal data in smart city applications. Centralized data repositories pose risks of data breaches, unauthorized access, and privacy violations, undermining citizen trust and compliance with data protection regulations. Federated learning mitigates these risks by keeping data localized and encrypted during the training process, but challenges remain in ensuring robust encryption schemes, access control mechanisms, and data anonymization techniques to safeguard privacy and prevent malicious attacks [29].

10.3.6 Regulatory Compliance

Federated learning in smart cities must follow data privacy, security, and ethics laws. Federated learning systems must comply with GDPR, CCPA, and sector-specific regulations, including data governance, consent management, and accountability. Policymakers, business stakeholders, and researchers must define standards, guidelines, and best practices for ethical and responsible federated learning in smart city infrastructure to meet legal requirements.

10.3.7 Resource Constraints

Edge devices in smart city environments often have limited computational resources, storage capacity, and battery life, posing challenges for federated learning implementation. These resource constraints can impact the efficiency and scalability of federated learning algorithms, hindering model training and inference tasks on edge devices. Moreover, edge devices may prioritize local tasks over federated learning computations, leading to inconsistencies in participation and data contribution. Addressing resource constraints requires optimizing model architectures, implementing lightweight algorithms, and prioritizing computational tasks based on-device capabilities and energy constraints [30].

10.3.8 Data Labeling and Annotation

Annotated data are essential for training supervised machine learning models in federated learning systems. However, data labeling and annotation processes can be resource-intensive, time-consuming, and costly, particularly in smart city environments where data may be unstructured or noisy. Moreover, labeling data for diverse applications, such as transportation, energy management, and public safety, requires domain expertise and contextual understanding. Developing efficient and scalable data labeling strategies, leveraging semi-supervised or unsupervised learning approaches, and incorporating human-in-the-loop annotation mechanisms are crucial for addressing data labeling challenges in federated learning for smart cities [31].

10.3.9 Interoperability and Standards

Interoperability and standards play a crucial role in enabling seamless integration and interoperability of federated learning systems with existing smart city infrastructure and technologies. However, disparate standards, protocols, and data formats across different domains and vendors can hinder interoperability and data exchange between federated learning platforms and legacy systems. Developing interoperability standards, data exchange protocols, and open-source frameworks for federated learning is essential to facilitate collaboration, data sharing, and innovation in smart city ecosystems. Moreover, ensuring compatibility with emerging technologies, such as 5G networks and edge computing platforms, is vital for enabling federated learning applications in smart cities.

10.3.10 Education and Awareness

Building awareness and capacity among stakeholders is crucial for the successful adoption and deployment of federated learning in smart cities. Many stakeholders, including city officials, policymakers, developers, and citizens, may lack awareness of federated learning concepts, benefits, and implications. Moreover, misconceptions or skepticism about the feasibility, effectiveness, and trustworthiness of federated learning may impede its adoption and acceptance. Investing in education, training, and outreach programs to raise awareness, build trust, and promote understanding of federated learning principles, applications, and best practices is essential for fostering a culture of data-driven innovation and collaboration in smart city environments.

10.3.11 Ethical and Social Implications

Federated learning raises ethical and social implications related to data governance, transparency, accountability, and societal impact. Concerns about data ownership, consent, and control may arise when individuals contribute data to federated learning systems without full awareness or understanding of how their data are used. Moreover, the potential for unintended consequences, such as reinforcing biases or perpetuating inequalities, requires careful consideration and mitigation strategies. Engaging stakeholders in ethical discussions, establishing governance frameworks, and conducting impact assessments are essential for addressing ethical and social implications and ensuring the responsible deployment of federated learning in smart city infrastructure.

In conclusion, addressing the challenges associated with federated learning is essential to unlock its transformative potential in shaping the future of smart city infrastructure. By overcoming barriers related to data heterogeneity, communication overhead, model synchronization, algorithmic bias, privacy, security, and regulatory compliance, federated learning can empower smart city stakeholders to derive actionable insights, deliver personalized services, enhance public safety, and promote sustainability in urban environments. Collaboration between researchers, policymakers, industry stakeholders, and communities is critical to developing innovative solutions and fostering equitable and inclusive urban development through federated learning.

10.4 PROPOSED WORK

In this section, we outline our proposed approach for leveraging federated learning to shape the future of smart city infrastructure. Our work aims to address key challenges and capitalize on opportunities to enhance the resilience, sustainability, and inclusivity of urban environments through decentralized, privacy-preserving, and efficient data analysis.

10.4.1 Data Collection and Preparation

Our proposed work begins with the collection and preparation of heterogeneous data sources from smart city environments. This includes data from sensors, IoT devices, social media feeds, and public records, covering various domains such as transportation, energy management, public safety, healthcare, and environmental monitoring. We develop data pipelines and preprocessing techniques to clean, normalize, and aggregate data, ensuring consistency and quality for subsequent analysis.

10.4.2 Federated Learning Framework Design

Next, we design a federated learning framework tailored to the unique characteristics and requirements of smart city infrastructure. Our framework incorporates privacy-preserving techniques, communication optimization strategies, and model synchronization mechanisms to address challenges related to data heterogeneity, communication overhead, and resource constraints. We leverage federated optimization algorithms, such as federated averaging and federated proximal algorithms, to improve efficiency and scalability while maintaining privacy and security.

10.4.3 Model Development and Training

We develop machine learning models tailored to specific smart city applications, including transportation optimization, energy consumption forecasting, public safety prediction, and environmental monitoring. These models leverage federated learning techniques to train on decentralized edge devices while preserving data privacy and security. We explore techniques for model personalization, fairness, and robustness to ensure equitable and reliable performance across diverse urban populations and use cases.

10.4.4 Evaluation and Validation

We conduct rigorous evaluation and validation of our federated learning framework and models using real-world datasets and simulation environments. We assess the accuracy, efficiency, and privacy-preserving capabilities of our approach compared to traditional centralized methods. We measure performance metrics such as model accuracy, convergence speed, communication overhead, and computational resources consumed to quantify the effectiveness and scalability of our proposed solution.

10.4.5 Deployment and Integration

Upon successful validation, we deploy and integrate our federated learning framework into existing smart city infrastructure and applications. This involves collaborating with city officials, policymakers, industry stakeholders, and community members to ensure seamless integration, interoperability, and compatibility with legacy systems and technologies. We provide training, support, and documentation to facilitate the adoption and usage of federated learning techniques among stakeholders.

10.4.6 Monitoring and Adaptation

We establish mechanisms for monitoring and adaptation to ensure the ongoing performance and relevance of our federated learning solution in dynamic smart city environments. This includes monitoring data quality, model performance, and user feedback to identify potential issues and opportunities for improvement. We employ techniques such as model retraining, adaptation to concept drift, and feedback loops to continuously optimize and refine our federated learning framework over time.

10.4.7 Ethical and Regulatory Compliance

Throughout the proposed work, we prioritize ethical and regulatory compliance to uphold privacy, fairness, transparency, and accountability in our federated learning approach. We adhere to data protection regulations such as GDPR and CCPA, implement privacy-preserving techniques such as differential privacy and federated encryption, and incorporate fairness-aware algorithms and bias mitigation strategies to mitigate ethical concerns and ensure equitable outcomes for all members of smart city populations.

10.4.8 Community Engagement and Collaboration

Community engagement and collaboration are integral to the success of our proposed work. We actively involve city residents, community organizations, and stakeholders in the development, implementation, and evaluation of our federated learning solution. We conduct outreach activities, workshops, and participatory design sessions to solicit input, gather feedback, and co-create solutions that address local needs and priorities. By fostering a sense of ownership, trust, and empowerment among community members, we aim to ensure the relevance, acceptance, and sustainability of our federated learning approach in smart city environments.

10.4.9 Capacity Building and Knowledge Sharing

We prioritize capacity building and knowledge sharing as part of our proposed work. We offer training programs, educational resources, and skill-building initiatives to empower city officials, policymakers, developers, and citizens with the knowledge and expertise needed to understand, adopt, and leverage federated learning techniques effectively. We organize seminars, webinars, and conferences to disseminate best practices, lessons learned, and case studies from our federated learning projects, fostering a culture of learning, collaboration, and innovation in smart city ecosystems.

10.4.10 Long-Term Sustainability and Scalability

Long-term sustainability and scalability are key considerations in our proposed work. We develop strategies and business models to ensure the financial viability and scalability of our federated learning solution beyond the initial deployment phase. This includes exploring funding opportunities, partnerships with industry stakeholders, and revenue-generation mechanisms to support ongoing maintenance, updates, and expansion of the federated learning infrastructure. Moreover, we design our solution with scalability in mind, considering factors such as data volume, user growth, and technological advancements to accommodate future needs and requirements of smart city environments.

10.4.11 Evaluation Metrics and Performance Indicators

We define clear evaluation metrics and performance indicators to assess the impact and effectiveness of our proposed work. These metrics include quantitative measures such as model accuracy, convergence speed, communication overhead, and computational resources consumed, as well as qualitative indicators such as user satisfaction, trust, and perceived value. By systematically measuring and evaluating the performance of our federated learning solution against predefined metrics and benchmarks, we ensure accountability, transparency, and continuous improvement throughout the project lifecycle.

10.4.12 Adaptation to Emerging Technologies and Trends

We remain vigilant and adaptable to emerging technologies and trends that may influence the trajectory of smart city infrastructure and federated learning. This includes monitoring advancements in areas such as edge computing, 5G networks, artificial intelligence, and blockchain technology to identify opportunities for synergies, integration, and innovation. By staying abreast of technological developments and evolving user needs, we can anticipate challenges, seize opportunities, and position our federated learning solution at the forefront of smart city innovation and transformation.

In conclusion, our proposed work encompasses a holistic approach to leveraging federated learning to shape the future of smart city infrastructure. By embracing community engagement, capacity building, long-term sustainability, and adaptation to emerging trends, we aim to create scalable, inclusive, and impactful solutions that empower cities to harness the collective intelligence of distributed data sources, drive innovation, and improve the quality of life for residents. Through collaboration, evaluation, and ongoing adaptation, we aspire to create a blueprint for resilient and thriving cities of the future, where federated learning serves as a catalyst for positive change and equitable urban development.

It aims to harness the transformative potential of federated learning to shape the future of smart city infrastructure. By addressing key challenges and capitalizing on opportunities, we seek to empower cities to leverage decentralized, privacy-preserving, and efficient data analysis to address urban challenges, drive innovation, and improve the quality of life for residents. Through collaboration, validation, and ongoing adaptation, we aim to create scalable, sustainable, and inclusive solutions that pave the way for resilient and thriving cities of the future.

10.5 WORK FLOW DIAGRAM

This diagram outlines the flow of data and processes within a smart city infrastructure utilizing federated learning. Data is collected from various sources such as sensors and IoT devices, processed and analyzed, and then used for model training and evaluation. Federated learning techniques ensure the privacy and security of the data while allowing for collaborative model training across devices. Finally, the trained global model is deployed back to the edge devices for real-time inference and decision-making within the smart city environment.

10.5.1 Data Collection

In smart city infrastructure, data are collected from various sources such as sensors, IoT devices, cameras, weather stations, etc. These data encompass a wide range of information including environmental data, traffic patterns, energy usage, and more.

10.5.2 Data Processing and Analysis

The gathered data are subjected to processing and analysis in order to derive significant insights. This stage often entails the use of edge computing, which refers to the processing of data in close proximity to its source in order to minimize delays and optimize bandwidth consumption. Cloud infrastructure may also be used for more demanding analytical jobs that are not feasible to be carried out at the edge.

10.5.3 Model Training and Evaluation

During the process of federated learning, this stage entails the training of machine learning models utilizing data that is decentralized. Federated learning techniques enable the training of models on several edge devices without the need to consolidate the data in a central location. Privacy-preserving methods, such as differential privacy, may be used to guarantee data privacy when training models.

10.5.4 Global Model

The models that have been trained on different edge devices are combined to form a global model. This global model represents the combined knowledge acquired from all participating devices while safeguarding the privacy of individual data.

10.5.5 Model Deployment

The global model is deployed back to the edge devices or IoT devices within the smart city infrastructure. This deployment ensures that the trained model is available for real-time inference and decision-making.

10.5.6 Real-Time Inference

With the deployed model, edge devices can make real-time decisions based on incoming data. These decisions could include optimizing traffic flow, controlling energy usage, detecting anomalies, and more. Real-time inference allows for immediate responses to events and situations within the smart city environment.

Overall, federated learning enables collaborative model training across distributed devices in a privacy-preserving manner, thereby shaping the future of smart city infrastructure by allowing for efficient data utilization and real-time decision-making while maintaining data privacy and security (Figure 10.1).

10.6 CONCLUSION

Federated learning presents a groundbreaking opportunity to shape the future of smart city infrastructure, offering a decentralized, privacy-preserving, and efficient approach to data analysis and machine learning. Through our exploration of federated learning applications, benefits, challenges, proposed work, and implications, it is evident that this technology holds immense promise for revolutionizing urban environments. As we conclude our examination, several key insights emerge. Firstly, federated learning offers a versatile solution for addressing the complex and diverse challenges faced by modern cities. From optimizing transportation systems and energy management to enhancing public safety and healthcare, federated learning enables cities to leverage distributed data sources and collaborative machine learning techniques to derive actionable insights and drive innovation. By decentralizing data analysis and preserving privacy, federated learning empowers cities to harness the collective intelligence of their citizens and infrastructure, paving the way for

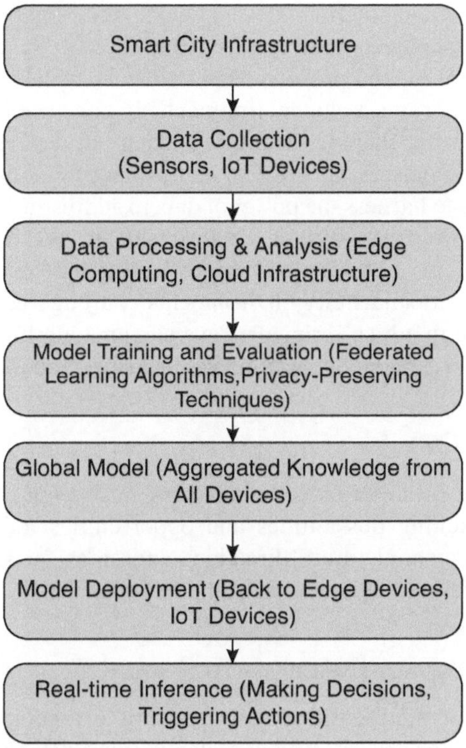

FIGURE 10.1 Workflow diagram.

more responsive, sustainable, and inclusive urban development. Secondly, while federated learning offers numerous benefits, it also poses significant challenges that must be addressed to realize its full potential. Data heterogeneity, communication overhead, model synchronization, algorithmic bias, privacy concerns, and regulatory compliance are among the key challenges that require careful consideration and mitigation strategies. By developing robust frameworks, algorithms, and governance mechanisms, cities can overcome these challenges and harness the transformative power of federated learning while safeguarding privacy, fairness, and transparency. Thirdly, our proposed work outlines a comprehensive approach to leveraging federated learning in smart city infrastructure. By focusing on data collection, framework design, model development, evaluation, deployment, and adaptation, we aim to create scalable, sustainable, and inclusive solutions that empower cities to address urban challenges and improve the quality of life for residents. Through community engagement, capacity building, and collaboration, we seek to foster a culture of innovation and cooperation that enables cities to leverage federated learning effectively and responsibly. Moreover, our proposed work emphasizes the importance of ethical considerations, long-term sustainability, and adaptation to emerging trends in shaping the future of smart city infrastructure. By prioritizing privacy, fairness, and accountability, we aim to ensure that federated learning benefits all members of society and promotes equitable urban development. Additionally, by staying abreast of technological advancements and evolving user needs, we can anticipate challenges, seize opportunities, and position federated learning as a catalyst for positive change and innovation in smart city environments. In conclusion, federated learning holds immense promise for shaping the future of smart city infrastructure. By decentralizing data analysis, preserving privacy, and promoting collaboration, federated learning enables cities to harness the power of data to address urban challenges, drive innovation, and improve the quality of life for residents. While challenges remain, our proposed work offers a pathway forward for leveraging federated learning effectively and responsibly in smart city environments. Through collaboration, innovation, and commitment to ethical principles, we can create a future where federated learning serves as a cornerstone of sustainable, inclusive, and thriving cities.

10.7 FUTURE SCOPE

As we look ahead, the future of federated learning in shaping smart city infrastructure is filled with exciting possibilities and opportunities for innovation. In this section, we explore potential future directions and areas for further research and development in the realm of federated learning for smart cities.

10.7.1 Advanced Privacy-Preserving Techniques

Future research in federated learning may focus on advancing privacy-preserving techniques to enhance the security and confidentiality of data in smart city environments. This could involve exploring novel encryption schemes, differential privacy mechanisms, and federated learning algorithms that provide stronger guarantees of privacy while maintaining model accuracy and utility. Additionally, research efforts may delve into developing federated learning frameworks that support multi-party computation and secure aggregation protocols to further safeguard sensitive data from unauthorized access or inference.

10.7.2 Edge Computing Integration

The integration of federated learning with edge computing technologies holds significant promise for enhancing the efficiency, scalability, and responsiveness of smart city infrastructure. Future research may explore ways to optimize model training and inference tasks on edge devices, leveraging their computational capabilities and proximity to data sources. This could involve developing lightweight federated learning algorithms, edge-aware model architectures, and dynamic resource allocation strategies to enable efficient and real-time data analysis at the network edge.

10.7.3 Dynamic Federation and Resource Management

Future research efforts may focus on developing dynamic federation and resource management strategies to adaptively allocate computational resources and data sources in federated learning systems. This could involve leveraging machine learning and optimization techniques to dynamically form federated learning cohorts based on data availability, device capabilities, and user preferences. Additionally, research may explore adaptive learning rate scheduling, model pruning, and federated ensemble methods to optimize resource utilization and improve convergence speed in federated learning scenarios with varying network conditions and device heterogeneity.

10.7.4 Interoperability and Standards Development

The establishment of interoperability standards and frameworks is essential for enabling seamless integration and collaboration between federated learning systems and existing smart city infrastructure. Future research may focus on developing open-source federated learning platforms, data exchange protocols, and interoperability frameworks that facilitate data sharing, model transfer, and collaboration among different cities, organizations, and stakeholders. This could involve standardizing data formats, communication protocols, and model serialization techniques to promote interoperability and scalability of federated learning solutions across diverse smart city ecosystems.

10.7.5 Ethical and Regulatory Guidelines

Future efforts in federated learning for smart cities may involve the development of ethical and regulatory guidelines to ensure responsible and equitable deployment of federated learning systems. This could include establishing principles for data governance, consent management, and algorithmic transparency to safeguard privacy, fairness, and accountability in federated learning applications. Moreover, research may focus on developing mechanisms for auditing, certification, and compliance verification to ensure that federated learning systems adhere to legal and ethical standards, such as GDPR, CCPA, and IEEE Ethically Aligned Design principles.

10.7.6 Citizen-Centric Federated Learning

Emphasizing citizen-centric approaches to federated learning is crucial for ensuring that smart city infrastructure serves the needs and preferences of residents while respecting their privacy and autonomy. Future research may explore participatory design methodologies, user-centric evaluation frameworks, and citizen engagement strategies to involve residents in the co-creation and validation of federated learning applications. This could involve developing user-friendly interfaces, personalized recommendation systems, and feedback mechanisms that empower citizens to actively contribute to and benefit from federated learning initiatives in their communities.

10.7.7 Cross-Domain Collaboration and Knowledge Sharing

Collaboration and knowledge sharing across different domains, disciplines, and stakeholders are essential for driving innovation and fostering interdisciplinary research in federated learning for smart cities. Future initiatives may involve establishing collaborative research networks, organizing interdisciplinary workshops, and facilitating knowledge exchange platforms that bring together researchers, practitioners, policymakers, and community members to share insights, best practices, and lessons learned from federated learning projects in diverse smart city contexts. This could lead to cross-pollination of ideas, accelerated innovation, and collective impact in shaping the future of smart city infrastructure through federated learning.

In conclusion, the future of federated learning in shaping smart city infrastructure is filled with opportunities for advancing privacy, efficiency, interoperability, ethics, citizen engagement, and collaboration. By embracing these future directions and addressing key research challenges, we can unlock the full potential of federated learning to create more resilient, sustainable, and inclusive cities for future generations. Through continued innovation, collaboration, and commitment to ethical principles, federated learning has the potential to revolutionize how we design, manage, and interact with smart city infrastructure, paving the way for a brighter and more equitable urban future.

REFERENCES

1. Li, Q., Zomaya, A. Y., & Zheng, W. (2023). Federated learning for mobile edge networks: A comprehensive survey. *IEEE Communications Surveys & Tutorials*, *22*(3), 2031–2063.
2. Konečný, J., McMahan, B., Yu, F. X., Richtárik, P., Suresh, A. T., & Bacon, D. (2023). Federated learning: Strategies for improving communication efficiency. *arXiv preprint arXiv:1610.05492.*
3. Bonawitz, K., Eichner, H., Grieskamp, W., Huba, D., Ingerman, A., Ivanov, V., Kiddon, C, Konečný, J., Mazzocchi, S., McMahan, H. B., Van Overveldt, T., Petrou, D., Ramage, D, & Roselander, J. (2019). Towards federated learning at scale: system design. arXiv preprint arXiv:1902.01046.
4. Verma, R. K. & Kishor, K. (2024). Image processing applications in agriculture with the help of AI. In: Khan M., Khan R., Praveen P., Verma A., & Panda M. (eds), *Infrastructure Possibilities and Human-Centered Approaches With Industry 5.0* (pp. 162–181). IGI Global, USA. https://doi.org/10.4018/979-8-3693-0782-3.ch010
5. McMahan, H. B., Moore, E., Ramage, D., Hampson, S., & Agüera y Arcas, B. (2022). Communication-efficient learning of deep networks from decentralized data. In: *Proceedings of the 20th International Conference on Artificial Intelligence and Statistics* (pp. 1273–1282). PMLR. , Google, Seattle, WA.
6. Kishor, K. (2023). Chapter 17 Application of quantum computing for digital forensic investigation. In: Yadav S. P., Singh R., Yadav V., Al-Turjman F., Kumar S. A. (eds) *Quantum-Safe Cryptography Algorithms and Approaches: Impacts of Quantum Computing on Cybersecurity* (pp. 231–248). De Gruyter, Berlin, Boston. https://doi.org/10.1515/9783110798159-017
7. Zhao, Y., Lu, H., Wu, Q., Li, Z., & Chen, H. H. (2022). Privacy-preserving federated learning: A comprehensive survey. *Future Generation Computer Systems*, *115*, 507–525.
8. Kishor K. (2022). Personalized federated learning. In: Yadav S. P., Bhati B. S., Mahato D. P., Kumar S. (eds) *Federated Learning for IoT Applications*. EAI/Springer Innovations in Communication and Computing. Springer, Cham. https://doi.org/10.1007/978-3-030-85559-8_3
9. Yang, Q., Liu, Y., Chen, T., & Tong, Y. (2022). Federated machine learning: Concept and applications. *ACM Transactions on Intelligent Systems and Technology (TIST)*, *10*(2), 1–19.

10. Kishor K., Sharma R., & Chhabra M. (2022) Student performance prediction using technology of machine learning. In: Sharma D. K., Peng S. L., Sharma R., Zaitsev D. A. (eds) *Micro-Electronics and Telecommunication Engineering.* Lecture Notes in Networks and Systems, vol. 373. Springer, Singapore. https://doi.org/10.1007/978-981-16-8721-1_53
11. Kishor, K., Nand, P., & Agarwal, P. (2017). Subnet based ad hoc network algorithm reducing energy consumption in MANET. *International Journal of Applied Engineering Research, 12*(22), 11796–11802.
12. Sheller, M. J., Konečný, J., McMahan, H. B., & Shroff, R. N. (2022). Federated learning in heterogeneous networks. *arXiv preprint arXiv:1912.04977.*
13. Kishor, K. (2022) Communication-efficient federated learning. In: Yadav S. P., Bhati B. S., Mahato D. P., Kumar S. (eds) *Federated Learning for IoT Applications.* EAI/Springer Innovations in Communication and Computing. Springer, Cham. https://doi.org/10.1007/978-3-030-85559-8_9
14. Wang, S., Tuor, T., Salonidis, T., Leung, K. K., Makaya, C., He, T., & Chan, K. K. (2022). Adaptive federated learning in resource constrained edge computing systems. *IEEE Journal on Selected Areas in Communications, 37*(6), 1205–1221.
15. Li, X., Huang, K., Yang, W., Liu, W., & Zhang, Z. (2021). Federated learning on non-IID data silos: A survey. *arXiv preprint arXiv:2103.10956.*
16. Zhou, Y., Huang, C., Lai, K. M., Liu, Q., & Yang, K. (2021). FADL: Fairness-aware decentralized federated learning. *arXiv preprint arXiv:2104.07349.*
17. Kishor, K., & Pandey D. (2022). Study and development of efficient air quality prediction system embedded with machine learning and IoT. In: Gupta D., Khanna A., Bhattacharyya S., Hassanien A. E., Anand S., Jaiswal A. (eds) *Proceeding International Conference on Innovative Computing and Communications.* Lecture Notes in Networks and Systems, vol. 471, Springer, Singapore. https://doi.org/10.1007/978-981-19-2535-1_24
18. Caruso, M. C., Russello, G., & Tuyls, K. (2019). On the design of a privacy-preserving platform for incentivising data sharing. *arXiv preprint arXiv:1902.08054.*
19. Ju, Chunhua, Qiuyang Gu, Gongxing Wu, and Shuangzhu Zhang. 2020. Local Differential Privacy Protection of High-Dimensional Perceptual Data by the Refined Bayes Network. Sensors 20(9), 2516. https://doi.org/10.3390/s20092516
20. Kishor, K., Saxena, N., & Pandey D. (2023). *Cloud-Based Intelligent Informative Engineering for Society 5.0,* 1st edition (pp. 1–234). Chapman and Hall/CRC, New York, ISBN: 9781003213895. https://doi.org/10.1201/9781003213895
21. Pham, H., Pagh, A., & Mai, Q. V. (2021). Federated learning: Challenges, methods, and future directions. *arXiv preprint arXiv:2104.11839.*
22. Dandil, B., & Lim, H. (2021). Securing federated learning systems: A survey. *arXiv preprint arXiv:2102.07689.*
23. Kishor, K., Sangal, S., & Shahroz (2023). Real-time traffic signs and lane line detection. In: Swaroop A., Kansal V., Fortino G., Hassanien A. E. (eds) *Proceedings of Fourth Doctoral Symposium on Computational Intelligence. DoSCI 2023.* Lecture Notes in Networks and Systems, vol. 726. Springer, Singapore. https://doi.org/10.1007/978-981-99-3716-5_71
24. Wang, X., Han, H., Chen, X., Jiang, C., Tian, Q., & Guo, B. (2021). OpenFL: An open-source framework for federated learning benchmark and development. *arXiv preprint arXiv:2104.11745.*
25. Kishor, K. (2023). Cloud computing in blockchain. In: Kishor K., Saxena, N., Pandey, D. (eds), *Cloud-based Intelligent Informative Engineering for Society 5.0,* 1st edition (pp. 79–105). Chapman and Hall/CRC, New York. ISBN: 9781003213895. https://doi.org/10.1201/9781003213895-5

26. Kishor, K., Nand, P., & Agarwal, P. (2018). Notice of retraction design adaptive subnetting hybrid gateway MANET protocol on the basis of dynamic TTL value adjustment. *Aptikom Journal on Computer Science and Information Technologies*, *3*(2), 59–65. https://doi.org/10.11591/APTIKOM.J.CSIT.115
27. Kishor, K., & Nand, P. (2023). "Wireless networks based in the cloud that support 5G." In: Kishor K., Saxena, N., Pandey, D. (eds), *Cloud-based Intelligent Informative Engineering for Society 5.0* , 1st edition (pp. 23–40). Chapman and Hall/CRC, New York. ISBN: 9781003213895. DOI: https://doi.org/10.1201/9781003213895-2
28. Jaggi, M., Smith, V., Takác, M., Terhorst, J., Krishnan, S., Hoffman, Michael I. Jordan & Uszkoreit, J. (2021). Communication-efficient distributed optimization using an approximate Newton-type method. *arXiv preprint arXiv:2102.06026.*
29. Kishor, K. (2023). Impact of cloud computing on entrepreneurship, cost, and security. In: Kishor K., Saxena, N., Pandey, D. (eds), *Cloud-based Intelligent Informative Engineering for Society 5.0* , 1st edition (pp. 171–191). CRC Press, New York. ISBN: 9781003213895. https://doi.org/10.1201/9781003213895-10
30. Smith, V., Chiang, K. N., Sanjabi, M., & Talwalkar, A. (2017). Federated multi-task learning. In: *Advances in Neural Information Processing Systems* (pp. 4425–4435). https://doi.org/10.48550/arXiv.1705.10467
31. Li, T., Sahu, A. K., Zaheer, M., Sanjabi, M., Talwalkar, A., & Smith, V. (2019). Federated optimization in heterogeneous networks. *arXiv preprint arXiv:1902.00146.*

11 Empowering Teaching Institutes

Integrating Federated Learning in the Internet of Things (IoT)

Debashree Chakravarty, Ipseeta Satpathy, Vishal Jain, B.C.M. Patnaik, and Sandeep Poddar

11.1 INTRODUCTION

The advent of the Internet of Things (IoT) has come up with new vistas in the sphere of education, offering unprecedented opportunities to reshape teaching methodologies and create dynamic, efficient, and privacy-respecting learning environments within teaching institutes. In recent times, the IoT has emerged to be the most transformative and adaptive technology apart from many. This paradigm shift in connectivity allows machines to execute complex tasks autonomously, revolutionizing various industries by creating a seamless network of communication between devices and the cloud. IoT essentially a collective network, thrives on the interconnectivity of devices and technologies, paving the way for innovative applications across diverse fields such as healthcare, agriculture, manufacturing, smart cities, and beyond.

As the world transitions from ancient methods to the IoT, teaching institutes have undergone a profound transformation. IoT in education implies to the interlinked network of devices and systems that enhance the teaching and learning experience. Smart classrooms equipped with interactive boards, connected devices, and data-driven analytics are becoming increasingly prevalent, promising a more immersive and engaging educational journey. IoT applicability in education spans various dimensions. In smart classrooms, IoT facilitates real-time communication between students and teachers, promotes interactive learning experiences, and offers personalized content based on individual progress. Wearable devices and smart learning applications further contribute to a more dynamic and personalized education system.

Integrating federated learning (FL) further enhances the educational landscape. FL, a decentralized machine learning approach, complements IoT in preserving privacy. It allows models to be trained across decentralized devices without exchanging raw data, addressing concerns related to centralized models. In the context of

DOI: 10.1201/9781003489368-11

education, FL enables the creation of personalized learning paths for students and provides real-time insights for educators without compromising sensitive information.

The proliferation of IoT devices in teaching institutes brings forth security and privacy challenges. Protecting sensitive student data, ensuring secure communication channels, and implementing stringent access controls become paramount. Educators and policymakers must work collaboratively to establish ethical guidelines and standards that safeguard user privacy while harnessing the benefits of IoT in teaching institutions.

The integration of IoT in education signifies a paradigm shift in the way we teach and learn. As technology continues to shape the educational landscape, understanding the implications, addressing challenges, and harnessing the potential of IoT in teaching institutes are essential endeavors. Keeping the right balance between innovation in technological and ethical acceptance paves the way for a lot of connected, personalized, and progressive educational ecosystems in the IoT era. This abstract intends to present a comprehensive overview of the paper, understanding the possible transformative effects of IoT technologies on traditional teaching methodologies and the overall educational ecosystem within institutes, to study the assessment of methods and technologies available for integrating (IoT) into teaching institutes while prioritizing privacy and data security. This involves investigating encryption protocols, secure data handling practices, and decentralized storage solutions. To understand the benefits of incorporating IoT into teaching institutes, including improved operational efficiency, enhanced teaching methodologies, and a secure learning environment. Explore the implications for educators, administrators, and students in embracing IoT for privacy-respecting learning experiences.

The meteoric evolution of technology, particularly the advent of the IoT, has introduced another age of potential outcomes across different areas. In the realm of education, the impact of IoT is proving to be transformative, presenting unparalleled opportunities to reshape traditional teaching methodologies. This paper delves into the potential of IoT, augmented by the integration of FL, to revolutionize teaching institutes, emphasizing the paramount importance of fostering privacy-respecting learning experiences.

As connectivity becomes more pervasive, machines gain the ability to autonomously execute complex tasks, creating seamless networks of communication between devices and the cloud. The IoT, as a collective network, thrives on this interconnectivity, offering innovative applications not only in education but also in fields such as healthcare, agriculture, manufacturing, and smart cities. FL, a decentralized machine learning approach, seamlessly integrates with IoT, providing an additional layer of privacy and autonomy in the training process, which is especially crucial in educational settings.

The synergy between IoT and FL becomes a catalyst for a paradigm shift in education. Local devices within teaching institutes become nodes in a decentralized network, allowing collaborative model training without compromising the privacy of sensitive data. This dynamic integration extends beyond traditional boundaries, creating smart classrooms where personalized learning experiences, real-time collaboration, and data-driven insights become intrinsic components of the educational ecosystem.

By exploring the potential of FL within the context of IoT in teaching institutes, this paper aims to elucidate the transformative impact on traditional educational methodologies. It navigates the intricate balance between technological advancements and the imperative need for privacy, emphasizing the collaborative potential of these innovations in reshaping the educational landscape. The subsequent sections delve into the key components, applications, and challenges, offering a comprehensive understanding of how the amalgamation of IoT and FL can redefine the contours of education, providing privacy-respecting, autonomous, and transformative learning experiences.

The integration of cutting-edge approaches is key to fostering privacy-respecting and transformative learning experiences. Among these innovations, FL emerges as a paradigm-shifting concept, particularly when seamlessly integrated into the fabric of the IoT. This introduction focuses on the pivotal role of FL, exploring how its decentralized nature harmonizes with IoT to empower teaching institutes with a transformative and privacy-respecting educational framework.

The Advent of FL: FL, a decentralized AI approach, has re-imagined the elements of cooperative model preparation. Not at all like customary unified models, FL empowers helping organizations to outfit the aggregate insight of dispersed gadgets without compromising the protection of delicate information. This interesting methodology involves sending model updates, as opposed to crude information, from nearby gadgets to a focal server, guaranteeing that the preparation cycle happens on the actual gadgets.

Integrating FL with IoT: The synergy between FL and the IoT amplifies the transformative potential within teaching institutes. As IoT establishes a network of interconnected devices, FL harmonizes seamlessly with this framework, extending its benefits to smart classrooms and educational ecosystems. This integration not only enhances the privacy-respecting nature of educational processes but also introduces novel opportunities for collaboration and personalized learning experiences.

Privacy as a Pillar of FL: Central to the integration is the emphasis on privacy. FL inherently preserves the confidentiality of individual data by confining the learning process to local devices. In the context of teaching institutes, this privacy-centric approach addresses concerns related to data security and ensures compliance with stringent privacy regulations. Consequently, educators can leverage the power of machine learning without compromising the trust and confidentiality of student and institutional data.

Transformative Learning Experiences: The amalgamation of FL and IoT heralds a new era of transformative learning experiences. Real-time collaboration, adaptive learning paths, and data-driven insights become integral components of this educational paradigm. Educators can tailor their approaches based on individual student progress, fostering a more responsive, dynamic, and personalized educational environment.

11.1.1 The Internet of Things

The IoT constitutes a network of physical objects, including cars, appliances, and household items. Imbued with software, sensors, electronics, actuators, and connectivity, these objects form a dynamic system (Letting and Mwikya, 2020). Each object is uniquely identifiable, housing an embedded computing system that facilitates seamless collaboration within the existing Internet infrastructure. The vision of the IoT has developed through the intermingling of advances, for example, remote association, continuous investigation, AI, sensors, and implanted frameworks.

The IoT is a significant strategic technology trend, rapidly transforming various aspects of daily life (Kishor et al., 2022a). Its ubiquity and promotion of intelligent as well as autonomous solutions distinguish IoT technologies from previous advances (Kishor et al., 2022b, 2023b). The foundation of this transformation lies in ubiquitous sensors and the ability to merge the physical as well as digital worlds. As a major paradigm shift, the IoT envisions integrating sensors into any object, utilizing machine-to-machine (M2M) interaction to integrate billions of devices into the Internet.

11.1.2 IoT in Higher Education

The expanding influence of IoT extends to varied industries, and higher education, specifically, universities, stands as a promising arena for IoT's impact (Kishor et al., 2023a, 2023c). Universities have the opportunity to lead in IoT technical development, foster innovative models, and address challenges related to Trust, Identity, Privacy, Protection, Safety, and Security (TIPPSS) relating to the IoT.

11.1.3 STEM Education and IoT Challenges

Recognizing the pivotal role of Science, Technology, Engineering, and Mathematics (STEM) in global economies, the fourth industrial revolution (4IR) introduced IoT as a key element (Kishor and Nand, 2023). However, there is a noticeable gap in higher education institutions offering STEM-related IoT courses. Research indicates that STEM students often lack designing experience and implementing IoT applications, leading to delayed entry into the workforce with minimal IoT exposure.

IoT, being a broad term, encompasses various methods of digital learning, such as E-Learning, M-Learning, and U-Learning (Kishor et al., 2017). These diverse approaches leverage electronic resources, handheld devices, and collaborative learning spaces, marking a shift toward pervasive information dissemination.

11.1.4 Studies on IoT in Education

Numerous studies have explored the application of IoT in education, shedding light on its potential benefits. From tracking student activity to swiftly addressing difficulties in study activities, IoT emerges as a tool with advantages and disadvantages in the teaching and learning landscape (Kishor and Saxena, 2023a).

Beyond the realm of teaching and learning, IoT finds applications in higher education institutions, from energy control and environmental monitoring to secure campus access and student health monitoring (Kishor and Sharma, 2022). The multifaceted use of IoT contributes to a comprehensive improvement in pedagogy and administrative decision-making.

In response to the evolving educational landscape, the Intelligence of Learning Things (IoLT) platform is proposed, presenting a mixed learning approach based on IoT (Kishor and Tyagi, 2023b). This platform offers a collaborative space for participants, fostering the exchange of ideas and the implementation of cutting-edge teaching methods.

11.1.5 Toward "Smart Universities" with IoT

Introducing the IoT in higher education holds the potential to revolutionize the traditional educational system (Kishorand Verma, 2023c). While notably IT&C corporations have initiated a project in the respective field, the establishment of a model for "smart universities" is still in progress. IoT platforms utilizing Cloud Computing services emerge as a promising technical solution for academia.

Exploring the synergy between IoT and big data, researchers argue for their positive effects on higher education (Chweya et al., 2019). The combination of these technologies supports intelligent connections, leading to more efficient learning environments. However, challenges related to security, transparency, and data volume require a thorough examination to grow the field of education.

The rapid expansion of the IoT presents both opportunities and challenges for education. Drawing parallels with Antoni Gaudí's architectural innovation, the paper explores the transformative potential of IoT in reshaping traditional teaching methodologies and fostering an interconnected, personalized, and secure learning environment. From its impact on higher education to diverse applications beyond teaching and learning, the IoT opens avenues for academic institutions to lead in technical development, innovate teaching models, and navigate the complexities of TIPPSS concerns. As academia explores the potential of IoT, the journey toward "smart universities" becomes an exciting opportunity for educators, researchers, and students to explore the creation of IoT systems, devices, applications, and services.

11.1.6 Theoretical Concept

The theoretical concept presented in the text revolves around the IoT and FL, particularly in the context of education and technology:

IoT: The primary theoretical framework is centered around the concept of the IoT. It refers to an individually addressed object of network or Internet technology to locate as well as communicate with each other with the use of devices. This concept is supported by references to seminal works in the field, such as the paper by Gubbi et al. (2013), establishing the foundational definition of IoT.

Technological Paradigm Shift: The text implies a paradigm shift in information and communication technology due to the rapid growth of the Internet, leading to the emergence of IoT. This shift is acknowledged as one of the hottest and most intriguing topics in contemporary technology discussions, as highlighted by Lee et al. (2013).

Educational Technology Advancements: Reference is made to the 2017 Horizon Report, a renowned study focusing on a step forward dimension education field and technological education. The inclusion of IoT as a technology projected to enable take-up within 2–3 years signifies a theoretical perspective on the evolving landscape of educational technology (Becker et al., 2023).

Practical Application of IoT Principles: The text emphasizes the importance of creating a learning climate that empowers the pragmatic use of hypothetical information inside the IoT field. This aligns with the practicality of understanding IoT principles for students and professionals in their future endeavors (Galarce-Miranda et al., 2022).

Multidisciplinary Work: There is recognition of the necessity of multidisciplinary work in understanding and implementing IoT principles. The capacity of current frameworks to incorporate has pointed out the significance of coordinated effort across disciplines (Laird and Bowen, 2016).

Enhanced Student Involvement: The development of educational materials using IoT principles is theorized to grow the involvement of students with content and improve learning settings. This suggests a pedagogical approach that integrates IoT concepts into educational materials to enhance engagement (Abdel-Basset et al., 2018).

Technological Advancements and Integration: The theoretical framework acknowledges technological advancements that enable the creation of less expensive but efficiently compact wireless systems for various gadgets. The integration of sensors, actuators, and communication tools is highlighted as components enabling frictionless connections within the IoT ecosystem (Du, 2012; Gubbi et al., 2013).

Low-Power Communications: The incorporation of low-power communications into IoT nodes is discussed, encompassing various protocols like ZigBee, Bluetooth, Wi-Fi, and NFC. This aligns with the theoretical understanding of the need for efficient communication in IoT devices.

Extension of Wireless Sensor Networks: The theoretical framework recognizes the extension of wireless sensor networks (WSN) to various gadgets, including books and wearable fitness trackers like FitBit, showcasing the broader impact of IoT on diverse applications (Gubbi et al., 2013).

The rapid expansion of industrial-scale IoT platforms and smart devices has led to a surge in data dimensions, fueling advancements in AI and machine learning research and applications (Lo et al., 2019). However, this growth in machine learning also amplifies concerns surrounding data privacy, as highlighted by the General Data Protection Regulation (GDPR), necessitating stringent data protection measures for compliance (Jobin et al., 2019). Machine learning systems often face "data hungriness issues" due to insufficient training data, accentuating the importance of finding solutions that deliver adequate data while respecting data owners' privacy.

In light of this test, Google proposed united learning in 2016. Combined learning, a variation of disseminated AI, permits model preparation on a decentralized organization of client gadgets. Significantly, unified learning empowers model preparation by utilizing privately gathered information without moving it outside the client gadgets. The interaction includes introducing a worldwide model on a focal server, broadcasting it to client gadgets for nearby preparation, and gathering privately prepared model boundaries, and collecting them to refresh worldwide model boundaries.

Fostering a unified learning framework requires a comprehensive methodology, joining a programming framework plan and AI information (Lo et al., 2021). In spite of having reference models for different frameworks like AI, enormous information, modern IoT, and edge figuring, there is at present no reference engineering for a start-to-finish unified learning framework. This paper tends to this hole by introducing an example situated federated learning reference architecture (FLRA) got from an efficient writing survey and modern prescribed procedures. The FLRA fills in as a plan rule, taking into account different quality credits and limitations for the turn of events and tasks of unified learning frameworks.

The theoretical concept is rooted in the transformative potential of IoT, encompassing technological advancements, educational implications, interdisciplinary collaboration, and the practical application of IoT principles. The referenced works by various authors contribute to building a comprehensive theoretical foundation for understanding and exploring the IoT paradigm.

11.2 OBJECTIVES OF THIS STUDY

- To understand the transformative impact of IoT technologies, including FL, as a modern trend in teaching and learning
- To study the privacy and security challenges associated with the proliferation of IoT devices, particularly in the context of FL, within teaching institutes.

11.3 THE MODERN TREND OF TRANSFORMATIVE IMPACT THROUGH IoT TECHNOLOGIES, INCLUDING FEDERATED LEARNING, IN TEACHING AND LEARNING ENVIRONMENTS

The current era witnesses an unprecedented transformation in the educational landscape, catalyzed by the integration of cutting-edge technologies. Among these influential trends, the advent of the IoT stands out as a beacon of innovation. In recent years, IoT technologies have emerged as a dynamic force, reshaping conventional teaching and learning methodologies and introducing a modern trend with immense potential for transformative impacts in the field of education. The educational sector, recognizing the power of connectivity and data-driven insights, is increasingly embracing IoT as a pivotal tool to revolutionize the teaching and learning experience.

At its core, IoT involves the seamless interconnection of devices and systems through the Internet, creating an expansive network that opens up innovative possibilities across diverse domains. This transformative impact is particularly pronounced in education, where IoT is not merely a technological addition but a catalyst for a more dynamic, efficient, and personalized learning environment. The integration of IoT in teaching and learning represents a departure from conventional methods, marking the inception of a modern trend characterized by enhanced connectivity and data-driven insights.

This technological evolution extends beyond the physical classroom, permeating virtual and remote learning environments. The introduction of wearable devices, smart applications, and IoT-driven educational platforms contributes to the creation of a holistic and interconnected educational ecosystem. These elements promise students personalized and adaptive learning experiences, transcending the limitations of traditional educational approaches.

As we delve into the exploration of this modern trend, it is crucial to recognize the synergy between IoT technologies and FL. FL, a decentralized machine learning approach, complements IoT in preserving privacy and facilitating collaborative model training without centralizing raw data. Together, these technologies form a potent combination, presenting unique opportunities and challenges for the educational sector. In navigating this transformative journey, educational institutions must strike a delicate balance between leveraging technological advancements for enhanced learning experiences and addressing the associated considerations of privacy, data security, and ethical practices. This presentation makes way for an extensive investigation of the modern trend of transformative impact through IoT technologies, including the intricate interplay with FL, in teaching and learning environments.

11.3.1 Federated Learning for IoT

The implementation of FL within IoT applications brings forth several key advantages. The distributed, collaborative, and privacy-preserving nature of FL addresses specific challenges encountered in the educational setting: (Zhang et al., 2022).

- **Preserving the Privacy of User Data**: In the ideal FL situation, each IoT gadget associated with the framework keeps up with the security of client information. During the united preparation process, crude information stays limited on the particular gadgets, and just the model updates are sent to the focal server. This approach limits the risk of individual information leakage, giving a robust security structure to delicate student data.
- **Improving Model Performance**: FL addresses limitations faced by individual IoT devices in terms of sufficient data for high-quality model learning. By collaboratively training a high-quality model, each IoT device benefits from the collective knowledge acquired from others' data, enhancing overall model performance. The periodic updating of local models within the FL framework allows for continuous improvement, surpassing what individual devices can achieve in isolation.
- **Flexible Scalability**: The appropriated idea of FL adds to adaptable adaptability, using calculation assets across multiple IoT devices in a parallel manner. This distributed learning approach is particularly advantageous as the hardware capabilities of edge devices increase. Incorporating all information to a server might strain the figuring assets at the edge and posture difficulties for remote correspondence organizations. By drawing in additional gadgets to join the FL structure, versatility is improved without forcing extra weights on a unified server. Moreover, FL minimizes the need for extensive transmission of raw IoT-collected data, reducing communication costs and improving scalability, especially in low-bandwidth IoT networks.

In essence, FL emerges as a powerful and adaptive mechanism within IoT-driven teaching institutes, ensuring the privacy of user data, enhancing model performance through collaboration, and providing a scalable solution that aligns with the evolving landscape of IoT technologies in education.

The landscape of education is experiencing a transformative wave, driven by the integration of the IoT and its advanced capabilities. A comprehensive analysis conducted by Rahmani et al. (2021) delves into the recent and future state of the IoT world, specifically in the sphere of education. The study explores the application of e-learning within a developed architectural framework, utilizing a range of IoT sensors such as cameras, microphones, and wearable devices to measure real-time physiological indicators like skin resistance, pulse rates, etc. This approach allows educational institutions to adapt their remote learning strategies, enhancing efficiency and maximizing resources while maintaining the integrity of their overall scholarly activities.

In a case study presented by Ahmed et al. (2022), Prairie View A&M University (PVAMU) serves as an example of integrating cutting-edge IoT technologies into the Computer Science (CS) program. The article outlines IoT learning modules designed to capture students' interest and seamlessly integrate them into existing CS curriculum courses. PVAMU's CS department has successfully implemented these modules, introducing a new project-based course focused on intelligent IoT technologies. The study also includes the results of an external review of the curriculum change, demonstrating encouraging effects on students' part of interest and knowledge in IoT across various courses and semesters. This research showcases how institutions can turn to good accounts for the potential of IoT to not only enhance traditional learning methods but also introduce innovative and practical approaches that prepare students for the complexities of a tech-driven world. As the integration of IoT in education continues to evolve, these studies contribute valuable insights into the positive outcomes and possibilities that emerge from this intersection of technology and learning.

The transformative influence of IoT technology is evident in the development of the University Intelligence Education Platform, as explored by Liu et al. (2021). This platform serves various functions, including online, control of attendance, instruction, and examination result inquiries. Through the creation of an IoT-based smart classroom architecture, the platform enhances the learning environment by incorporating IoT technology to manage instructional tasks. The intelligent education information's function module and cloud service layer are developed using IoT technology, resulting in a robust IoT-based intelligent education platform. Test results validate the platform's efficacy, demonstrating higher throughput, reduced application delay, and improved classroom assignment results, emphasizing its reliability and application effectiveness.

Smart Learning Environments (SLEs) find application in the governance processes of smart cities, as investigated by Setiawan et al. (2022). The researchers propose the IoT Virtual e-Learning System (IoT-Ve-LS) paradigm as a smart e-learning tool (SeT). This paradigm functions as a vast network, crucially connecting e-learners to Online Tutors (OTs). IoT-Ve-LS aims to secure a structure for accepting, storing, and distributing data in a cloud database. The report emphasizes enhancing intrinsically motivated online learning, demonstrating the necessity and impact of IoT in contemporary SLEs. The IoT-Ve-LS introduces real-time scenarios that address current challenges in educational development, offering another development in showing natural frameworks for shrewd urban communities later on.

The impact of technology on education is further explored by Shrestha and Furqan (2020), emphasizing the role of IoT in shaping how people live, work, learn, and play. IoT involves various devices and technologies working in tandem, transforming the educational landscape. Online IoT tools like Kahoot, along with communication platforms like Google Docs and Telegram, streamline tasks such as grading and information dissemination, providing efficiency for both teachers and students. The study highlights how IoT devices contribute to improved staff and student management, enhanced security, and resource conservation in institutions. Overall, these studies underscore the multifaceted applications of IoT in education, from intelligent classroom platforms to SLEs and beyond.

11.3.2 Ubiquitous Learning (U-Learning): Transforming Education with IoT

Ubiquitous Learning (U-Learning) has arised as a transformative educational paradigm characterized by interoperability, pervasiveness, and seamlessness. This architecture connects, integrates, and shares essential elements of resources of learning including collaborative learning, content and services, ultimately enhancing acquisition skills, even for participants with less self-efficacy (Tham and Verhulsdonck, 2023). The advantages of U-Learning, as highlighted by Villamizar et al. (2023), encompass various aspects.

Ubiquitous Learning Environment: The persistent need for communication and learning among students, teachers, and academics is addressed by U-Learning. It offers a learning climate that is relevantly mindful, recreates genuine situations, and takes into consideration vivid encounters. The learner's location is continually sensed by the system, providing relevant information based on environmental situations, and operates independently of network changes while the operator is in motion. Additionally, the U-Learning system adapts to the platform used by the learner, ensures data permanence, and offers real-time streaming for better service quality.

Challenges in U-Learning: However, challenges accompany the U-Learning concept, as noted by Shapsough and Zualkernan (2018). These challenges include the potential expense of ubiquitous equipment, misuse of technologies by instructors and students, difficulties in knowledge accumulation, and concerns about participant anonymity leading to diminished self-awareness and potential security risks.

IoT stands the possibility to alter schooling, by improving functional proficiency, growth opportunities, wellbeing, and security. As education evolves, embracing IoT becomes crucial for staying competitive and providing high-quality education. The advancement of IoT technology has shifted the emphasis from traditional memorization-based learning to a student-centered approach (Bucea-Manea-Ţoniş et al., 2022).

Benefits of Integrating IoT in U-Learning: The integration of IoT in U-Learning brings numerous advantages:

- **Improved Resource Management**: Streamlining monitoring procedures for school infrastructure, security systems, and cost reduction.
- **Monitoring and Controlling Utility**: Ensuring secure environments, tracking school bus movements, and enabling remote access to teaching materials.
- **Real-time Data Collection**: Processing large volumes of data for safety monitoring, student progress tracking, and identifying areas for improvement.
- **Effectiveness and Efficiency**: Transforming the traditional system into a U-Learning model to boost efficiency and personalized learning experiences.
- **Addressed Safety Concerns**: Utilizing IoT security applications to enhance safety, encourage healthy behaviors, and simplify facility management.

The integration of IoT in U-Learning offers a dynamic and flexible educational structure with potential benefits and challenges. Addressing security concerns, establishing pedagogical support, and ensuring technological advancements are aligned as essential for realizing the full potential of this transformative educational approach.

11.3.3 Navigating Challenges in Integrating IoT for Enhanced Teaching and Learning

The integration of the IoT into education presents both opportunities and challenges, as highlighted by Margianti and Mutiara (2015). While the IoT can enhance students' training skills by providing solutions to previously unsolvable problems and enabling the tracking of academic behavior through wearable technology, implementing IoT in education requires addressing several challenges. One significant challenge is the need for educational institutions to adapt their curricula to include IoT courses. This adaptation is crucial to equip graduates with the skills necessary for working on diverse IoT projects. Additionally, ensuring that the benefits of IoT are incorporated into established curricula, providing orientation sessions for staff, offering development opportunities for teachers professionally, and bringing issues to light among understudies about IoT applications are fundamental stages.

The nascent stage of the IoT introduces challenges such as wireless coverage, sensors that are expensive and battery life, as noted by Zhang et al. (2020). Engineers and developers working on IoT applications must tackle these challenges to enhance user accessibility. Despite the potential of Mobile Learning (M-Learning) such applications like augmented reality and learning analytics are such were their adoption requires further development and research. Security and privacy concerns emerge as significant barriers to IoT adoption in education. Implementation of IoT in education in future efforts should prioritize addressing these concerns to ensure the effectiveness and ethical use of IoT technologies.

Cybersecurity risks associated with the collaboration of IoT in education are emphasized, including the potential for cyberattacks, system failures, and data theft (Mahmoud, 2015). The interconnected nature of IoT devices increases vulnerability to cyber threats, potentially rendering educational institutions useless in the face of targeted attacks. Information security concerns incorporate start to finish security, verification, and information privacy, requiring dependable and secure IoT applications for instructive settings. The substantial amount of information created by IoT gadgets highlights the significance of tending to versatility issues, particularly in information examination for instructive purposes.

The potential for dehumanization, where autonomous systems reduce human interaction, poses ethical concerns in educational settings (Kassab, 2019). While modern technology significantly impacts daily lives, striking a balance between technological advancement and ethical considerations is crucial. Operational protocols based on IoT technologies for services and applications should prioritize ethical considerations to avoid dehumanization.

A few extra difficulties are related to the IoT in educating and learning, each requiring cautious thought and key arrangements. The underlying expenses of getting and sending IoT gadgets can be huge for destitute instructive establishments. Looking for subsidizing through awards, joint efforts with tech organizations, or taxpayer-supported initiatives is suggested. The incorporation of IoT arrangements with obsolete foundation might introduce troubles, and tending to this requires updating network equipment and quality of service (QoS) and load-adjusting procedures.

Issues related to IoT device breakdowns, data privacy, compliance with regulations, unequal access, and scalability necessitate comprehensive maintenance plans, clearly defined data use and privacy rules, adherence to relevant regulations, and strategies to ensure equal access for all students. Collaborative efforts with community organizations and careful planning for scalability from the beginning can address some of these challenges. Overall, addressing these challenges requires a holistic approach, including stakeholder cooperation and a commitment to data security and privacy. When appropriately implemented, the IoT has the potential to significantly enhance teaching and learning in educational institutions.

Security and Privacy Challenges: Despite the advantages, integrating IoT into education presents security and privacy challenges (Saeed et al., 2021). Standards for security and privacy must be established to protect valuable data, and collaborative efforts are essential to identify and mitigate potential business risks associated with data breaches.

11.4 SECURING THE FUTURE: PRIVACY PREVENTION IN IoT-DRIVEN TEACHING INSTITUTES THROUGH FEDERATED LEARNING

The expansion of IoT devices in teaching institutes brings forth security and privacy challenges that require careful consideration and strategic solutions. As educational institutions increasingly adopt IoT technologies to enhance teaching and learning experiences, protecting sensitive student data becomes paramount. One of the primary concerns revolves around safeguarding student information. With the deployment of various IoT devices, such as smart classrooms equipped with interactive boards and connected devices, there is a significant amount of data generated.

This includes student interactions, progress, and potentially sensitive personal information. Guaranteeing strong safety efforts is fundamental to forestall unapproved access and potential information breaks.

Implementing stringent access controls is crucial in mitigating security risks. Educational institutions need to establish clear policies and protocols governing who has access to IoT-generated data. This involves restricting access only to authorized personnel, such as teachers, administrators, and relevant staff. Access control mechanisms should be regularly updated and monitored to adapt to evolving security needs.

Furthermore, educators and policymakers must work collaboratively to lay out moral rules and guidelines for the mindful utilization of IoT in educational establishments. This involves addressing not only technical security aspects but also the ethical considerations surrounding data privacy. Institutions should prioritize transparency and communication with students, ensuring they are aware of how their data are being used and protected. As the integration of IoT in education continues to advance, the focus on security and privacy challenges should remain at the forefront. Proactive measures, robust access controls, and ethical considerations will contribute to creating a secure and privacy-respecting learning environment for students in teaching institutes. FL will collectively contribute to the creation of a secure and privacy-respecting learning environment for students in teaching institutes.

11.4.1 Federated Learning in Education: Addressing Privacy Challenges and Revolutionizing Learning Analytics

The integration of FL in the field of education has garnered significant attention, primarily driven by the imperative to address privacy challenges while revolutionizing learning analytics. A review of the existing literature provides insights into the evolution, applications, challenges, and potential solutions associated with FL in educational contexts.

Evolution of FL in Education: The inception of FL can be traced back to Google's pioneering work, presenting a decentralized approach to machine learning. In education, where data privacy is paramount, FL has emerged as a promising solution. Early works by Lavaur et al. (2022) laid the foundation for FL's potential to train unified prediction models across multiple devices without compromising individual privacy. This evolutionary shift has resonated with educational institutions seeking to harness data-driven insights while respecting privacy norms.

Applications in Learning Analytics: FL's applications in learning analytics are diverse and impactful. Educational institutions have adopted FL to tailor learning experiences by analyzing data across decentralized devices. A study by Kairouz et al. (2021) outlines the basic training flow of FL in education, emphasizing the iterative process of model updates from local datasets. The technology enables the construction of global models trained on multiple data sources, fostering a collaborative and adaptive learning environment.

Privacy Challenges: Privacy challenges are inherent in traditional data-centric approaches, especially concerning sensitive educational information. FL addresses these challenges by allowing participants to upload model gradients or weights derived from their local data, mitigating the risk of exposing sensitive information. However, the literature acknowledges potential concerns such as communication efficiency, model robustness, and the need for secure aggregation protocols (Fereidooni et al., 2021)

Revolutionizing Learning Analytics: The integration of FL in education marks a paradigm shift in learning analytics. By decentralizing the training process, FL not only preserves individual privacy but also enhances the efficiency and security of learning analytics. This revolution is particularly significant in the era of stringent data protection regulations, such as the GDPR (2018), which emphasizes individual control over personal data.

Potential Solutions and Future Directions: Literature on FL in education also explores potential solutions to its challenges. Secure aggregation protocols (Fereidooni et al., 2021) advancements in communication efficiency, and robustness mechanisms are areas of ongoing research. As FL continues to evolve, future directions include exploring collaborative learning scenarios between educational organizations, ensuring scalability, and addressing the specific needs of diverse learning environments.

FL is an AI approach that empowers model preparation across decentralized edge gadgets without the need to move crude information to a focal server. With regards to showing organizations with IoT gadgets, FL allows the training of machine learning models on data generated by these devices while keeping the data localized.

11.4.2 Key Components of Federated Learning in IoT-Driven Teaching Institutes

- **Decentralized Model Training:** Instead of consolidating data in a central server, FL allows each IoT device, such as interactive boards and connected devices in smart classrooms, to locally train a model.
- **Model Updates:** After local training, each device generates model updates, representing the changes made to the model based on its local data. These updates are typically in the form of model gradients or weights.
- **Secure Aggregation:** The model updates from each device are securely aggregated without exposing raw data. This ensures that the central server receives only aggregated information, preserving the privacy of individual device data.
- **Global Model Update:** The focal server consolidates the collected model updates to make a worldwide model that embodies experiences from every single decentralized gadget. This global model is then used for further iterations or analysis.

11.4.3 Privacy and Security Benefits

- **Data Localization:** FL allows sensitive data to remain localized on IoT devices, reducing the risk of data breaches or unauthorized access. This is particularly crucial in educational settings where student data privacy is a top priority.

- **Secure Collaboration:** The secure aggregation process ensures that even during the model update phase, individual contributions from IoT devices are protected. This collaborative approach enhances the overall security of the learning analytics process.
- **Ethical Considerations:** FL aligns with ethical considerations by minimizing the exposure of individual data. The approach prioritizes privacy, transparency, and user awareness, essential aspects in educational institutions.

11.4.4 Implementation Challenges and Considerations

- **Communication Efficiency:** Efficient communication between devices and the central server is essential for timely model updates. Bandwidth limitations or connectivity issues may pose challenges.
- **Model Robustness:** Ensuring the robustness of the global model, considering diverse data sources from different devices, is an ongoing consideration in FL implementations.
- **Regulatory Compliance:** Adhering to data protection regulations, including educational privacy laws, is crucial. Compliance with regulations such as GDPR ensures that privacy is balanced throughout the FL process.

FL in IoT-driven teaching institutes introduces a secure and privacy-respecting approach to harness the potential of IoT devices for educational advancements. This collaborative and decentralized model training methodology aligns with the objectives of "Securing the Future," contributing to enhanced privacy prevention in the ever-evolving landscape of educational technology

11.4.5 Privacy Prevention

Ensuring privacy in the context of the IoT within educational settings is a critical consideration, with specific attention to safeguarding sensitive student data and maintaining stringent access controls. This privacy preservation effort involves various layers within the IoT architecture tailored for educational institutions.

Privacy-Preserving in IoT: This section delves into privacy preservation in the IoT, focusing on a three-layer structure of the IoT stack (Seliem et al., 2018; Safaei Yaraziz et al., 2023; Sciancalepore et al., 2022; Sen and Dasgupta, 2023).

The Privacy Protection in IoT Gadget Layer: The IoT gadget layer is especially powerless to assaults, requiring the execution of strong safety efforts. Access control, confirmation, information encryption, and cryptography innovation are critical components in strengthening this layer. For example, specific RFID sticking and a nonlinear key calculation can improve access control and information encryption, individually. The IPSec convention and cryptography innovation assume essential parts in giving confirmation, security assurance, privacy, genuineness, and information trustworthiness. Furthermore, advanced marks and hash values add to getting correspondence conventions actually.

Protection Safeguarding in Stage/Framework Layer: The stage/foundation layer includes smart information handling and goes up against security challenges connected with information classification and respectability. Existing organization conventions might demonstrate lacking for M2M correspondence in asset-restricted conditions, possibly blocking associations between machines. Security weaknesses emerging from the heterogeneity of these organizations can adversely affect network security, interoperability, and coordination. To resolve these issues, new security instruments customized for IoT conditions are basic. These incorporate start to finish verification and key arrangement instruments, network virtualization, and the reception of IPv6 as standard organization layer conventions to help inborn safety efforts.

Protection Safeguarding in Application Layer: The application layer is assorted, with various applications focusing on different spaces, each having extraordinary information assortment prerequisites. Thus, security contemplations and necessities at the application layer contrast from the past two layers, including both non-specialized perspectives like protection mindfulness and security of the executives and specialized angles like cryptography and key arrangements. The consolidation of symmetric and deviated cryptosystems and certificate move innovation becomes fundamental in tending to these different security needs. This nuanced approach guarantees a thorough and versatile security structure across the different applications inside the IoT environment.

A part of IoT arrangements using out-of-date frameworks may present difficulties; addressing them calls for upgrading network hardware and using load-adjusting and quality-of-service techniques.

Privacy Preservation in Ubiquitous Learning Environments: Privacy concerns within Ubiquitous Learning Environments (ULE) necessitate a strategic approach to mitigate security risks and safeguard sensitive data. The multifaceted nature of ULE introduces several potential vulnerabilities that demand attention.

Unauthorized Access Risks: One essential security risk in U-learning is the danger of unapproved admittance to delicate information, especially when students associate with learning materials through open Wi-Fi organizations or other unstable associations. Relieving this hazard implies a coordinated exertion from instructors and students to utilize secure organizations and powerful verification measures, like two-factor confirmation. By guaranteeing secure associations, the probability of unapproved access is essentially diminished.

Information Breaks and Cyberattacks: The potential for information breaks or cyberattacks on learning stages and cell phones is a relevant concern. Such occurrences can prompt the robbery of individual data, monetary information, or licensed innovation. To counter this gamble, associations should execute rigid security conventions, including encryption and firewalls, to invigorate their frameworks and safeguard important information. Hearty safety efforts act as an impressive guard against outside dangers, cultivating a protected learning climate.

Social Designing Attacks: ULE is helpless to social designing assaults, wherein cybercriminals endeavor to maneuver students toward unveiling delicate data through strategies like phishing tricks. Anticipation methodologies include instructing students on recognizing and staying away from dubious messages or messages. Associations ought to organize approaches and strategies to identify and answer social designing occurrences. By advancing online protection mindfulness, the instructive local area can on the whole shield against tricky strategies.

While U-Learning presents a bunch of benefits, tending to security gambles is central. Executing powerful safety efforts, teaching students on prescribed procedures, and cultivating a culture of network protection are fundamental to guaranteeing the viability and security of U-Learning drives. In the more extensive setting of the Web of Things (IoT) innovation in training, focusing on network protection and protection becomes basic to lay out a completely safe learning climate. An extensive, multifaceted methodology enveloping training, gadget design, access controls, and checking frameworks engages instructive foundations to tackle the upsides of IoT while limiting related gambles. The essential utilization of IoT can improve the general security stance of instructive foundations by checking and controlling access and delicate information.

The landscape of education is going through a significant change, propelled by the integration of IoT technologies into teaching institutes. This shift presents an array of opportunities to revolutionize traditional learning methodologies, creating dynamic, interconnected educational environments. However, as we delve into this era of IoT-driven teaching institutes, it is crucial to recognize and address the security and privacy challenges that accompany this technological revolution.

Embracing the IoT Revolution in Education: The advent of IoT in teaching institutes signifies a paradigm shift, promising enhanced connectivity, interactive learning experiences, and unprecedented insights into student performance. Connected classrooms, smart devices, and data-driven analytics offer the potential to reshape education positively. Yet, this transformative journey is not without its hurdles.

The Duality of IoT in Education: Promise and Peril: While IoT opens the door to a world of possibilities in education, it brings forth a duality that demands careful navigation. On one hand, there is the promise of improved operational efficiency, personalized learning experiences, and a technologically advanced educational ecosystem. On the other, there are challenges related to security breaches, privacy infringements, and ethical considerations that require immediate attention.

Security Concerns in the IoT-Empowered Educational Landscape: As educational institutions embrace IoT technologies, a pressing concern emerges—the safeguarding of sensitive student data. The vast amount of information generated by IoT devices, ranging from academic performance metrics to personal preferences, demands robust security measures. This challenge necessitates a closer look at stringent access controls, encryption protocols, and comprehensive strategies to prevent unauthorized access and data breaches.

Ethical Considerations in the IoT Era of Education: The intersection of technology and education raises ethical questions that cannot be ignored. Striking the right balance between leveraging IoT for educational enhancement and upholding ethical standards becomes paramount. Addressing concerns about data use, surveillance, and responsible information handling requires the establishment of clear guidelines and adherence to privacy regulations.

Proactive Measures for a Secure Educational Future: In this era of IoT-driven teaching institutes, proactive measures are indispensable to mitigate risks and secure the future of education. From addressing network overburdening and potential disruptions to implementing QoS and robust maintenance plans, educational leaders and technology developers must collaboratively design strategies that prioritize security and privacy.

Preserving Student Privacy Amidst the IoT Revolution: Preserving student privacy becomes a focal point in the IoT revolution. Establishing a framework for responsible data use, ensuring transparency in data collection, and limiting information usage solely for educational purposes are essential steps. Adhering to regulations such as COPPA or GDPR becomes imperative to protect student data from unauthorized access and usage.

The increasing prevalence of IoT devices in teaching institutes presents a transformative landscape for education but it also introduces critical security and privacy summons that demand meticulous examination. This study deep into the intricate web of issues associated with the proliferation of IoT devices in educational settings, with a specific focus on safeguarding sensitive student data and the imperative need for stringent access controls.

Protecting Sensitive Student Data: As educational institutions embrace IoT technologies to enrich the learning experience, the sheer volume of data generated, including student interactions and personal information, raises concerns about data protection. The study scrutinizes the measures required to fortify the security infrastructure and shield student data from potential breaches. This involves exploring encryption protocols, secure data handling practices, and decentralized storage solutions to guarantee the classification and respectability of delicate data.

Stringent Access Controls: A critical aspect of addressing security challenges involves the implementation of stringent access controls. The study advocates for clear policies and protocols governing access to IoT-generated data within educational institutions. By restricting access to authorized personnel such as teachers, administrators, and relevant staff, institutions can fortify their defenses against unauthorized access and potential misuse of sensitive information. Regular updates and monitoring of access control mechanisms are emphasized to adapt to evolving security needs.

Ethical Guidelines and Standards: Beyond technical security measures, the study underscores the importance of establishing ethical guidelines and standards for responsible IoT use in teaching institutes. Integration between educators and policymakers is crucial to developing a framework that ensures transparent communication with students. This involves creating awareness about how their data are utilized, fostering a culture of trust, and prioritizing ethical considerations surrounding data privacy.

11.5 CONCLUSION

In conclusion, the integration of IoT technologies, coupled with the implementation of FL, represents a transformative and pivotal trend in the realm of teaching and learning. The advent of IoT has revolutionized traditional educational methodologies, fostering dynamic, efficient, and privacy-respecting learning environments within teaching institutes. As the educational landscape undergoes a profound transformation, it is crucial to recognize the multifaceted implications, challenges, and potential benefits associated with this technological evolution. The interconnectedness facilitated by IoT devices, from smart classrooms to wearable devices, has ushered in a new era of interactive and personalized learning experiences. This paradigm shift has the potential to significantly enhance operational efficiency and teaching methodologies, providing a more engaging and immersive educational journey for students and educators alike. However, with these opportunities come security and privacy challenges, particularly concerning the vast proliferation of IoT devices in teaching institutes.

This paper focuses on the significance of FL as a complementary and essential element in this educational landscape. FL, with its decentralized machine learning approach, addresses critical concerns related to privacy and security. By permitting models to be prepared across decentralized gadgets without trading crude information, FL ensures that sensitive information remains localized, mitigating risks associated with centralized models. In the sphere of education, FL empowers the creation of personalized learning paths for students and provides real-time insights for educators, all while upholding the paramount importance of safeguarding sensitive information. As educational institutions grapple with the challenges and opportunities presented by IoT and FL, a balanced approach that harmonizes technological innovation with ethical considerations is imperative. Educators, administrators, and policymakers must collaborate to establish robust ethical guidelines and standards that prioritize user privacy while harnessing the benefits of IoT in teaching institutions. The dynamic, personalized, and progressive educational ecosystem envisioned in the IoT era can be fully realized through the integration of FL, which not only preserves privacy but also contributes to the security and efficiency of the learning environment.

In essence, the importance of FL in education lies in its ability to reconcile the advancements in technology with the ethical imperatives of user privacy, ultimately shaping a future where transformative, connected, and personalized education becomes a reality.

REFERENCES

Abdel-Basset, M.; Manogaran, G.; Mohamed, M.; Rushdy, E. Internet of Things in Smart Education Environment: Supportive Framework in the Decision-Making Process. *Concurrency and Computation Practice and Experience* **2018**, *31*, e4515.

Ahmed, I.A.G.; Bellam, K.; Yang, Y.; Preuss, M. Integrating IoT Technologies into the CS Curriculum at PVAMU: A Case Study. *Education Sciences* **2022**, *12*, 840.

Aldowah, H.; Ul Rehman, S.; Ghazal, S.; Naufal Umar, I. Internet of Things in Higher Education: A Study on Future Learning. *Journal of Physics: Conference Series* **2017**, *892*, 012017.

Al-Emran, M.; Alkhoudary, Y.A.; Mezhuyev, V.; Al-Emran, M. Students and educators attitudes towards the use of M-Learning: Gender and smartphone ownership differences. *International Journal of Interactive Mobile Technologies* **2019**, *13*, 127–135.

Almufarreh, A.; Arshad, M. Promising Emerging Technologies for Teaching and Learning: Recent Developments and Future Challenges. *Sustainability* **2023**, *15*, 6917.

Bagheri, M.; Movahed, S.H. The Effect of the Internet of Things (IoT) on Education Business Model. In *Proceedings of the 2016 12th International Conference on Signal-Image Technology & Internet-Based Systems (SITIS)*, Naples, Italy, 28 November–1 December **2016**.

Becker, S.A.; Cummins, M.; Davis, A.; Freeman, A.; Hall, C.G.; Ananthanarayanan, V. NMC Horizon Report: 2017 Higher Education Edition. https://www.learntechlib.org/p/174879/ (Accessed on 27 May **2023**).

Bucea-Manea-Ţoniş, R.; Vasile, L.; Stănescu, R.; Moanţă, A. Creating IoT-Enriched Learner-Centered Environments in Sports Science Higher Education during the Pandemic. *Sustainability* **2022**, *14*, 4339.

Chweya, R.; Ibrahim, O. Internet of Things (IoT) Implementation in Learning Institutions: A Systematic Literature Review. *Pertanika Journal of Science & Technology* **2021**, *29*, 471–517.

Fereidooni, H.; Marchal, S.; Miettinen, M.; Mirhoseini, A.; Möllering, H.; Nguyen, T.D.; Rieger, P.; Sadeghi, A.R.; Schneider, T.; Yalame, H.; Zeitouni, S. "SAFELearn: Secure Aggregation for private FEderated Learning. In *2021 IEEE Security and Privacy Workshops (SPW)*, San Francisco, CA, USA, 2021, pp. 56–62. https://doi.org/10.1109/SPW53761.**2021**.00017

Galarce-Miranda, C.; Gormaz-Lobos, D.; Hortsch, H. An Analysis of Students' Perceptions of the Educational Use of ICTs and Educational Technologies During the Online Learning. *International Journal of Engineering Pedagogy (iJEP)* **2022**, *12*, 62–74.

Gubbi, J.; Buyya, R.; Marusic, S.; Palaniswami, M. Internet of Things (IoT): A Vision, Architectural Elements, and Future Directions. *Future Generation Computer Systems* **2013**, *29*, 1645–1660.

Jobin, A., Ienca, M., Vayena, E. The Global Landscape of AI Ethics Guidelines. *Nature Machine Intelligence* **2019** *1*(9), 389–399.

Kairouz, P.; McMahan, H.B.; Avent, B.; Bellet, A.; Bennis, M.; Bhagoji, A.N.; Bonawitz, K.; Charles, Z.; Cormode, G.; Cummings, R.; D'Oliveira, R.G. Advances and Open Problems in Federated Learning. *Foundations and Trends(r) in Machine Learning* **2021**, *14*(1–2), 1–210.

Kishor, K. Personalized Federated Learning. In Yadav S.P., Bhati B.S., Mahato D.P., Kumar S. (eds) *Federated Learning for IoT Applications*. EAI/Springer Innovations in Communication and Computing. Springer, Cham, **2022a**. https://doi.org/10.1007/978-3-030-85559-8_3

Kishor, K. Communication-Efficient Federated Learning. In Yadav S.P., Bhati B.S., Mahato D.P., Kumar S. (eds) *Federated Learning for IoT Applications*. EAI/Springer Innovations in Communication and Computing. Springer, Cham, **2022b**. https://doi.org/10.1007/978-3-030-85559-8_9

Kishor, K. Impact of Cloud Computing on Entrepreneurship, Cost, and Security. In Kishor K., Saxena, N., Pandey, D. (eds), *Cloud-based Intelligent Informative Engineering for Society 5.0* , 1st edition, pp. 171–191. CRC Press, New York, **2023a**. ISBN: 9781003213895. https://doi.org/10.1201/9781003213895-10

Kishor, K. Cloud Computing in Blockchain. In Kishor K., Saxena, N., Pandey, D. (eds), *Cloud-based Intelligent Informative Engineering for Society 5.0*, 1st edition, pp. 79–105. Chapman and Hall/CRC, New York, **2023b**. ISBN: 9781003213895. https://doi.org/10.1201/9781003213895-5

Kishor, K. Chapter 12 Review and Significance of Cryptography and Machine Learning in Quantum Computing. In Yadav, S.P., Singh, R., Yadav, V., Al-Turjman, F., Kumar, S.A. (eds) *Quantum-Safe Cryptography Algorithms and Approaches: Impacts of Quantum Computing on Cybersecurity*, pp. 159–176, De Gruyter, Berlin, Boston, **2023c**. https://doi.org/10.1515/9783110798159-012

Kishor, K.; Nand, P. Wireless Networks Based in the Cloud That Support 5G. In *Cloud-Based Intelligent Informative Engineering for Society 5.0*, 1st edition, pp. 23–40. Chapman and Hall/CRC, New York, **2023**. ISBN: 9781003213895. https://doi.org/10.1201/9781003213895-2

Kishor, K.; Nand, P.; Agarwal, P. Subnet Based Ad Hoc Network Algorithm Reducing Energy Consumption in MANET. *International Journal of Applied Engineering Research* **2017**, *12*(22), 11796–11802.

Kishor, K.; Saxena, N.; Pandey, D. *Cloud-Based Intelligent Informative Engineering for Society 5.0*, 1st edition, pp. 1–234. Chapman and Hall/CRC, New York, **2023a**. ISBN: 9781003213895. https://doi.org/10.1201/9781003213895

Kishor, K.; Sharma, R.; Chhabra, M. Student Performance Prediction Using Technology of Machine Learning. In Sharma D.K., Peng S.L., Sharma R., Zaitsev D.A. (eds) *Micro-Electronics and Telecommunication Engineering*. Lecture Notes in Networks and Systems, vol. 373. Springer, Singapore, **2022**. https://doi.org/10.1007/978-981-16-8721-1_53

Kishor, K.; Tyagi, R.; Bhati, R.; Rai, B.K. Develop Model for Recognition of Handwritten Equation Using Machine Learning. In Mahapatra, R.P., Peddoju, S.K., Roy, S., Parwekar, P. (eds) Proceedings of International Conference on Recent Trends in Computing. Lecture Notes in Networks and Systems, vol. 600. Springer, Singapore, **2023b**. https://doi.org/10.1007/978-981-19-8825-7_23

Kishor, K.; Verma, R.K. Cloud Computing-Based Smart Agriculture. In Sharma A., Chanderwal N., Khan R. (eds) *Convergence of Cloud Computing, AI, and Agricultural Science*, pp. 120–136. IGI Global, Hershey, Pennsylvania, USA, **2023c**. https://doi.org/10.4018/979-8-3693-0200-2.ch006

Khan, M.Z.; Alhazmi, O.H.; Javed, M.A.; Ghandorh, H.; Aloufi, K.S. Reliable Internet of Things: Challenges and future trends. *Electronics* **2021**, *10*, 2377.

Laird, L.; Bowen, N. A New Software Engineering Undergraduate Program Supporting the Internet of Things (IoT) and Cyber-Physical Systems (CPS). In *Proceedings of the 2016 ASEE Annual Conference & Exposition*, New Orleans, LA, USA, 26 June–28 August **2016**.

Lavaur, L.; Pahl, M. -O.; Busnel Y.; Autrel F. The Evolution of Federated Learning-Based Intrusion Detection and Mitigation: A Survey. *IEEE Transactions on Network and Service Management* **2022**, *19*(3), 2309–2332. https://doi.org/10.1109/TNSM.2022.3177512

Letting, N.; Mwikya, J. Internet of Things (IoT) and Quality of Higher Education in Kenya; a Literature Review. *International Journal of Management and Leadership Studies* **2020**, *2*(1), 14–26. https://core.ac.uk/download/pdf/286894462.pdf

Liu, J.; Wang, C.; Xiao, X. Internet of Things (IoT) Technology for the Development of Intelligent Decision Support Education Platform. *Scientific Programming* **2021**, *2021*, 6482088.

Lo, S.K.; Lu, Q.; Zhu, L.; Paik, H.Y.; Xu, X.; Wang, C. Architectural Patterns for the Design of Federated Learning Systems. arXiv preprint arXiv:2101.02373, **2021**.

Margianti, E.; Mutiara, A. Application of Cloud Computing in Education. In *Proceedings of the 1st International Joint Conference Indonesia-Malaysia-Bangladesh-Ireland (IJCIMBI)*, Banda Aceh, Indonesia, 27–28 April **2015**.

Njeru, A.M.; Omar, M.S.; Yi, S.; Paracha, S.; Wannous, M. Using IoT technology to Improve Online Education through Data Mining. In *Proceedings of the 2017 International Conference on Applied System Innovation (ICASI)*, Sapporo, Japan, 13–17 May **2017**.

Rahmani, A.M.; Ali Naqvi, R.; Hussain Malik, M.; Malik, T.S.; Sadrishojaei, M.; Hosseinzadeh, M.; Al-Musawi, A. E-Learning Development Based on Internet of Things and Blockchain Technology During COVID-19 Pandemic. *Mathematics* **2021**, *9*, 3151.

Safaei Yaraziz, M.; Jalili, A.; Gheisari, M.; Liu, Y. Recent Trends Towards Privacy-Preservation in Internet of Things, Its Challenges and Future Directions. *IET Circuits Devices & Systems* **2023**, *17*, 53–61.

Satu, M.S.; Roy, S.; Akhter, F.; Whaiduzzaman, M. IoLT: An IoT Based Collaborative Blended Learning Platform in Higher Education. In *Proceedings of the 2018 International Conference on Innovation in Engineering and Technology (ICIET)*, Dhaka, Bangladesh, 27–28 December **2018**.

Sciancalepore, S.; Zannone, N. PICO: Privacy-Preserving Access Control in IoT Scenarios through Incomplete Information. In *Proceedings of the 37th ACM/SIGAPP Symposium on Applied Computing, Virtual Conference*, 25–29 April **2022**, pp. 147–156.

Setiawan, R.; Devadass, M.M.V.; Rajan, R.; Sharma, D.K.; Singh, N.P.; Amarendra, K.; Ganga, R.K.R.; Manoharan, R.R.; Subramaniyaswamy, V.; Sengan, S. IoT Based Virtual E-Learning System for Sustainable Development of Smart Cities. *Journal of Grid Computing* **2022**, *20*, 24.

Shapsough, S.; Zualkernan, I. A Generic IoT Architecture for Ubiquitous Context-Aware Assessments. In *Proceedings of the 10th International Conference on Education Technology and Computers*, Matsue, Japan, 9–11 April 2018; Association for Computing Machinery, New York, NY, USA, **2018**, pp. 69–73.

Shrestha, S.K.; Furqan, F. IoT for Smart Learning/Education. In Proceedings of the 2020 5th International Conference on Innovative Technologies in Intelligent Systems and Industrial Applications (CITISIA), Sydney, Australia, 25–27 November **2020**.

Stankovic, J.A. Research Directions for the Internet of Things. *IEEE Internet Things Journal* **2014**, *1*, 3–9.

Tham, J.C.K.; Verhulsdonck, G. Smart Education in Smart Cities: Layered Implications for Networked and Ubiquitous Learning. *IEEE Transactions on Technology and Society* **2023**, *4*, 87–95.

Villamizar, S.B.C.; Ruiz, L.L.V.; Suarez, A.A.G. Perspectives of Ubiquitous Learning in Educational Contexts. *Journal of Positive Psychology and Wellbeing* **2023**, *7*, 86–90.

Zhang, T.; Gao, L.; He C.; Zhang M.; Krishnamachari B.; Avestimehr, A.S. Federated Learning for the Internet of Things: Applications, Challenges, and Opportunities. *IEEE Internet of Things Magazine* **2022**, *5*(1), 24–29. https://doi.org/10.1109/IOTM.004.2100182

12 A Critical Role for Federated Learning in IoT

Kaushal Kishor, Angeles Quezada, and Satya Prakash Yadav

12.1 INTRODUCTION

Artificial intelligence (AI) is enhancing the utilization of intelligent services and applications, leading to the widespread integration of the Internet of Things (IoT) into every aspect of our lives. Conventional AI systems often use centralized data collecting and processing. Given the vast scale of IoT networks and the challenges surrounding data privacy, this method is unlikely to be effective in practical situations. Federated learning (FL) is an AI method that promotes collaboration and decentralization in the development of intelligent IoT applications. AI may be trained on dispersed IoT devices without the need to share data. This article explores the emergence of FL in IoT networks. The study commences with a concise overview of the advancements in FL and the IoT, followed by an analysis of their integration. We assess the IoT services provided by FL. We provide services for IoT data interchange, unloading, caching, attack detection, geolocation, mobile crowdsensing, privacy, and security. The next step is to use FL in large-scale IoT applications such as smart healthcare, smart transportation, unmanned aerial vehicles (UAVs), smart cities, and smart industries. This study also emphasizes significant discoveries on FL-IoT services and applications. To conclude, we will address present challenges and forthcoming investigations in this swiftly expanding domain.

In recent years, the IoT has made significant progress by enabling the connection of diverse objects to the Internet via sensing and processing mechanisms [1]. AI technologies, such as deep learning (DL), have been extensively used to train data models in order to enhance the understanding of IoT data. Intelligent IoT applications, such as those in the fields of smart healthcare, transportation, and cities, may be created [2]. Typically, AI operations are carried out on cloud servers or data centers to facilitate data learning and modeling [3]. This method is disadvantageous due to the growing volume of IoT data. Due to network capacity and latency issues, the increasing growth of IoT data at the network edge may delay data transit to faraway servers [4]. External servers for AI training create privacy risks, including data breaches, since training data may include user addresses and preferences [5]. Thus, advanced AI technologies are needed to improve intelligent IoT networks and apps' efficacy and privacy. Intelligent and privacy-enhanced IoT systems should use FL. FL uses a central server to synchronize devices and collect data without sharing datasets [6]. Intelligent IoT networks may leverage several IoT devices as workers to communicate

DOI: 10.1201/9781003489368-12

with a server for neural network training. To be more precise, the aggregator constructs a comprehensive model with adjustable learning parameters. Each worker acquires the most recent version from the aggregator, computes its model update using its own dataset, sometimes using stochastic gradient descent (SGD), and transmits the local update to the aggregator. Next, the aggregator consolidates all modifications made to the models in order to generate an updated global model. To enhance training and safeguard user privacy, the aggregator may use a network of workers' computational capabilities. The aggregator transmits the comprehensive update to the local workers, who calculate their individual updates until the global training process is finished. FL's innovative operating paradigm enables it to provide several advantages for IoT. Some of the advantages include:

- **Data Privacy Enhancement**: In the context of FL, the training process at the aggregator does not need access to raw data. This minimizes the quantity of confidential user information shared with other entities, safeguarding data confidentiality. The privacy protection feature of FL makes it well-suited for developing intelligent and secure IoT devices since it addresses stringent data privacy standards such as the GDPR.
- FL minimizes communication latency by eliminating the need to transmit IoT data to the server, hence lowering data offloading. It preserves the spectrum and reduces power consumption during data training.
- **Improved Learning Quality**: FL speeds up the process of training models and enhances their accuracy. FL may use a network of IoT devices to get access to a diverse range of computer resources and statistics. On the other hand, AI approaches that are centralized and have restricted access to data and computational capacity may exhibit subpar performance.

12.1.1 FL's Distributed Learning Improves Intelligent Network Scalability

FL is being evaluated as a potential location for IoT applications, such as smart healthcare, smart transportation, and UAVs, due to the advantages it offers. FL permits the use of machine learning models without the need to provide patient data to medical institutions, thereby enabling the development of intelligent health services [7,8]. FL allows hospitals to locally train the AI model and send the acquired parameters to the aggregator for global calculation while ensuring that healthcare information remains confidential. FL promotes cooperation among institutions to expedite the process of diagnosing and treating medical conditions, all while ensuring the privacy of users. FL has shown intelligent vehicular services by integrating vehicles with roadside devices for data analysis and machine learning [9]. The services provided encompass autonomous driving, prediction of road safety, precise vehicle identification, and improvement of privacy. The recent success of FL-IoT applications makes it an opportune moment to highlight this crucial research issue. Section A: Comparison and Our Contributions. Emerging technologies such as FL and IoT have led to a surge in the examination of relevant research. The study investigated technological challenges related to FL systems, including data privacy and network attacks. The study conducted an analysis of FL in mobile edge networks [10,11]. The primary subjects of discussion are the problems of implementing FL and the optimization of FL in edge networks. A recent research has analyzed the potential of FL's wireless network in several aspects. Fog radio access networks use FL and investigate methods for mitigating accuracy loss and compressing models [12]. Explores FL's role in 5G, specifically focusing on its roles in edge computing, spectrum sharing, and core network administration. The integration of FL with 6G wireless applications faced significant obstacles, as seen in [13]. There is currently a lack of a thorough evaluation of FL-IoT applications, ranging from intelligent transport to smart urban areas [14–16]. To address these restrictions, we suggest studying the integration of FL in IoT networks. Our examination of FL applications for IoT data sharing, offloading, caching, attack detection, geolocation, mobile crowdsensing, privacy, and security is at the forefront of current research. The primary benefit of FL lies in its use in intelligent healthcare, transportation, UAVs, smart cities, and smart industries. Additionally, included are significant survey results [17]. In conclusion, we address significant research challenges and promising opportunities in the field of FL-IoT.

The contributions to the above section are highlighted as follows.

- We analyze the usage of FL in IoT networks in detail. The paper begins with an overview of FL and IoT advances and then examines their integration goals.
- FL is used in IoT services, including data exchange, caching, attack detection, geolocation, mobile crowdsensing, and privacy and security.
- We fully evaluate FL's potential in IoT applications, such as healthcare, transportation, UAVs, cities, and industries, focusing on communication and networking difficulties.

12.2 FEDERATED LEARNING FOR SERVICES RELATED TO IoT

12.2.1 FL as IoT Data-Sharing Alternative

The IoT depends on the flow of data to fulfill specific user requirements in various applications. FL suggests giving more priority to the exchange of information rather than exchanging IoT data. This method allows for the creation of intelligent IoT networks that have very low delay and prioritize privacy. The study offers a cooperative data-sharing platform specifically designed for industrial IoT applications. This architectural design facilitates the safe and efficient transmission of data for decentralized entities such as data owners and requestors. FL is a solution that tackles the issue of restricted resources for users of the IoT. This approach uses requester queries to predict data-sharing requests. It guarantees accurate computation and retrieval of the correct responses to these queries in order to facilitate sharing. FL uses a blockchain [18] to safeguard information exchange from outside attackers. By using immutable data blocks maintained by all parties, this link enables transparency and data ownership without a central authority. Requests and data transferred on the blockchain are immutable. This functionality greatly reduces data leakage and network security issues. Simulation studies using real-time datasets demonstrate the efficiency of FL-based sharing. The findings demonstrate precise acquisition of knowledge and enhanced protection. The paper [19] introduces a federated tensor mining system designed specifically for industrial IoT. This system uses FL to amalgamate data from several sources. This enables the extraction of information using tensor-based techniques while still guaranteeing security. A centralized server facilitates the sharing of homomorphism-encrypted data for tensor mining across many factories. The server compiles encrypted data into a tensor, while local factories store the original data. This strategy ensures the protection of data privacy throughout the process of tensor mining. Eavesdroppers and attackers have the ability to intercept encrypted data from the centralized server, but they do not possess the decryption key.

FL facilitates the decentralized sharing of data across automobile networks. A specialized asynchronous federated data-sharing system for the Internet of Vehicles (IoV) was created in [20]. Every vehicle functions as a client in the FL system and shares data with a centralized base station (BS) aggregate server. Cars that need traffic prediction or route selection may obtain mobile base stations (MBS) data. MBS employs a worldwide model that relies on automobile data [21]. This approach utilizes computational methods to transform the act of sharing into a problem of actor-critic reinforcement learning. This approach analyses both detrimental and innocuous node behaviors. In order to make educated judgments about sharing data, we choose the most efficient nodes that have reduced costs for sharing. Blockchain enhances the security and reliability of data. This process verifies the authenticity of model parameters and stores them securely in unchangeable ledgers [22].

In addition, this paper [23] introduces an asynchronous FL system for the purpose of sharing resources in vehicular networks. Gradient descent training now incorporates local differential privacy to safeguard local updates, ensuring secure and resilient FL exchange. Decentralized FL models include vehicle model updates at distant MBSs in order to reduce communication costs and address security concerns that are seen in centralized FL systems. The distributed FL technique outperforms centralized FL alternatives in terms of accuracy and privacy preservation in actual dataset experiments. In the context of hierarchical blockchain vehicular networks, Ref. [24] provides information transmission via FL.

A cloud server collects non-independent and identically distributed (IID) data from edge devices in heterogeneous federated learning (HFL) systems [25]. Heterogeneous federated learning (HFL) employs a federated swapping model that minimizes the utilization of shared data points. This approach effectively mitigates the adverse impact of non-IID data on weight divergence, which is often seen in conventional FL. Semi-supervised learning is used to anticipate video analysis on edge devices. FL enhances image classification accuracy by 3.8% and object identification accuracy by 1.1% compared to FedAvg when used to real-world video data. Connecting to a distant cloud might cause delays in FL methods.

The technique shown in [26,27] offers a cost-effective optimization strategy for coordinating edge devices with the cloud, resulting in reduced communication time. This enhances the efficiency of scheduling shared data, managing admissions, and adjusting correctness. The simulation results demonstrate that the suggested method has the potential to reduce latency and enhance privacy in many network scenarios. The study investigates a mobile IoT FL system that utilizes cloud servers and mobile devices as learning clients. A mobile cloud video recommendation system headquartered in FL is being investigated. Each cloud utilizes non-IID datasets such as user profiles and movie textual information to train a local FL approach. This is done by using a dual-convolution probabilistic matrix factorization model. The federated recommendation approach integrates cloud data with computed local updates to compensate for network delay [28,29].

12.3 FL FOR IOT DATA OFFLOADING AND CACHING OPTIMIZATION

FL has excelled in IoT data offloading and caching in various applications in addition to data sharing.

12.3.1 FL-Based IoT Offloading Optimization

IoT data loading in FL was recently investigated. Reference [30] describes deep reinforcement learning-based FL data loading for IoT networks. The FL concept is utilized to develop an intelligent loading strategy that reduces expenses for a variety of task probabilities by using various IoT devices as distributed reinforcement learning (DRL) agents. Distributing learning across agents and centralizing implementation decreases IoT network load while increasing data privacy. The FL approach synchronizes IoT devices to increase learning model training accuracy. Improve learning accuracy while taking into account latency and energy consumption. FL clients (mobile devices) improve local loading via Lagrangian Dual problems. In contrast to heterogeneity-unaware equal task allocation, the FL technique increases loading performance by increasing accuracy [31]. FL and SVM anticipate user relationships, reducing task computation and transmission energy. FL enables mobile users to create SVM models from their own datasets and then communicate local modifications to a global model depending on user association and task data size. As a result, user input aids the BS in determining ideal user associations for loading, conserves energy, and improves learning accuracy [32].

The authors of [33] suggest integrating video data into federated video analytics systems via FL. MEC servers use NNs to modify frame sizes throughout the network and regulate resizing. MEC server and edge devices create an FL algorithm to improve edge network video loading in terms of accuracy and latency. An FL-based vehicle network loading approach was investigated [33]. FL offers a cooperative learning architecture that allows vehicles and RSUs to share a learning model while mitigating user privacy concerns about direct data loading. This novel technique may reduce agent selection and loading task data exchange learning costs. In [34], FL analyses fog computing network data. FL enables mobile devices to improve loading based on resource constraints and accuracy without sharing raw data with the fog server, ensuring privacy. To decrease model errors and learning costs, mobile devices may transmit model changes. FL-based operation improves network resource utilization and local model learning accuracy when employing MNIST datasets containing hundreds of images.

12.3.2 FL-Optimized IoT Data Caching

Edge servers may store mobile IoT data [35]. FL can assist in developing innovative caching solutions to address IoT networks' exponential mobile data growth. Users may be skeptical of third-party servers and reticent to supply sensitive data for caching, therefore FL may alleviate privacy concerns in traditional learning systems. FL assists in developing proactive edge computing data caching solutions that do not need user data access in order to anticipate the most popular files [36]. FL enables mobile users to download the neural network (NN) model from an edge server for local training. They may continue transmitting the learned model to the server for aggregation. Server-based data file recommendations leverage latent features to match users and files. FL-enabled caching efficiency outperforms Oracle, random, and Thompson sampling because of improved cache capacity and communication cycles. This learning system safeguards privacy, conserves spectrum resources, and enhances learning by reducing unbalanced data and poor communication with each client. FL offers DRL training for data caching and content substitution [37]. User equipment (UEs) learn a shared model and save raw data locally, while a BS cloud learns a global model by aggregating local updates. FL users utilize a Markov decision process to replace content.

FL and blockchain technology are being used to develop powerful learning algorithms for edge computing content caching [38]. Edge servers employ blockchain smart contracts to determine access and credit eligibility during content loading and caching. IoT devices utilize their own datasets to learn user and file properties for local training, which improves the cache hit rate. They then communicate gradient changes to the edge server to determine file popularity. This method compresses gradients on devices to decrease FL data. Cache efficiency is much greater in Movie Lens dataset simulations than in FL-based methods. FL is used to anticipate content popularity in [39] and [40]. Proactive edge caching at BSs may decrease traffic while increasing user satisfaction by selectively storing and prioritizing vital data. FL is an excellent method for measuring popularity without disclosing user information or preferences. The NN may increase cache hit rates by training a preference-weighted popularity model on local devices based on previous request data. FL combines local updates on a central edge server to predict popular files for caching, maximizing data utilization for all participants [41,42].

12.3.3 FL for Internet of Things Localization

Smart gadgets use localization technology to provide location-based services. Wireless signals and CSI may be used to find target things. Conventional AI/machine learning (ML)-based localization systems are affected by changes in the mobile environment and limitations on privacy due to the centralized data processing structure for object localization [43]. FL provides state-of-the-art privacy-enhancing geolocation services. A house interior localization technique that protects privacy utilizes FL [44]. Mobile users can create AI models by using beacon signal strength and tagged locations. The central server constructs a worldwide multilayer perceptron model by using local updates to accurately predict localization. A simulation of fingerprinting using the UJIIndoorLoc database RSS data [45] demonstrates enhanced localization accuracy and privacy in various inside environments. According to the study [46], FL has the potential to address the challenges of central AI server task learning and privacy in indoor localization systems. FL and centralized indoor localization techniques decrease the need for collecting fingerprints and lower network costs, all while maintaining privacy.

Decentralized location determination in a building is accomplished by using a large number of mobile devices. Crowdsourced data are unlabeled, whereas fingerprint data are marked. Each device runs a DL model and uploads the findings to a central server, which then generates a statistical localization model. FL excels in lowering communication costs, protecting privacy, and distributing fingerprint data.

FL [47] is used in the WiFi network localization technique. Mobile devices may use WiFi data to generate DNN models and deep autoencoder noise reduction signatures. A central server builds a global model from local weights. Homomorphism encryption protects the communication channel when unloading. A laboratory corridor experiment demonstrated reliable and secure location estimation. The Federated Localization (FedLoc) architecture is utilized in IoT networks to provide exact localization [48]. FL lowers location prediction bias and safeguards privacy by training models on a large number of users' local fingerprint data. GPS-based indoor localization, inertial sensor target tracking, and BS wireless traffic prediction are all available in FL [49].

12.3.4 FL for Internet of Things Mobile Crowdsensing

Mobile crowdsensing leverages ubiquitous mobile devices to detect and gather data from physical settings for end users. This method is viable because of the IoT. Intelligent mobile crowdsensing systems use conventional AI and ML frameworks [50] for model training and data processing. Nevertheless, conventional crowdsensing techniques sometimes need direct access to user data, thereby jeopardizing privacy. Given the enormous amount of data generated by IoT devices, it is not feasible for a single server to process all the sensory input. FL may expedite the process of training crowdsensing models. Recent research using FL has shown potential.

12.3.5 FL-Based IoT Privacy and Security Methods

IoT devices on IoT networks raise new security and privacy concerns. These gadgets monitor user activity and gather personal information. Several AI/ML algorithms discover and detect IoT network threats and privacy bottlenecks, hence addressing security and privacy concerns. Traditional systems' centralized IoT data collecting and public data-sharing jeopardize user privacy. FL seems to have possibilities for IoT privacy and security. FL secures automobile IoT data privacy through two-phase mitigation, including intelligent data translation and collaborative data leakage detection [51]. The vehicular FL technique lets players (e.g., automobiles) train models locally using their own data without a curator, protecting data privacy more than earlier systems. To assure data usefulness after transformation, many vehicles utilize cooperative mapping to convert raw data from many sources into trained data models. The trained model provides accurate resource allocation data without providing raw data, which improves privacy. To preserve vehicular IoT network privacy, FL is used with local differential privacy to disrupt car gradients while preserving gradient utility [52]. This stops attackers from deducing the original data from the modified gradient.

12.4 FOR APPLICATIONS RELATING TO THE IoT

12.4.1 The FL for Intelligent Healthcare

AI has been extensively used in the field of smart healthcare to evaluate health data and enhance services, namely in the area of sickness diagnosis via the use of intelligent imaging. Conventional AI models pose privacy problems since they rely on learning from public data stored in the cloud or data center. HIPAA specifically governs the handling of confidential healthcare information, which sets it apart from other sectors [53]. In intricate healthcare environments, where hospitals and insurance firms use healthcare databases for the purpose of data analysis and processing, simply eliminating metadata such as patient information is insufficient to safeguard patient privacy. Conventional AI solutions that rely on a central server for data processing are not appropriate for use in the healthcare industry. FL may provide privacy-conscious analysis without the need for data exchange by presenting alternative choices.

FL has been extensively studied for its capacity to provide adaptable and privacy-enhancing electronic health record (EHR) management in healthcare operations. A cooperative learning approach for EHRs is presented in reference [53]. The FL technique utilizes many hospitals and a cloud server. Each hospital uses EHRs and a cloud server to operate a NN. A data perturbation strategy is used to safeguard the parameters of an FL model. This technique involves perturbing the data used for training, which helps prevent attacks on the model's memory while it is learning.

The decentralized and privacy-preserving nature of FL has the potential to improve the delivery of medical services by promoting secure collaboration in healthcare. The paper [54] discusses the interaction of medical IoT devices via FL. Every device operates a NN on a powerful server to identify arrhythmia using electrocardiogram information. When applied to 64 IoT devices in a practical arrhythmia detection system, FL demonstrates lower communication overhead and less loss of accuracy compared to Federated Averaging (FedAvg).

12.4.2 FL for Smart Transportation

AI/ML techniques have recently been used to enhance intelligent transport systems (ITSs) via the centralized acquisition of vehicle data at the data center [55]. Nevertheless, this method entails the transfer of information in secure environments, which might potentially give rise to privacy issues. FL is a technique that allows the deployment of AI capabilities at the network edge. This process involves the collaboration of several entities, such as autos, to collectively train AI models that are then distributed globally. This approach obviates the need for massive data transmission and guarantees the safeguarding of user privacy. This research focuses on the use of FL in the domain of smart transportation. The primary objective of our analysis is to examine the use of FL in two distinct domains: vehicular traffic planning and vehicular resource management. Various designs using fuzzy logic (FL) have been proposed to enhance vehicular traffic planning, a vital component of ITS for forecasting traffic patterns and managing cars to reduce congestion. For example, FL is considered a viable alternative to traditional centralized ML algorithms for traffic prediction problems. The process entails executing ML models directly on edge devices, such as autos, using their own datasets that include information on road layout, traffic flow, and weather. FL can provide resource management techniques for vehicle-to-vehicle (V2V) networks, as well as traffic planning. A technique using FL is introduced in [56] to enhance ultra-reliable low-latency communication (URLLC) in automobiles. This approach enables each vehicle to get information on a generalized Pareto distribution (GPD) of network queues, particularly in the context of power management and resource allocation. Crucially, these data are acquired without revealing the specific samples of the queue length.

12.5 FL FOR UNMANNED AERIAL VEHICLES

UAVs are vital for the transportation of goods, surveillance of natural disasters, and execution of military operations. Their adaptability and reliable connectivity make them essential in 5G/6G wireless networks. AI and ML have been used to enhance ground BS UAV operations in order to enhance UAV networks. The mentioned tasks include trajectory planning, power management, and target detection [57]. The quick movement and high altitude of UAVs pose challenges for maintaining continuous connectivity with BSs in dynamic aerial environments. Therefore, centralized AI/ML solutions for UAV operations may not be optimal, particularly when transmitting substantial volumes of data over aerial connections. FL enhances the learning capabilities of intelligent UAV networks. This enables several UAVs to communicate with one another without transmitting unprocessed data to BSs, while yet maintaining data confidentiality. This inquiry focuses on the analysis of Free Space Optics (FSO) in UAV communications and network management. Multiple studies have investigated the impact of FL in enhancing UAV communications. FL offers a decentralized method for managing the routing of large UAV networks, as described in reference [58]. Each UAV operates a NN and transmits the model parameters to a central unit in order to generate a global model. This approach provides a more precise estimation of the population density function of UAVs. FL mitigates the transmission of data over the air, hence decreasing communication delays and addressing privacy concerns. FL would also use the full processing capacity of UAVs to accelerate the training of global models. The simulations demonstrate that FL algorithms have the potential to decrease transmission time, minimize drone motion energy, and mitigate collision risks in turbulent scenarios.

12.6 FL FOR SMART CITY

Smart city ecosystems use intelligent gadgets, advanced infrastructure, monitoring systems, and communication technologies to provide urban populations with food, water, and energy. AI/ML technologies are used to process large amounts of data from sensors, devices, and people in real-time to develop smart cities. The majority of AI-based smart city solutions have a centralized learning architecture hosted on a cloud server, which is not capable of scaling to accommodate the increasing number of smart devices. Decentralized smart city applications get advantages from the anonymity and reduced connection latency provided by FL. FL may assist in the management of data and grids for smart cities. FL is an ideal location for incorporating distributed AI capabilities into intelligent data management systems for smart cities due to its decentralization and emphasis on privacy. The implementation findings demonstrate high accuracy and little testing loss while analyzing unlabeled smart city data. FL, in [59], manages data streams from widespread IoT devices functioning as FL clients, allowing them to do local learning without disclosing their data to others. This approach tackles the challenges of handling smart city data from various streams and devices, as well as resolving concerns related to user privacy. Smart urban communication, social economy sharing, social activity monitoring, and global citizen networking will revolutionize smart cities. Smart Grid FL provides power to homes and businesses via electrical networks, benefiting industrial systems and industries. Intelligent solutions that enhance privacy in FL contribute to the safety of smart grids. FL enables the development of predictive models in edge data centers. These centers use a recurrent NN to predict future energy requirements based on data on client energy usage. Information is collected at a cloud server in order to generate a worldwide prediction model [60].

Client privacy is protected by not storing energy choices and dwelling addresses on the cloud. The prediction model has the potential to surpass centralized learning on a single server by coordinating data centers across metropolitan areas [61].

12.7 FL FOR SMART INDUSTRY

Intelligent manufacturing uses AI methods, such as ML and DL, to analyze vast quantities of data generated by industrial equipment. The data are used for process modeling, monitoring, prediction, and control throughout all stages of production [62]. AI operations depend on training data, which must be exchanged across organizations and manufacturers. Transferring large amounts of data across industrial networks for AI purposes is inefficient because of issues over user privacy. FL (FL) can serve as a viable solution for industrial system intelligence, eliminating the need for data exchange or sacrificing privacy. The use of robotics and Industry 4.0 in the state of FL: Robotics is essential for the automated and programmable manufacture of automotive industrial systems. Real-time data processing and data privacy pose significant issues for robotic systems. FL might address these problems by equipping robots with cognitive capabilities instead of depending on a server for data processing. FL may be used to include AI models in on-site robots for data learning in robotics, eliminating the delay caused by network transmission [63]. Every robot contributes gradient parameter updates to construct a cloud model while keeping its raw data undisclosed. This technique ensures privacy by using differential privacy. During communication cycles, the server can accumulate a substantial amount of robot data in order to construct a robust learning model. Local robots exhibit superior efficiency and accuracy in learning imitation as compared to centralized learning systems. Fog/edge computing, which is an extension of cloud computing, enables federated robots to interact rapidly [64]. FL can facilitate the implementation of decentralized intelligent industrial 4.0 systems. The paper [65] presents the implementation of FL with privacy protection. This design allows several mobile users to collaboratively develop a shared AI model. The model is constructed by encrypting local gradients using homomorphism cipher text and a distributed Gaussian method on a cloud server. The existence of local variations and common parameters in FL communication cycles may possibly result in privacy violations; yet, our method has the capacity to alleviate these concerns. To increase compound TCP efficiency in industrial 4.0 WiFi networks, the authors recommend employing FL to coordinate WiFi upload and download dynamics and decrease wireless data losses across numerous access points. Each access point regression analyses QoS data samples and updates learning parameters to feed a central server. An integrated FL-blockchain architecture was developed for industrial IoT [66,67]. FL employs stationary and dynamic data to spread learning on local IoT devices. The immutable records and smart contracts of blockchain may validate FL learning interactions. An effective FL implementation in Industrial Edge-Based IoT Networks: Industrial edge computing is popular in IoT networks for its fast processing and storage [68]. Edge computing in the Industrial Internet of Things (IIoT) may speed up decision-making and save network traffic. Smart and private edge-based IIoT applications are possible using FL technology. The distributed learning capacities of edge devices, IIoT network training federation, communication, and network resource concerns must be considered [69].

Efficient utilization of network resources is necessary for achieving effective FL training in distributed edge IoT networks. The concept of [70] revolves around a unified challenge that aims to maximize computation and transmission in edge IoT systems based on FL. The objective is to minimize the energy consumption of local processing and wireless transmission. This issue should be approached in an iterative manner. The approach is straightforward and produces novel closed-form solutions for optimizing time, bandwidth, power, computing frequency, and learning accuracy at each iteration. Within FL-based edge networks, the model owner is responsible for allocating worker energy and selecting global model transmission channels. Efforts should be made to limit energy use while maximizing the success of transmission. The study presented in Ref. [71] utilizes DRL as a solution to address these challenges. This approach allows the model owner to choose the most efficient energy and channels without requiring any specialized knowledge in networking. The model owner formulates a stochastic optimization problem to maximize global model transmissions while simultaneously reducing energy and channel expenses. The paper also investigates a method for managing robot resources using an IIoT-based FL methodology [72]. This study examines the limitations of bandwidth, processing capacity, and battery life. Previous FL systems operated on the assumption of abundant resources, thereby disregarding these constraints. The communication cycle assesses the resources used for model training and the trust scores of the robot in order to determine client training. Robots that experience delays in updating their models may see a change in their trust score in the subsequent training cycle. This modification has the potential to decrease the occurrence of the straggler effect in robotic-based FL systems.

12.8 CHALLENGES AND DIRECTIONS IN THE RESEARCH INDUSTRY

12.8.1 Concerns Regarding Safety and Confidentiality in FL

FL may protect dispersed IoT systems' privacy by not sharing raw data during learning. Both learning clients and servers pose security and privacy risks to FL [73]. An attacker may change data characteristics or inject a false subset to add backdoors to the model and change local clients' training objectives. Backdoor poisoning involves contaminating clients' local model alterations before servers load them. The central server can manipulate weight gradients, which multiply learning layer and feature mistakes, to compromise aggregated local updates. This modification lets attackers steal gradient training data in a few rounds [74]. Thus, globally updating may expose personal attributes of locally utilized training data, violating user privacy. IoT networks must keep user preferences and home addresses across FL-applied intelligent domains, making this problem critical. Unprotected FL hinders security and privacy in intelligent IoT systems, making it difficult to engage people for collaborative data training.

Given current circumstances, safe FL systems must develop new threat mitigation measures. Differential privacy or dummy perturbation methods can protect training records from unauthorized access using composition theorems. Artificial noise, such as Gaussian noise, is added to NN gradients in [75], to protect training data and personal information while assuring convergence. This strategy will prevent the server or malicious users from learning about user samples from received messages, regardless of auxiliary information or attack. A confidential and privacy-focused FL system for industrial data mining is described in a study. Refer to [76] for system specifics. Reducing server-participant parameters reduces privacy risks without affecting model accuracy. A proxy server, which acts as a middleman between the server and all participants, maintains strong privacy by using Gaussian differential privacy on shared parameters. In addition, [77] proposes a robust aggregation method to maximize server-mediated, unauthenticated network security in FL systems. Encrypting local updates and facilitating key exchange between users and the server with a double-masking structure ensures equitable data verification and key agreement during aggregation. FL–IoT privacy protection research requires innovative methods to protect clients (like IoT devices) and communication (like wireless server–client connections).

12.8.2 Communication and Learning Convergence Issues of FL–IoT

Challenges related to the integration of FL–IoT result in communication and learning convergence concerns. The presence of unbalanced, non-independent, and identically distributed (non-IID) data in both the uplink and downlink training of FL exposes communications to vulnerabilities. The variation in client training data volumes and dispersion is caused by differences in sensor settings. Due to the exponential increase in the number of clients, it is not feasible to perform direct client-server updates because of network congestion. The convergence of FL algorithms in IoT systems is uncertain because of the limited training capacity and the complexity of data in IoT devices. FedAvg, a widely used approach in FL, implies that all clients of IoT devices participate in every communication cycle for FL training [78]. Nevertheless, the occurrence of device connectivity issues and battery drain often renders this assumption unworkable. FL approaches that use first-order gradient descent to reduce the loss function exhibit sluggish convergence [79].

There are several FL performance solutions available. Reference [80] provides a highly efficient communication protocol that compresses both uplink and downlink communications. This protocol is capable of handling a larger number of clients and effectively distributing data. Scarification, ternarization, error accumulation, and optimal Golomb encoding may enhance uplink compression and accelerate global server parallel training while maintaining learning convergence. Furthermore, Ref. [81] presents FetchSGD, an FL approach for optimizing IoT networks. This technique constructs accurate models to enhance communication efficiency. During each communication cycle, clients compute a gradient based on their local data and then use Count Sketch to compress it before transmitting it to the central aggregator. The aggregator is responsible for maintaining the momentum and error accumulation count drawings, as well as the round weight update. Therefore, the proposed approach reduces communication in every iteration while still maintaining the required quality criteria for federated training. The momentum FL architecture enhances the convergence of algorithms [82]. The central server uses momentum gradient descent (MGD) to minimize the loss function with optimum learning parameters, resulting in reduced processing requirements compared to first-order gradient descent-based FL systems.

12.8.3 Resource Management in FL-IoT

FL-IoT employs scalable data parallelism and on-device training at IoT devices before consolidating learning parameters at a worldwide server. Training data necessitates the allocation of storage and CPU resources from all IoT devices to provide synchronized server updates. Regrettably, IoT devices that have limited computing capacity may cause a delay in the aggregate of parameters that are synchronized with a server. Training DL models such as DNN on IoT devices may provide challenges due to limitations in CPU frequency and energy, particularly when dealing with picture and audio data [83]. Edge devices possess more computational capacity for large-scale AI training compared to IoT devices. Therefore, enhancing single-device AI/ML models may reduce the computational load on IoT devices.

12.8.4 Potential AI Learning Functions That May Be Applied to IoT Sensors

Implementing AI algorithms on IoT devices might provide challenges. Limitations in hardware, memory, and power hinder certain IoT sensors from being able to train an AI model [84]. Advanced ML algorithms need a substantial amount of memory and computational power for tasks such as training models, storing parameters, and training variables. Training fundamental image classification models such as ResNet-50 [85] requires a significant amount of CPU processing and many terabytes of RAM. The exorbitant expense associated with transmitting AI training data is a significant obstacle to implementing on-device FL. The exchange of models between IoT sensors and the server leads to higher communication expenses as the size of the model grows [86]. This would discourage IoT sensors from building large FL algorithm models on devices. Energy management in FL–IoT, particularly in lightweight mobile devices with limited battery capacity, is a significant issue of concern. Deliver comprehensive and effective training by using equipment available locally, while maintaining high standards of quality, in order to save energy.

12.9 CONCLUSION

FL is a novel AI technology that has gained popularity because of its ability to deliver scalable, privacy-preserving services and applications for the IoT. This article looks at how FL may improve IoT networks. It does this by conducting a thorough survey and participating in in-depth conversations that include the most current discoveries in the field. The lack of a comprehensive evaluation of FL's usage in the IoT context is the impetus for this research. To narrow this gap, we began by reviewing the most recent breakthroughs in the IoT and FL, as well as providing practical suggestions on how to integrate them. Subsequently, we conducted considerable research on FL applications in key IoT services. We especially investigated its applications in mobile crowdsensing, attack detection, geolocation, data unloading and caching, IoT privacy and security, and geolocation. Since then, we have been looking at the most current advances in integrated FL-IoT applications in a variety of key use case areas, including smart cities, smart transportation, smart healthcare, smart industrial, and UAVs. Numerous significant findings from the complete survey have been summarized and examined. In conclusion, we have successfully identified numerous significant research challenges and prospective areas for further investigation. This discovery is expected to spark further interest in this new subject, motivating more research efforts targeted at fully realizing FL–IoT.

REFERENCES

1. M. Mohammadi and A. Al-Fuqaha, "Enabling cognitive smart cities using big data and machine learning: Approaches and challenges," *IEEE Commun. Mag.*, vol. 56, no. 2, pp. 94–101, Feb. 2018.
2. K. Kishor "Communication-efficient federated learning." In: Yadav S. P., Bhati B. S., Mahato D. P., Kumar S. (eds) *Federated Learning for IoT Applications*. EAI/Springer Innovations in Communication and Computing. Springer, Cham, 2022. https://doi.org/10.1007/978-3-030-85559-8_9
3. Y. Sun, M. Peng, Y. Zhou, Y. Huang, and S. Mao, "Application of machine learning in wireless networks: Key techniques and open issues," *IEEE Commun. Surveys Tutor.*, vol. 21, no. 4, pp. 3072–3108, 4th Quart., 2019.

4. K. Kishor "Personalized federated learning." In: Yadav S. P., Bhati B. S., Mahato D. P., Kumar S. (eds) *Federated Learning for IoT Applications.* EAI/Springer Innovations in Communication and Computing. Springer, Cham, 2022. https://doi.org/10.1007/978-3-030-85559-8_3
5. J. Konecný, H. B. McMahan, F. X. Yu, P. Richtarik, A. T. Suresh, and D. Bacon, "Federated learning: Strategies for improving communication efficiency," Oct. 2017. [Online]. Available: arXiv:1610.05492.
6. K. Kishor and D. Pandey, "Study and development of efficient air quality prediction system embedded with machine learning and IoT." In: Gupta D., Khanna A., Bhattacharyya S., Hassanien A. E., Anand S., Jaiswal A. (eds) *Proceeding International Conference on Innovative Computing and Communications.* Lecture Notes in Networks and Systems, vol. 471, Springer, Singapore, 2022. https://doi.org/10.1007/978-981-19-2535-1_24
7. M. J. Sheller, G. A. Reina, B. Edwards, J. Martin, and S. Bakas, "Multi-institutional deep learning modeling without sharing patient data: A feasibility study on brain tumor segmentation." In: Crimi A., Bakas S., Kuijf H., Keyvan F., Reyes M., and van Walsum T. (eds) *Brainlesion: Glioma, Multiple Sclerosis, Stroke and Traumatic Brain Injuries.* Lecture Notes in Computer Science, Springer International, Cham, Switzerland, 2019, pp. 92–104.
8. K. Kishor, "Chapter 17 Application of quantum computing for digital forensic investigation." In: Yadav S. P., Singh R., Yadav V., Al-Turjman F., Kumar S. A. (eds) *Quantum-Safe Cryptography Algorithms and Approaches: Impacts of Quantum Computing on Cybersecurity.* De Gruyter, Berlin, Boston, 2023, pp. 231–248. https://doi.org/10.1515/9783110798159-017
9. Y. Liu, J. J. Q. Yu, J. Kang, D. Niyato, and S. Zhang, "Privacy- preserving traffic flow prediction: A federated learning approach," *IEEE Internet Things J.*, vol. 7, no. 8, pp. 7751–7763, Aug. 2020.
10. Z. Li, V. Sharma, and S. P. Mohanty, "Preserving data privacy via federated learning: Challenges and solutions," *IEEE Consum. Electron. Mag.*, vol. 9, no. 3, pp. 8–16, May 2020.
11. W. Y. B. Lim et al., "Federated learning in mobile edge networks: A comprehensive survey," *IEEE Commun. Surveys Tutor.*, vol. 22, no. 3, pp. 2031–2063, 3rd Quart., 2020.
12. K. Kishor and P. Nand "Wireless networks based in the cloud that support 5G." In: Kishor K., Saxena, N., Pandey, D. (eds), *Cloud-based Intelligent Informative Engineering for Society 5.0*, 1st edition. Chapman and Hall/CRC, New York, 2023, pp. 23–40. ISBN: 9781003213895. https://doi.org/10.1201/9781003213895-2
13. Y. Liu, X. Yuan, Z. Xiong, J. Kang, X. Wang, and D. Niyato, "Federated learning for 6G communications: Challenges, methods, and future directions," Jul. 2020. [Online]. Available: arXiv:2006.02931.
14. Z. Du, C. Wu, T. Yoshinaga, K.-L. A. Yau, Y. Ji, and J. Li, "Federated learning for vehicular Internet of Things: Recent advances and open issues," *IEEE Open J. Comput. Soc.*, vol. 1, pp. 45–61, 2020.
15. J. Xu and F. Wang, "Federated learning for healthcare informatics," 2019. [Online]. Available: arXiv:1911.06270.
16. B. Brik, A. Ksentini, and M. Bouaziz, "Federated learning for UAVs- enabled wireless networks: Use cases, challenges, and open problems," *IEEE Access*, vol. 8, pp. 53841–53849, 2020.
17. Y. Lu, X. Huang, Y. Dai, S. Maharjan, and Y. Zhang, "Blockchain and federated learning for privacy-preserved data sharing in industrial IoT," *IEEE Trans. Ind. Informat.*, vol. 16, no. 6, pp. 4177–4186, Jun. 2020.

18. D. C. Nguyen, P. N. Pathirana, M. Ding, and A. Seneviratne, "Blockchain for 5G and beyond networks: A state of the art survey," *J. Netw. Comput. Appl.*, vol. 166, Sep. 2020, Article no. 102693.
19. L. Kong, X.-Y. Liu, H. Sheng, P. Zeng, and G. Chen, "Federated tensor mining for secure industrial Internet of Things," *IEEE Trans. Ind. Informat.*, vol. 16, no. 3, pp. 2144–2153, Mar. 2020.
20. Y. Lu, X. Huang, K. Zhang, S. Maharjan, and Y. Zhang, "Blockchain empowered asynchronous federated learning for secure data sharing in Internet of Vehicles," *IEEE Trans. Veh. Technol.*, vol. 69, no. 4, pp. 4298–4311, Apr. 2020.
21. K. Kishor, P. Singh, and R. Vashishta, "Develop model for malicious traffic detection using deep learning." In: Sharma D. K., Peng S. L., Sharma R., Jeon G. (eds) *Micro-Electronics and Telecommunication Engineering*. Lecture Notes in Networks and Systems, vol. 617. Springer, Singapore, 2023. https://doi.org/10.1007/978-981-19-9512-5_8
22. K. Kishor, "Cloud computing in blockchain." In: Kishor K., Saxena, N., Pandey, D. (eds), *Cloud-based Intelligent Informative Engineering for Society 5.0*, 1st edition. Chapman and Hall/CRC, New York, 2023, pp. 79–105. ISBN: 9781003213895 https://doi.org/10.1201/9781003213895-5
23. Y. Lu, X. Huang, Y. Dai, S. Maharjan, and Y. Zhang, "Differentially private asynchronous federated learning for mobile edge computing in urban informatics," *IEEE Trans. Ind. Informat.*, vol. 16, no. 3, pp. 2134–2143, Mar. 2020.
24. T.-C. Chiu, Y.-Y. Shih, A.-C. Pang, C.-S. Wang, W. Weng, and C.-T. Chou, "Semi-supervised distributed learning with non-IID data for AIoT service platform," *IEEE Internet Things J.*, vol. 7, no. 10, pp. 9266–9277, Oct. 2020.
25. Z. Zhou, S. Yang, L. J. Pu, and S. Yu, "CEFL: Online admission control, data scheduling and accuracy tuning for cost-efficient federated learning across edge nodes," *IEEE Internet Things J.*, vol. 7, no. 10, pp. 9341–9356, Oct. 2020.
26. Kishor, K. Using a half cheetah habitat for random augmentation computing. *Multimed Tools Appl* (2024). https://doi.org/10.1007/s11042-024-19084-0
27. K. Kishor, N. Saxena, and D. Pandey (eds), *Cloud-Based Intelligent Informative Engineering for Society 5.0*, 1st edition. Chapman and Hall/CRC, New York, 2023, pp. 1–234. ISBN: 9781003213895. https://doi.org/10.1201/9781003213895
28. S. Duan, D. Zhang, Y. Wang, L. Li, and Y. Zhang, "JointRec: A deep-learning-based joint cloud video recommendation framework for mobile IoT," *IEEE Internet Things J.*, vol. 7, no. 3, pp. 1655–1666, Mar. 2020.
29. K. Kishor, "Impact of cloud computing on entrepreneurship, cost, and security." In: Kishor K., Saxena, N., Pandey, D. (eds), *Cloud-based Intelligent Informative Engineering for Society 5.0*, 1st edition. CRC Press, New York, 2023, pp. 171–191. ISBN: 9781003213895. https://doi.org/10.1201/9781003213895-10
30. J. Ren, H. Wang, T. Hou, S. Zheng, and C. Tang, "Federated learning-based computation offloading optimization in edge computing-supported Internet of Things," *IEEE Access*, vol. 7, pp. 69194–69201, 2019.
31. T. Tuor, S. Wang, T. Salonidis, B. J. Ko, and K. K. Leung, "Demo abstract: Distributed machine learning at resource-limited edge nodes," In *Proc. Conf. Comput. Commun. Workshops (INFOCOM WKSHPS),* Apr. 2018, pp. 1–2. IEEE, Washington, DC, DOI: 10.1109/SMARTCOMP.2019.00030
32. K. Kishor, R. Sharma, and M. Chhabra "Student performance prediction using technology of machine learning. In: Sharma D. K., Peng S. L., Sharma R., Zaitsev D. A. (eds) *Micro-Electronics and Telecommunication Engineering.* Lecture Notes in Networks and Systems, vol. 373. Springer, Singapore, 2022. https://doi.org/10.1007/978-981-16-8721-1_53
33. Y. Deng, T. Han, and N. Ansari, "FedVision: Federated video analytics with edge computing," *IEEE Open J. Comput. Soc.*, vol. 1, pp. 62–72, 2020.

34. J. Cao, K. Zhang, F. Wu, and S. Leng, "Learning cooperation schemes for mobile edge computing empowered Internet of Vehicles," In *Proc. IEEE Wireless Commun. Netw. Conf. (WCNC)*, Seoul, South Korea, May 2020, pp. 1–6.
35. S. Wang, X. Zhang, Y. Zhang, L. Wang, J. Yang, and W. Wang, "A survey on mobile edge networks: Convergence of computing, caching and communications," *IEEE Access*, vol. 5, pp. 6757–6779, 2017.
36. Z. Yu et al., "Federated learning based proactive content caching in edge computing," In *Proc. IEEE Global Commun. Conf. (GLOBECOM)*, Abu Dhabi, UAE, Dec. 2018, pp. 1–6.
37. X. Wang, C. Wang, X. Li, V. C. M. Leung, and T. Taleb, "Federated deep reinforcement learning for Internet of Things with decentralized cooperative edge caching," *IEEE Internet Things J.*, vol. 7, no. 10, pp. 9441–9455, Oct. 2020.
38. L. Cui et al., "CREAT: Blockchain-assisted compression algorithm of federated learning for content caching in edge com- puting," *IEEE Internet Things J.*, Aug. 5, 2020, https://doi.org/10.1109/JIOT.2020.3014370
39. K. Qi and C. Yang, "Popularity prediction with federated learning for proactive caching at wireless edge," In *Proc. IEEE Wireless Commun. Netw. Conf. (WCNC)*, Seoul, South Korea, May 2020, pp. 1–6.
40. Y. Wu, Y. Jiang, M. Bennis, F. Zheng, X. Gao, and X. You, "Content popularity prediction in fog radio access networks: A federated learning based approach," In *Proc. Int. Conf. Commun. (ICC)*, Dublin, Ireland, Jun. 2020, pp. 1–6.
41. K. Kishor, P. Nand, and P. Agarwal, "Secure and efficient subnet routing protocol for MANET," *Indian Journal of Public Health*, vol. 9, no. 12, p. 200, 2018, https://doi.org/10.5958/0976-5506.2018.01830.2
42. K. Kishor, P. Nand, and P. Agarwal, "Notice of retraction design adaptive subnetting hybrid gateway MANET protocol on the basis of dynamic TTL value adjustment," *Aptikom Journal on Computer Science and Information Technologies*, vol. 3, no. 2, pp. 59–65, 2018, https://doi.org/10.11591/APTIKOM.J.CSIT.115
43. L. Chen et al., "Robustness, security and privacy in location-based services for future IoT: A survey," *IEEE Access*, vol. 5, pp. 8956–8977, 2017.
44. B. S. Ciftler, A. Albaseer, N. Lasla, and M. Abdallah, "Federated learning for RSS fingerprint-based localization: A privacy-preserving crowdsourcing method," In *Proc. Int. Wireless Commun. Mobile Comput. (IWCMC)*, Limassol, Cyprus, Jun. 2020, pp. 2112–2117.
45. J. Torres-Sospedra et al., "UJIIndoorLoc: A new multi-building and multi-floor database for WLAN fingerprint-based indoor localization problems," In *Proc. Int. Conf. Indoor Position. Indoor Navig. (IPIN)*, Busan, South Korea, Oct. 2014, pp. 261–270.
46. W. Li, C. Zhang, and Y. Tanaka, "Pseudo label-driven federated learning based decentralized indoor localization via mobile crowdsourcing," *IEEE Sensors J.*, vol. 20, no. 19, pp. 11556–11565, Oct. 2020.
47. Y. Liu, H. Li, J. Xiao, and H. Jin, "FLoc: Fingerprint-based indoor localization system under a federated learning updating framework," In *Proc. Int. Conf. Mobile Ad-Hoc Sensor Netw. (MSN)*, Shenzhen, China, Dec. 2019, pp. 113–118.
48. F. Yin et al., "FedLoc: Federated learning framework for data-driven cooperative localization and location data processing," May 2020. [Online]. Available: arXiv: 2003.03697.
49. K. Kishor, P. Nand, and P. Agarwal, "Subnet based ad hoc network algorithm reducing energy consumption in MANET," *International Journal of Applied Engineering Research*, vol. 12, no. 22, pp. 11796–11802, 2017.
50. Z. Zhou, H. Liao, B. Gu, K. M. S. Huq, S. Mumtaz, and J. Rodriguez, "Robust mobile crowd sensing: When deep learning meets edge computing," *IEEE Netw.*, vol. 32, no. 4, pp. 54–60, Jul. 2018.

51. Y. Lu, X. Huang, Y. Dai, S. Maharjan, and Y. Zhang, "Federated learning for data privacy preservation in vehicular cyber-physical systems," *IEEE Netw.*, vol. 34, no. 3, pp. 50–56, May 2020.
52. Y. Zhao et al., "Local differential privacy based federated learning for Internet of Things," *IEEE Internet Things J.*, Nov. 10, 2020, https://doi.org/10.1109/JIOT.2020.3037194
53. B. C. Drolet, J. S. Marwaha, B. Hyatt, P. E. Blazar, and S. D. Lifchez, "Electronic communication of protected health information: Privacy, security, and HIPAA compliance," *J. Hand Surg.*, vol. 42, no. 6, pp. 411–416, Jun. 2017.
54. B. Yuan, S. Ge, and W. Xing, "A federated learning framework for healthcare IoT devices," May 2020. [Online]. Available: arXiv: 2005.05083.
55. H. Ye, L. Liang, G. Y. Li, J. Kim, L. Lu, and M. Wu, "Machine learning for vehicular networks: Recent advances and application examples," *IEEE Veh. Technol. Mag.*, vol. 13, no. 2, pp. 94–101, Jun. 2018.
56. A. M. Elbir and S. Coleri, "Federated learning for vehicular networks," Jun. 2020. [Online]. Available: arXiv: 2006.01412.
57. K. Kishor, R. Rani, A. K. Rai, and V. Sharma, "3D application development using unity real time platform." In: Swaroop A., Kansal V., Fortino G., Hassanien A. E. (eds) *Proceedings of Fourth Doctoral Symposium on Computational Intelligence. DoSCI 2023.* Lecture Notes in Networks and Systems, vol. 726. Springer, Singapore, 2023. https://doi.org/10.1007/978-981-99-3716-5_54
58. H. Shiri, J. Park, and M. Bennis, "Communication-efficient massive UAV online path control: Federated learning meets mean-field game theory," Aug. 2020. [Online]. Available: arXiv: 2003.04451.
59. D. R. Mukhametov, "Ubiquitous computing and distributed machine learning in smart cities," In *Proc. Wave Electron. Appl. Inf. Telecommun. Syst. (WECONF)*, St. Petersburg, Russia, Jun. 2020, pp. 1–5.
60. S. Perez, J. Perez, P. Arroba, R. Blanco, J. L. Ayala, and J. M. Moya, "Predictive GPU-based ADAS management in energy-conscious smart cities," In *Proc. IEEE Int. Smart Cities Conf. (ISC2)*, Oct. 2019, pp. 349–354.
61. A. Taik and S. Cherkaoui, "Electrical load forecasting using edge computing and federated learning," In *Proc. Int. Conf. Commun. (ICC)*, Dublin, Ireland, Jun. 2020, pp. 1–6.
62. K. Kishor and R. K. Verma, "Cloud computing-based smart agriculture." In: Sharma A., Chanderwal N., Khan R. (eds) *Convergence of Cloud Computing, AI, and Agricultural Science.* IGI Global, Hershey, Pennsylvania, USA, 2023, pp. 120–136. https://doi.org/10.4018/979-8-3693-0200-2.ch006
63. W. Zhou, Y. Li, S. Chen, and B. Ding, "Real-time data processing architecture for multi-robots based on differential federated learning," In *Proc. IEEE SmartWorld Ubiquitous Intell. Comput. Adv. Trust. Comput. Scalable Comput. Commun. Cloud Big Data Comput. Internet People Smart City Innov. (SmartWorld/SCALCOM/UIC/ATC/CBDCom/IOP/SCI)*, Guangzhou, China, Oct. 2018, pp. 462–471.
64. A. K. Tanwani, N. Mor, J. Kubiatowicz, J. E. Gonzalez, and K. Goldberg, "A fog robotics approach to deep robot learning: Application to object recognition and grasp planning in surface decluttering," In *Proc. Int. Conf. Robot. Autom. (ICRA)*, Montreal, QC, Canada, May 2019, pp. 4559–4566.
65. M. Hao, H. Li, X. Luo, G. Xu, H. Yang, and S. Liu, "Efficient and privacy-enhanced federated learning for industrial artificial intelligence," *IEEE Trans. Ind. Informat.*, vol. 16, no. 10, pp. 6532–6542, Oct. 2020.
66. S. R. Pokhrel and S. Singh, "Compound-TCP performance for industry 4.0 WiFi: A cognitive federated learning approach," *IEEE Trans. Ind. Informat.*, vol. 17, no. 3, pp. 2143–2151, Mar. 2021.

67. Kishor, K., Shukla, A., Thakur, A. (2024). Vehicle Classification and License Number Plate Detection Using Deep Learning. In: Sharma, D.K., Peng, SL., Sharma, R., Jeon, G. (eds) Micro-Electronics and Telecommunication Engineering. ICMETE 2023. Lecture Notes in Networks and Systems, vol 894. Springer, Singapore. https://doi.org/10.1007/978-981-99-9562-2_5
68. K. Kishor, S. Sangal, and Shahroz, "Real-time traffic signs and lane line detection." In: Swaroop A., Kansal V., Fortino G., Hassanien A. E. (eds) *Proceedings of Fourth Doctoral Symposium on Computational Intelligence. DoSCI 2023.* Lecture Notes in Networks and Systems, vol. 726. Springer, Singapore, 2023. https://doi.org/10.1007/978-981-99-3716-5_71
69. S. Gupta, S. Tyagi, and K. Kishor, "Study and development of self sanitizing smart elevator." In: Gupta D., Polkowski Z., Khanna A., Bhattacharyya S., Castillo O. (eds) *Proceedings of Data Analytics and Management.* Lecture Notes on Data Engineering and Communications Technologies, vol. 90. Springer, Singapore, 2022. https://doi.org/10.1007/978-981-16-6289-8_15
70. K. Kaushal. "Chapter 12 Review and significance of cryptography and machine learning in quantum computing." In: Yadav S. P., Singh R., Yadav V., Al-Turjman F., Kumar S. A. (eds) *Quantum-Safe Cryptography Algorithms and Approaches: Impacts of Quantum Computing on Cybersecurity.* De Gruyter, Berlin, Boston, 2023, pp. 159–176. https://doi.org/10.1515/9783110798159-012
71. H. T. Nguyen, N. C. Luong, J. Zhao, C. Yuen, and D. Niyato, "Resource allocation in mobility-aware federated learning networks: A deep rein- forcement learning approach," In *Proc. IEEE 6th World Forum Internet Things (WF-IoT)*, New Orleans, LA, Jun. 2020, pp. 1–6.
72. A. Imteaj and M. H. Amini, "FedAR: Activity and resource-aware federated learning model for distributed mobile robots," Jan. 2021. [Online]. Available: arXiv: 2101.03705.
73. L. Lyu, H. Yu, and Q. Yang, "Threats to federated learning: A survey," Mar. 2020. [Online]. Available: arXiv:2003.02133.
74. B. Zhao, K. Fan, K. Yang, Z. Wang, H. Li, and Y. Yang, "Anonymous and privacy-preserving federated learning with industrial big data," *IEEE Trans. Ind. Informat.*, Jan. 15, 2021, https://doi.org/10.1109/TII.2021.3052183
75. K. Bonawitz et al., "Practical secure aggregation for federated learning on user-held data," Nov. 2016. [Online]. Available: arXiv:1611.04482.
76. R. K. Verma and K. Kishor, "Image processing applications in agriculture with the help of AI." In: Khan M. A., Khan R., Praveen P., Verma A., Panda M. (eds) *Infrastructure Possibilities and Human-Centered Approaches With Industry 5.0.* IGI Global, Hershey, Pennsylvania, USA, 2024, pp. 162–181. https://doi.org/10.4018/979-8-3693-0782-3.ch010
77. A. Nilsson, S. Smith, G. Ulm, E. Gustavsson, and M. Jirstrand, "A performance evaluation of federated learning algorithms," In *Proceedings of the. 2nd Workshop Distributed Infrastructre for Deep Learning (DIDL)*, Dec. 2018, Association for Computing Machinery, New York, pp. 1–8. https://doi.org/10.1145/3286490.3286559
78. T. Li, A. K. Sahu, M. Zaheer, M. Sanjabi, A. Talwalkar, and V. Smith, "Federated optimization in heterogeneous networks," Dec. 2018. [Online]. Available: arxiv.abs/1812.06127.
79. F. Sattler, S. Wiedemann, K.-R. Muller, and W. Samek, "Robust and communication-efficient federated learning from non-i.i.d. data," *IEEE Trans. Neural Netw. Learn. Syst.*, vol. 31, no. 9, pp. 3400–3413, Sep. 2020.
80. D. Rothchild, A. Panda, E. Ullah, N. Ivkin, I. Stoica, V. Braverman, J. Gonzalez, and R. Arora, "Fetchsgd: Communication-efficient federated learning with sketching," *Int. Conf. on Machine Learning (ICML).* Online: PMLR, Jul. 2020, pp. 8253–8265.

81. W. Liu, L. Chen, Y. Chen, and W. Zhang, "Accelerating federated learning via momentum gradient descent," *IEEE Trans. Parallel Distrib. Syst.*, vol. 31, no. 8, pp. 1754–1766, Aug. 2020.
82. M. Tsukada, M. Kondo, and H. Matsutani, "A neural network-based on-device learning anomaly detector for edge devices," *IEEE Trans. Comput.*, vol. 69, no. 7, pp. 1027–1044, Jul. 2020.
83. J. Chauhan, S. Seneviratne, Y. Hu, A. Misra, A. Seneviratne, and Y. Lee, "Breathing-based authentication on resource-constrained IoT devices using recurrent neural networks," *Computer*, vol. 51, no. 5, pp. 60–67, May 2018.
84. S. Dhar, J. Guo, J. Liu, S. Tripathi, U. Kurup, and M. Shah, "On-device machine learning: An algorithms and learning theory perspective," 2019. [Online]. Available: arXiv:1911.00623.
85. E. Rezende, G. Ruppert, T. Carvalho, F. Ramos, and P. De Geus, "Malicious software classification using transfer learning of ResNet-50 deep neural network," In *Proc. 16th IEEE Int. Conf. Mach. Learn. Appl. (ICMLA)*, Cancun, 2017, pp. 1011–1014.
86. P. Kairouz et al., "Advances and open problems in federated learning," 2019. [Online]. Available: arXiv:1912.04977.

Index